BIOLOGY for Engineers

Vidya Rajesh, *ME*, *PhD*

Professor, Department of Biological Sciences

BITS Pilani—Hyderabad Campus, Jawahar Nagar

Cengage

Australia • Brazil • Canada • India • Mexico • Singapore • United Kingdom • United States

Biology for Engineers
Vidya Rajesh

For product information and technology assistance, contact us at
asia.infoindia@cengage.com

For permission to use material from this text or product, submit all requests online at
www.cengage.com/permissions

ISBN-13: 978-93-6053-626-8
ISBN-10: 93-6053-626-1

Cengage Learning India Pvt. Ltd.
T-303, 3rd Floor
DAV Complex, Plot 3
Mayur Vihar Phase 1
Delhi 110091

Cengage is a leading provider of customized learning solutions with employees residing in nearly 40 different countries and sales in more than 125 countries around the world. Find your local representative at: **www.cengage.com.**

Cengage is committed to creating inclusive and diverse products, including perspectives that represent the diversity of people across all backgrounds, identities, opinions, and experiences.
If you have any feedback related to this, please contact us at **support.cengage.com.**

To learn more about Cengage products, platforms and services, visit **www.cengage.co.in.**

Printed by:
IBH Print & Bind, Pvt. Ltd.
Plot No. F-411, Sector-63, Noida, (U.P.) -201301

Printed in India
Print Number: 02 Print Year: 2026

To

my parents and teachers who taught me

and

my students from whom I continue to learn

Preface

Biology is the study of complicated things that have the appearance of having been designed with a purpose.

—Richard Dawkins

This textbook is the culmination of my more than two decades of teaching experience at BITS Pilani, especially the first-year course BIO F111 General Biology to the cream of Indian students pursuing their education at the premier institute. Over the years, the pedagogical challenges in making a biology course valuable and interesting to a reluctant technical brain, thrilled to explore physical sciences and engineering but resistant to biology, spurred me to innovate and present the biological concepts in unique ways. A holistic delivery of content by introducing stimulating examples, new modern conceptual twists, and real-life applications helped to keep the knowledge relevant and engaging to engineering students. Constant efforts were made to erase the perceptual bias of biology being a subject of too much theory with rote learning and numerous diagrams. These measures ensured that in my classes, the learning objective of this course proved to be useful, both, in the professional and personal life decisions of students.

The insights gained in the classroom, constant student feedback about the value and clarity of content, along with the detailed subject insights from my scrawled lecture notes, regularly updated relevant examples and applications from the class presentations, discussions, and expert insights obtained from fellow colleagues and instructors, all these have guided me in conceptualizing and presenting the contents of this book. The design of the content and ways of explaining the concepts have been further enhanced by the references to available global textbooks and reference books.

Today, in this interconnected world, irrespective of their vocation and field of study, it is becoming increasingly important for all individuals to understand the nature of life sciences and the fundamentals of biological concepts, to navigate the new-age challenges and understand and appreciate the modern biological research and its outcomes and implications on their lives. Non-biologists, academicians, cross-disciplinary researchers, and anyone with an interest and aim to learn biology even without a prior background, all, will benefit from this book. In the present Indian context, this textbook will be of great significance as technical institutes across the country now recommend a foundation-level biology course for all engineering students, where this book for engineers and 'non-majors' would be appropriate to be adapted as the textbook for learning the different aspects of biology in a simplified manner.

This textbook introduces and acts as a foundational resource to learn about living systems, their properties, basic biochemical and physiological processes, and to gain knowledge on biological systems vis-à-vis their nature, behavior, and functioning. A special emphasis on human physiology and anatomy will give the students insights to various aspects of the structure and function of their own bodily systems. The readers will get introduced to modern-day developments of genetics and recombinant DNA technology and their applications in daily life. The content has also been designed to enable the student to understand the intricate relationship that living organisms have with their environment and become aware of the influence of biology in almost every aspect of their lives.

Organization of the Book

The book contains 20 *chapters* distributed into 3 *units*, covering all the facets of life sciences with the right mix and balance between classical and modern biology. *Unit* I contains Chapters 1 to 9 and deals with the intricacies of life and life processes at the cellular level. *Unit* II contains Chapters 10 to 15 and deals with human biology and functional systems at the organism level. *Unit* III contains Chapters 16 to 20 and deals with all the aspects influencing living organisms and their life and society.

The contents of each chapter are designed into manageable sections for an easy and interesting read. Numerous pedagogical aids have been incorporated to support learning.

The *learning objectives* and *chapter takeaways* give direction to the learning; a *Prelude* to each chapter sets the context; interspersed *experimental insights* present important historical scientific developments and *interesting facts* break the monotony and help to retain the interest of the readers; clear and detailed *figures* and *diagrams* complement the text; chapter-end *summary* and *exercises* help the readers to refresh and evaluate the learning; and, finally additional and diverse reading resources are provided to augment.

Due care has been taken to relate the principles of biology with other science and engineering disciplines, wherever possible, and provide experimental and practical insights, as applicable, to offer and ensure an overall holistic experience to each learner.

After reading the book, the student will gain a complete understanding of the basic concepts of biology, synthesize information from other relevant sources, and apply the knowledge gained to design new and better solutions to present-day challenges.

I hope that the readers find this book stimulating, enjoyable, and useful. I would be grateful to receive any comments or suggestions from the readers for the improvement of future editions of this textbook.

—Vidya Rajesh

Acknowledgements

At the outset, I thank my institution, BITS Pilani, for giving me the opportunity to teach biology to my esteemed UG students and for bestowing on me the foresight about the critical role of biology in the future world of interdisciplinary and multidisciplinary research and innovation. The pioneering introduction of the General Biology course in the curriculum for all engineering students almost four decades back along with offering dual degrees of science and engineering is testimony to this foresight. The BITS ecosystem played a great role in my evolution as a teacher and researcher. The experience has provided me the background and confidence to design this textbook.

Teaching biology at all the three tiers of education—UG, PG, and PhD—along with every conversation with intellectuals in the classroom, corridor, seminar hall, and chamber of the institute over the past two decades, both, have been invaluable resources for my growth and depth of scientific knowledge.

I am equally grateful to all my curiosity-driven students, intellectually supportive colleagues, and mentors, who all have shaped my understanding and knowledge of the subject. I acknowledge the role of every scholar I have interacted with and every book and article that I have read in this long journey for being instrumental in chiseling me as a biologist.

I acknowledge the support of my biologist friends, Dr Monica and Dr Kasturi, who patiently reviewed the chapters of the book and provided their insights on the technical content, its presentation, and the overall flow and understanding therein. Their commitment and support over such short notice is truly appreciated.

This project would not have been completed without the support and encouragement of my family—my parents, sister Soumya, husband Rajesh, and son Adarsh, who have constantly stood by me in every endeavour of my life.

I also thank my publishers Cengage Learning India and the entire team for pursuing me, constantly motivating me to take up as well as complete this assignment, and providing all the support required in this journey. They have been instrumental in coordinating on all the aspects of this book through useful discussions on scheduling timelines, content review and mapping, sourcing figures and drawings, and overall contributing towards ensuring accuracy and the best presentation of each chapter. Their inputs, suggestions and contributions have been invaluable in the book's production.

There are many people without whose help or influence this book would not have been possible. Thanking all others as well who have directly or indirectly inspired, influenced, or supported this book!

—Vidya Rajesh

Brief Contents

Contents

PART I HOW LIFE WORKS!

CHAPTER 1

The Study of Life

LEARNING OBJECTIVES

After reading this chapter, you will be able to:

- Understand what biology is and its relevance in a multidisciplinary problem-solving approach.
- Know the outcomes resulting from the synergy between biology and the other disciplines.
- Establish why engineers and students not majoring in biology should also have a working knowledge of the principles of biology.

CHAPTER TAKEAWAY **Basic knowledge of biology is critical for all.**

This chapter establishes what biology is and reinforces its relevance to all students, irrespective of their vocation or discipline.

PRELUDE

The COVID-19 global experience suddenly threw into our lives a plethora of new terms in daily colloquial use—Corona, pandemic, RT-PCR, variants, antibody, vaccine, immunity, quarantine, outbreak, hypoxemia, herd immunity, R naught (R0), etc. ....to name a few. In a sense, all humans on the planet Earth started using a new vocabulary consisting of these biological terms at least once every day. The pandemic also brought to the fore the value to professionals dealing with the virus and disease management—doctors, researchers, lab personnel, quarantine helpers, nurses, life science teachers, science communicators, medical news reporters, policy makers, public health workers, etc. Overall, the pandemic brought back the understanding that fundamentals of 'life and living,' such as safety, hygiene, diet, medical support for treatment, new scientific discoveries, and uncompromised research methodologies, were extremely valuable and non-negotiable. The pandemic was one clear example where the power as well as limitations of science were exposed and recognized. Thus, irrespective of occupation, a working knowledge of biology became inevitable.

What Is Biology!

Biology is the science that deals with the study of life; 'bio' means '-life' and '-logy' means 'a study or branch of knowledge'. The natural world around us is a multi-faceted complex puzzle that human beings have been trying to make sense of for over millions of years. The pieces of the puzzle have been put together in an iterative manner with observation, followed by testing and analysis, over the years; however, the more the pieces fit together by critical thinking, the bigger the puzzle becomes.

The ocean of the unknowns, coupled with interconnectedness and sense of order with the right mix of simplicity and complexity in the biological world, provides a curious model for scientific

exploration (Figure 1.1). The study of biology and understanding life helps in decoding the layers of shared and unshared features of natural systems.

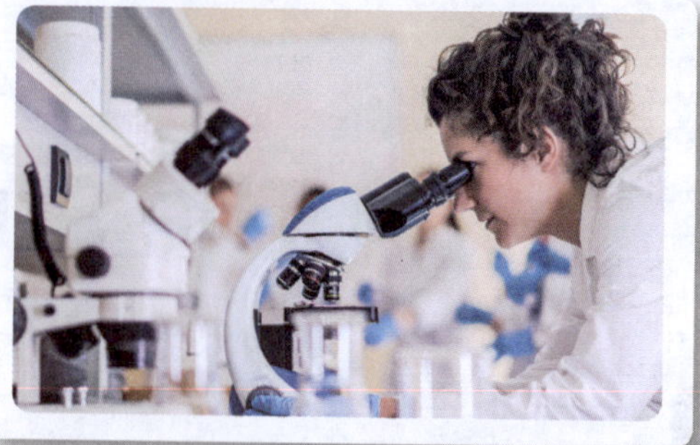

Fig. 1.1 A biology student/researcher attempting to understand the complexities of biological world

Interesting Fact 1.1

Some facts about living systems

- No machine is as energy-efficient as a cell.
- No network is as efficient as the neural network in human beings.
- No robot is as mechanically perfect as the human body.

Why Study Biology!

As a science, biology has roots in ancient civilization, although modern biology is much more recent. Biology is a fascinating, vast, and eclectic field composed of many specialized disciplines that study the structure, function, growth, distribution, evolution, or other features of living organisms (Figure 1.2).

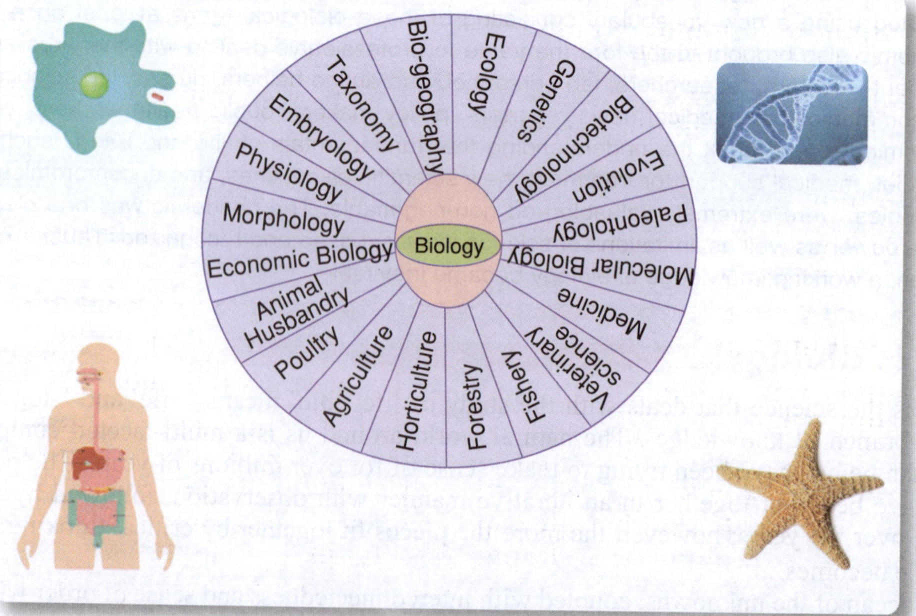

Fig. 1.2 Branches of biology

Over the centuries, biology has progressed and made significant contributions to mankind in genetics, taxonomy, evolution, health and disease, nutrition, environment, etc. Discoveries in Biology have had ground-breaking impact, and some of the major ones, over the past, are listed below (Table 1.1):

Table 1.1 Some Major Discoveries in Biology

1.	Classification of living organisms—modern taxonomy by Carl Linnaeus	18th Century
2.	Discovery of microscope and observation of cells by Anton von Leeuwenhoek and Robert Hooke	18th Century
3.	Theory of evolution by Charles Darwin	19th Century
4.	Principles of heredity by Gregor Mendel	19th Century
5.	Discovery of the antibiotic Penicillin by Alexander Fleming	20th Century
6.	Structure of DNA by Watson and Crick	20th Century
7.	Cloning of Dolly by Ian Wilmut	20th Century
8.	Whole genome sequencing and Human Genome Project	21st Century
9.	Developments in stem cell research	21st Century
10.	CRISPR-Cas-based genome editing	21st Century

The journey continues, and the 21st Century is termed the 'Century of Biology' due to the revolutionary developments and contributions leading to path-breaking technologies and innovations. Problem-solving approaches using multidisciplinary science cutting across discipline boundaries provide unique insights and practical solutions to the many challenges facing mankind.

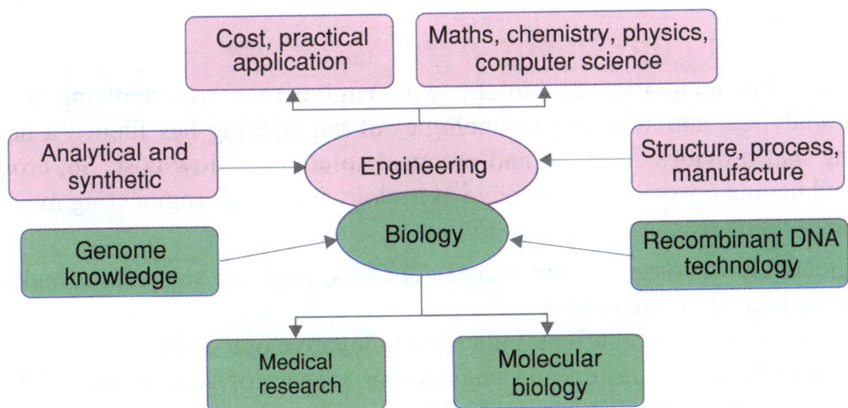

Fig. 1.3 Overlapping attributes from biology and engineering that help in creative solutions

Non-conventional research at the intersection of two or more disciplines is popular and will lead to exciting outcomes and, definitely, seems to be a norm, going forward. Biology is at the forefront along with the other pillars of science, such as Chemistry, Mathematics, and Physics, in developing technologies of the future. New interdisciplinary and multidisciplinary fields have emerged gradually in the new era of knowledge creation, which have synergistic and close associations with biology such as biophysics, biochemistry, material science, bioinformatics, nanotechnology, synthetic biology, biomechanics, systems and computational biology, biosensors, bioelectronics, biochips, bio-robotics, etc. (Figure 1.3).

Study of biological systems has inspired many innovations in day-to-day life. Aeroplanes inspired by bird flight and cameras with the human eye are some of the most obvious ones (Figure 1.4)!

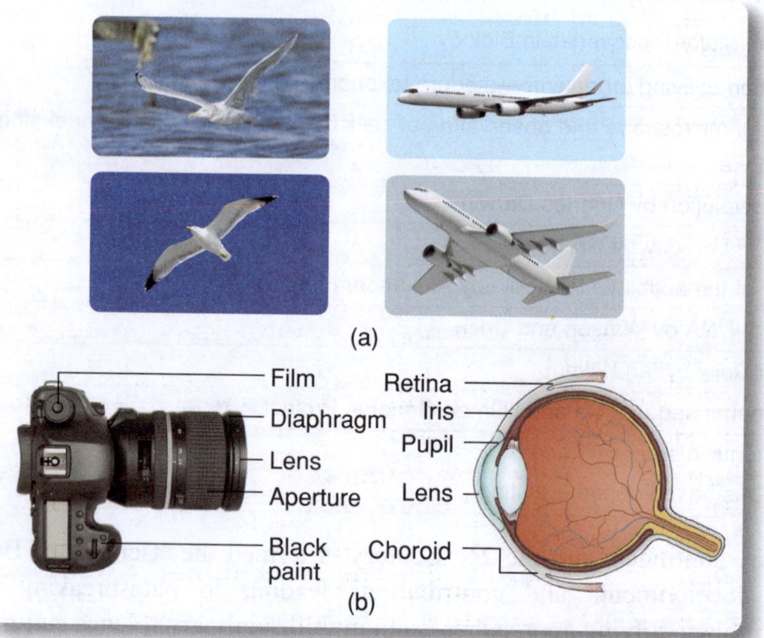

Fig. 1.4 Bird flight and human eye in designing of (a) aeroplane and (b) camera, respectively

Importance of Biology in Engineering

It is evident that integration of biology with engineering, i.e., applying the concepts of biology with analytical and synthetic technologies of engineering has fuelled a new domain of bioengineering that provides sensitive and practical solutions at low cost. So, cross-pollination of concepts and having a working knowledge of biology is key for engineering majors in the new world because of many reasons that are as follows:

(a) To effectively interact and understand concerns, problem statements, and for providing solutions related to biosystems
(b) To improvise existing solutions and create innovative outcomes
(c) To contribute to the design and engineering aspects of sustainable solutions to grand challenges pertaining to health, food, climate, and environment
(d) To become aware of new biological developments (genetically modified foods, superbugs, designer babies, emerging diseases, biological warfare) and their implications as future parents and informed citizens
(e) To understand ethical issues and incorrect application of biological principles, which can have damaging consequences

Knowledge of biology is useful to other disciplines, and other disciplines also contribute to applications in biology! These are undeniable truths. So, the useful, amazing, ever-evolving, educative, and game-changing story of biology is a critical need for all. It is relevant, but is it daunting?

Innovative outcomes by synergistic association between biology and other disciplines

Events in life sciences have revolutionized agriculture, industry, medicine, and environment protection, and found solutions to many challenging problems. Innovations in other disciplines using the concepts associated with biology have changed the world in the 20th and 21st centuries. Marriage between the attributes of engineering and science has led to interdisciplinary outcomes involving two or more disciplines, as highlighted in Table 1.2.

Table 1.2 Examples of Innovative Products and Processes Fuelled by Interdisciplinary Approaches Involving Biological Principles and Knowledge

Innovation	Major Discipline	Inspiration and Synergistic Insights from Biology
Evolution of prosthetics: replacement limbs, artificial retina, etc.	Material Science, Chemical Engineering, Mechanical Engineering	From inanimate mimics to regeneration from biological living components using molecular biology
Synthetically designed sensors to detect metal ions	Chemistry, Material Science	Screening for zinc and other metal ions in blood—inexpensive, non-invasive methods
Artificial muscles for robotic movement	Robotics, Mechanical Engineering	Mimicking muscle biology
Super microscopes for 3D single-cell visualization of ex-vivo animal organs	Imaging, Mechanical Engineering and Design	Real-life observations facilitate health research by studying living components.
Airplane, Velcro, Shinkansen bullet train, cameras	Design Engineering	Bio-inspired design—flight of birds, sticking of dogs hair, kingfisher beak, human eye, etc.
Biomimetic systems, Efficient flexible robots—robotic cheetah, robotic snail	Robotics	Bio-inspired design— understanding physical principles of living systems and applying to engineering design
Artificial neural networks, Hyper organs such as ultrasensitive nose, sound navigation and identification	Computer Science	Processing set-ups inspired by the human brain
Bio-instrumentation, Point of care devices, Miniaturization of gadgets for health monitoring	Electronics and Instrumentation, Mechanical photonics, 3D printing and imaging	Biologically relevant diagnostic home devices to monitor health parameters such as blood pressure, oxygen sensing, glucose monitoring, etc.
Creation of large databases for genome sequencing and computational analysis, Machine learning approaches, Artificial intelligence (AI)	Computer Science, Bioinformatics, Computational biology	Comparative genomics— sequence, structural, and functional analysis of genes and proteins and biological networks between and across organisms; bio-inspired cognition and intelligence

[*Contd*]

[*Contd*]

Design of protection systems and ecological enhancement, Constructed wetlands, Biogas production, Bio-concrete for sustainable structures, etc.	Civil Engineering, Natural Resource Engineering	Knowledge of ecological aspects, biotic and abiotic contributors and their niche roles such as energy dynamics and bio-geocycles
Bio-MEMS (Biological micro-electro-mechanical systems)	Electrical and Electronics Engineering (EEE)	Living cells and tissues incorporated into micro-devices for applications such as micro-fluidic devices and organs on a chip
Sustainable production of chemicals, materials, pharmaceuticals, and fuels from renewable sources by fermentation, using living cells and/or their enzymes	Chemical Engineering, Pharmacy	Micro-organisms and enzymes used for the production of food supplements (such as vitamins), bio-based plastics (such as polylactic acid) and energy carriers (such as biogas and bio-ethanol)
Predictive modelling of infection spread and disease dynamics during pandemic	Mathematics, Public Health	COVID–19 pandemic saw the emergence of multiple statistical models predicting impact and outcome.

Interesting Fact 1.2

Is biology difficult? Let's look at it using an example.

Example: Pneumonoultramicroscopicsilicovolcanoconiosis

- (*pneu-*) or (*pneumo-*), which means related to lung
- *ultra*, meaning extreme
- *microscopic*, meaning small
- (*silico-*), which refers to silicon
- (*volcano-*), which refers to the mineral particles that make up a volcano
- (*coni-*), a derivative of the Greek word *konis* meaning dust
- (*-osis*), which means affected with

Lungs are affected with extremely small (*ultramicroscopic*) silicon (*silico-*) and volcanic (*volcano-*) dust (*coni-*) particles. Thus, Pneumonoultramicroscopicsilicovolcanoconiosis is a disease of the lungs resulting from the inhalation of very fine silicate or quartz dust. That was not so difficult, now. Was it?

Future Directions for Biology as a Science

While great strides have been made already, and we have acquired a lot of knowledge about how individual bits and individual organisms work, the shape of biology as a science in the future continues to excite.

The focus would be on tackling big problems of society such as population control, ecological imbalances and climate change, global disease management and personalized medicine, emerging new and drug-resistant pathogens, alternative-energy bio-resources, understanding systems biology for emotional, psychological and social well-being, biotechnological research for sustainable food production, medicines and fuels, space explorations, and so on (Figure 1.5). Collaborative efforts across disciplines to capture new possibilities and manage risks could have far-reaching impacts and bring meaningful changes across societies and economies.

(a)

(b)

(c)

(d)

(e)

(f)

Fig. 1.5 Possibilities and impacts in future

To sustain and address these challenges, human resources trained in biology would be crucial. Understanding the fundamental principles of life sciences will help to manage day-to-day

challenges in future, as biological innovations will influence our everyday lives and the choices we may need to make. Knowledge of biology will be key to handle some of the future possibilities and making relevant life choices like:

- Choosing genetically modified food with superior benefits.
- Understanding the impact of radiation on health and well-being.
- The reality of superbugs and how to tackle the same.
- Would human organs be available over the counter like automobile spare parts.
- Reproductive technologies and the advancements, including the possibility of human cloning in future.

These have a direct impact on how we and our forthcoming generations would live.

Awareness of biology for future citizens and parents is also mandatory in the new global environment for the greater public good, to make and understand public policies on issues such as vaccine hesitancy, radiation impact, and social and emotional impacts of fast-paced developments like metaverse, drug resistance, and biological-waste hazard. Knowledge of biology will be the key!

Experimental Insight 1.1 **Bio-inspired Design**

Why bio-mimicry and bio-inspired design fuelled innovations still have a long way to go?

Despite initial success in the bio-inspired design products already available, practically, there are multiple aspects where possibilities of improvisation exist going forward. For example, even though robotics and digital computers have made great strides and are as good as they are in playing chess, working spreadsheets, and solving complex maths equations, they still cannot tackle fine and complex activities that the human beings do.

If you send the digital computer to play football, rushing across multiple positions in the field, passing the ball to the next strategic partner, and focusing on opponents at the same time, these are monumental tasks for the computerized machine. Similarly, the human eye can capture and process information at speeds higher than the best cameras. The luminescence, colour, and motion are also captured well without any blurring of images. The best cameras still have a long way to go in replicating that. Parallel processing does solve this challenge to a certain extent, but the robotic networks doing so are still primitive, with huge and complex setups and less effectiveness.

Hence, the living systems set benchmarks for scientific innovations to attain the same progressively.

Tips

Best learning outcomes in biology! Curiosity is the best teacher; so understanding the fundamental concepts through personal (self) concept maps, asking questions, being imaginative, ready to experiment, and analyse are the best ways to learn biology. Teaching biology with objective rationale, real-life examples, connecting theory to applications, and audio-visual illustrations of concepts enhance the learning outcomes, even for new entrants.

SUMMARY

- Biology is the study of living systems.
- It is a fascinating, vast field with multiple sub-specializations.
- Many interdisciplinary innovations take a leaf out of the book of biology.
- Bio-inspired design and cross-pollination of concepts between biology and other disciplines have led to many useful products in the past century.
- Knowledge of biology for engineers and non-majors has several benefits.
- To sustain and address the big problems and challenges of society in future, trained manpower with knowledge of biology is crucial.

EXERCISES

A. Objective-Type Questions

1. Which of the following is the most recent discovery of biology?
   (a) Penicillin (b) Dolly—the sheep
   (c) Human Genome Project (d) Theory of evolution
2. Which of the following would be a product of Bio-MEMS?
   (a) Aeroplane (b) Flexible robots
   (c) Microfluidic devices: organ-on-a-chip (d) Biogas production
3. Structure of DNA was elucidated by
   (a) Gregor Mendel (b) Watson and Crick (c) Charles Darwin (d) Ian Wilmut
4. Which term below best captures the meaning of 'Pneumo'?
   (a) Silica (b) Sheep (c) Lung (d) Chip
5. Which engineer has a role in the design of prosthetics for human use?
   (a) Computer science (b) Civil engineer
   (c) Ocean engineer (d) Material scientist
6. Velcro is an example of bio-inspired design. True or False?

B. Short-Answer Questions

1. Give two futuristic challenges in the next century that will require the knowledge of biology?
2. Justify with 3–4 reasons why teaching biology as an ability-enhancing course to students of all disciplines is a must?

C. Self-Study Questions

1. Conduct a google search for articles and YouTube videos to know more and understand the interdisciplinary innovations depicted in Table 1.1.
2. Search and select TED talks that outline the modern innovations having a life science–engineering basis.
3. Why was vaccine hesitancy seen during the pandemic (COVID-19)?
4. If GM foods are in the market, would you be willing to consume?
5. How was predictive modelling helpful in the decision-making process for administrators during the pandemic (COVID-19)?

FURTHER READINGS

1. The Bio Revolution: Innovations transforming economies, societies, and our lives. McKinsey Global Institute, May 2020.

2. Biology for Engineers: A case study of cross-disciplinary teaching and learning, prepared for the *International Journal of Engineering Education*, 2006.

USEFUL WEBLINKS

1. https://www.who.int/emergencies/diseases/novel-coronavirus-2019
2. https://leverageedu.com/blog/branches-of-biology/
3. https://www.livescience.com/44549-what-is-biology.html

VIDEO LINKS

1. https://www.youtube.com/watch?v=zGQk6jwdosw

CHAPTER 2

Life as We Know It

LEARNING OBJECTIVES

After reading this chapter, you will be able to:

- Visualize how life originated on the Earth
- Browse the theories proposed on the origin of life and how they were proved/disproved
- View the historical perspective and timelines of life on Earth
- See what it means to be alive and how living things differ from non-living things
- Know what metabolic, generative, responsive, and control processes are
- Understand why the cell is a fundamental unit of living organisms

CHAPTER TAKEAWAY What *life* means and how it evolved?

This chapter presents interesting insights into how life evolved and to know what living means. This chapter presents interesting insights into how life evolved and to know what 'living' means by defining the basic characteristics of life.

PRELUDE

Based on our present knowledge about the solar system, the Earth is by far the most (or, only) habitable planet that supports life. The diversity of living forms that we see today and the driving climate change have propelled questions on the influence of the environment on the sustenance of life and the adaptations of life forms. The influence of the sun, oxygen, and greenhouse gases (GHGs) have all contributed to this—the present-day Earth. Theories of origin, simulation experiments, computer models of biogeochemical cycles, and fossil and geological field data from rocks have fuelled various hypotheses, thus, contributing to our understanding of the evolution of the present-day planet. Before a biologist embarks on the journey further deep, some primitive and fundamental questions such as, 'What is life?,' 'When did life begin?,' 'legal definitions of life and death,' etc., should be clear. 'When we transplant an organ from a dead body, is the organ still living?,' 'Does an individual die before every cell of his body has died?' etc., are some curious modern-day questions. Similarly, also interesting to note is how one of the major pathogens causing diseases—the viruses—are not even alive in the true sense but, still, end up creating havoc, the recent COVID pandemic being the case in point!

Evolution of Life on Earth

Origin of life on Earth is an intriguing as well as a complex story. Even to the uninitiated, the dinosaurs, and their extinction, have been fascinating, to say the least. Similarly, understanding life on Earth, starting from its origin to the evolved diversity of the present day, has intrigued philosophers, religious leaders, and scientists alike from time immemorial.

If we ask a simple question—from where did life originate on the Earth? At the very basic, there will be four possibilities in very simplistic terms:

(i) It was created and implanted on the Earth (Divine/religious creation theory).
(ii) It existed in some other 'place', and it came/moved to Earth (Panspermia theory).
(iii) It is formed spontaneously from inanimate objects.
(iv) Environmental triggers created and maintained life.

Several theories on these lines have been proposed and debated, as follows (Figure 2.1):

(i) The ancient religious doctrines believed the 'creation' theory that life was created by God, and genesis accounts were believed.
(ii) The Panspermia theory that life came from outer space gained support as astronomers showed events of meteorites striking primitive Earth and bringing life with it from another universe. Modern research has not yet found life elsewhere, and there is clearly a lack of concrete evidence from space explorations. Also, the possibility of life forms surviving long journeys to Earth and proving their colonization remains a question mark in the absence of clear concrete evidence. Moreover, this still does not answer the query about the origin of life either on Earth or elsewhere.

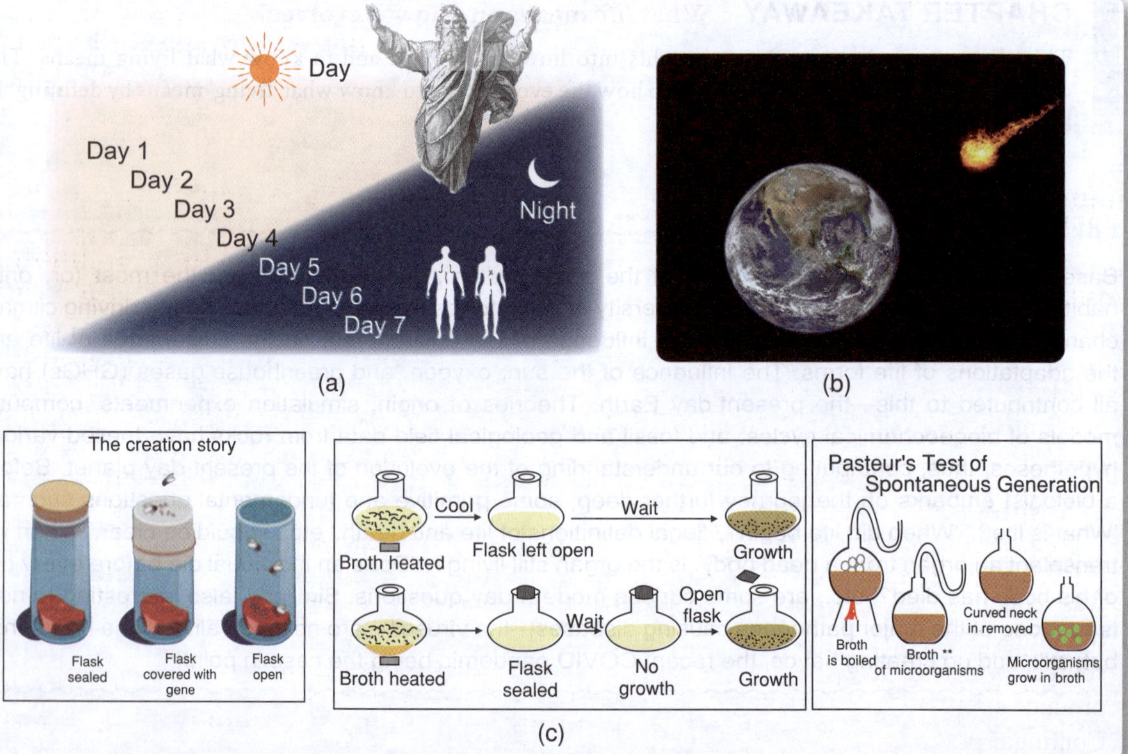

Fig. 2.1 Theories on the origin of life: (a) Creation theory; (b) Panspermia theory; (c) Spontaneous generation

(iii) The theory of *spontaneous generation* argued that the formation of life spontaneously from inanimate objects is called 'abiogenesis'. Spoilage of food by unknown entities, production of flies and maggots around food, animal waste and garbage, appearance of frogs and toads in ponds—such observations fuelled this theory of spontaneous generation further.

Experimentation to rationalize this theory began in the 17th Century, with scientists providing evidence to prove or disprove the theory. Francesco Redi in 1668 showed the appearance of maggots only in open meat jars, not in closed ones. He disproved spontaneous generation, claiming that maggots appeared from the eggs of flies that were attracted to the meat.

This observation was contested by John Needham in 1745, who showed the presence of life in a sealed broth proposing that life was created from non-life, thus proving spontaneous generation. These contradictory observations led to a debate on the theory of spontaneous generation, which was finally disproved in the 18th and 19th centuries by the experiments of Lazzaro Spallanzani and Louis Pasteur. Spallanzani argued that the appearance of life in the bottle is due to microbes entering via air. Pasteur, in 1862, conclusively sealed Spallanzani's observation by doing experiments open to air, in bottles with swan necks (S-curved necks). No life was seen in these flasks even after a year, as the shape of the bottle prevented access to airborne microbes because they were trapped due to gravity in the S-curve. He also showed that if the neck is broken, airborne particles lead to microbes in the broth. These observations sealed the debate on spontaneous generation and proved that microbes were airborne and did not appear spontaneously from inanimate objects (Figure 2.2).

Fig. 2.2 The proponents of the theories on the origin of life

Biochemical Evolution and Origin of The First Cell

In the 20th Century, the modern hypothesis that living cells originated in a warm, primitive ocean from inorganic molecules emerged. Haeckel, Oparin, Haldane, Miller, and Urey were the scientists from the 20th Century who contributed to and provided evidence in support of this theory.

Our planet Earth was formed 4.6 billion years ago. Gradual, extensive chemical evolution pre-dated the formation of the first 'cell'. In this *primordial soup theory*, the primitive atmosphere was reducing with very low oxygen levels and abundant energy and gases released from sunlight, lightning, meteorite impact, cosmic irradiation, radioactive material, volcanoes, and geothermal sources, which helped to synthesize stable organic molecules in the liquid water of ancient oceans to produce a diverse range of organic compounds in the pre-biotic (pre, before; biotic, living) period. Continual progressive steps of polymerization then led to the accumulation of macromolecules,

which resulted in the formation of the first living 'protocell' (Figure 2.3). This entire biochemical evolution in the primitive oceans of the Earth dates to approximately 3.7 billion years ago (Table 2.1).

(a) (b)

Fig. 2.3 Biochemical evolution: Macromolecules and cells

Table 2.1 Historical Perspective of Life on Earth

S. No.	Timeline	Event
1.	4500 mn years ago	Planet Earth is formed.
2.	4000 mn years ago	The earliest life as self-replicating RNA is formed.
3.	3700 mn years ago	The existence of cells resembling prokaryotes
4.	3500 mn years ago	Bacteria develop primitive forms of photosynthesis.
5.	2100 mn years ago	More complex eukaryotic cells appear.
6.	600–1000 mn years ago	Multicellular organisms appear.
7.	475–500 mn years ago	Primitive plants move onto land.
8.	450 mn years ago	Arthropods—the first animals to move onto land.
9.	400 mn years ago	The first insects evolve.
10.	150 mn years ago	Giant dinosaurs were common at this time.
11.	65 mn years ago	Major extinctions occur.
12.	50–20 mn years ago	Early reptiles, fish, flowers, and modern mammals evolve.
13.	3 mn years ago	The bipedal Australopithecines (early hominins) evolve in the Savannas of Africa.
14.	195 thousand years ago	Omo 1, Omo 2—the earliest known *Homo sapiens* appear—archaic (Omo river—Ethiopia)
15.	100 thousand years ago	Advanced transitions to modern humans (*Homo sapiens*) appear in Africa.
16.	1 AD	Human population becomes 150 million.
17.	1835 AD	Human population becomes 1 billion.
18.	1969 AD	Humans walk on the moon.
19.	2022 AD	Human population becomes 7.97 billion.

Experimental Insight 2.1 Miller–Urey Experiment

Miller–Urey experiment, in 1953, simulated the early environment of primitive Earth in a warmed flask of water with reducing gases such as methane (CH_4), ammonia (NH_3), and hydrogen (H_2). Simulated lightning by electric sparks and repeated rounds of heating and cooling (for condensation and rains) for one week generated simple organic amino acids in the liquid pool from inorganic raw materials. This established that building blocks of life can be produced without requiring life to synthesize them. Gradually, diversity of these organic molecules increased and proteins, nucleic acids, nucleoproteins, protoribosomes, and protoviruses started to appear on the Earth (Figure 2.4).

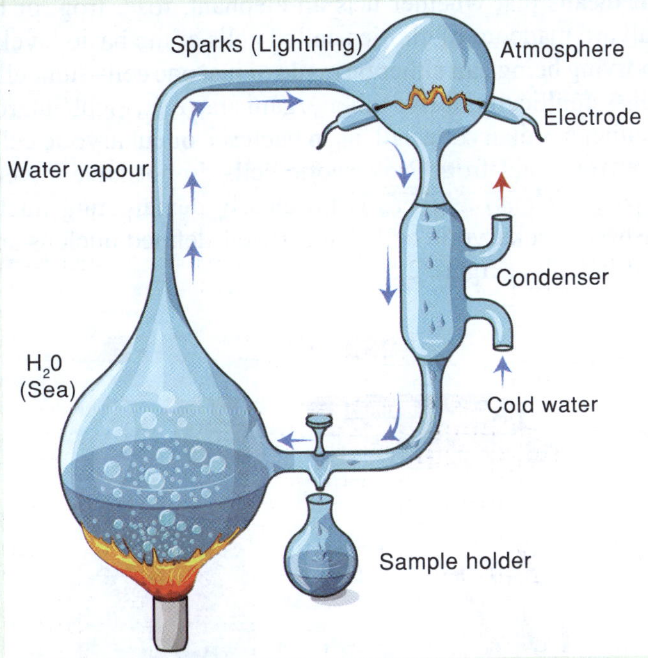

Fig. 2.4 Miller–Urey experiment

The first living cell would then have been formed as aggregates of nucleoproteins surrounded by nutrient shells (Orgel concept) or the mixture of macromolecules aggregated as large colloidal particles (microsphere coacervates) and then surrounded by surface membrane of phospholipids (Oparin concept). The first living cells were called the protobionts or protocells. These protocells then evolved into prokaryotes, eukaryotes, and all forms of living beings.

Characteristics of Living Organisms

Picture the following list of objects—rock, clock, car, frog, cloud, money, rose, elephant, fish, etc.

Now, if you are asked to classify these as living and non-living, without blinking an eye, you will identify clock, car, cloud, and money as non-living and frog, rose, elephant, and fish as living. If we need to define the reason, the simplest would be that living beings are alive and have life. But then, the next question would be, what does it mean to be alive?

As biology is the study of *life and the living*, it becomes imperative to understand what living organisms are. So, what are the characteristics or the common threads of fundamental principles that define living beings and distinguish them from the non-living? When we say the early protobiont was the first living cell, what were the attributes and structures essential to call it living?

Ability to manipulate energy and matter through a common fundamental and structural unit called a *cell* is unique to living systems. Living things have five characteristics that are not found in non-living things.

1. **Unique structural organization** All living beings are made up of cell(s) as the fundamental structural and functional unit. Cells have a limiting membrane with various structures/organelles formed from various biomolecules inside them that perform different functions. It means that whether it is an elephant, rose, frog, or bacterium, or any other organsim, all are made up of entities called cells at the basic level. What differs is that an organism or living being can either be made of just one cell—unicellular (e.g., bacteria)—or of many cells—multicellular like other organisms (e.g., elephant, rose, frog). Similarly, the cell can be either a prokaryotic cell, as in bacteria, or eukaryotic cell, as in larger organisms like elephant, rose, and frog. Prokaryotic cells do not have a well-defined nucleus (pro, before; karyon, nucleus) as there is no clearly demarcating nuclear membrane and are primitive; whereas eukaryotic cells have a well-defined nucleus and are better evolved in structure and function (Figure 2.5).

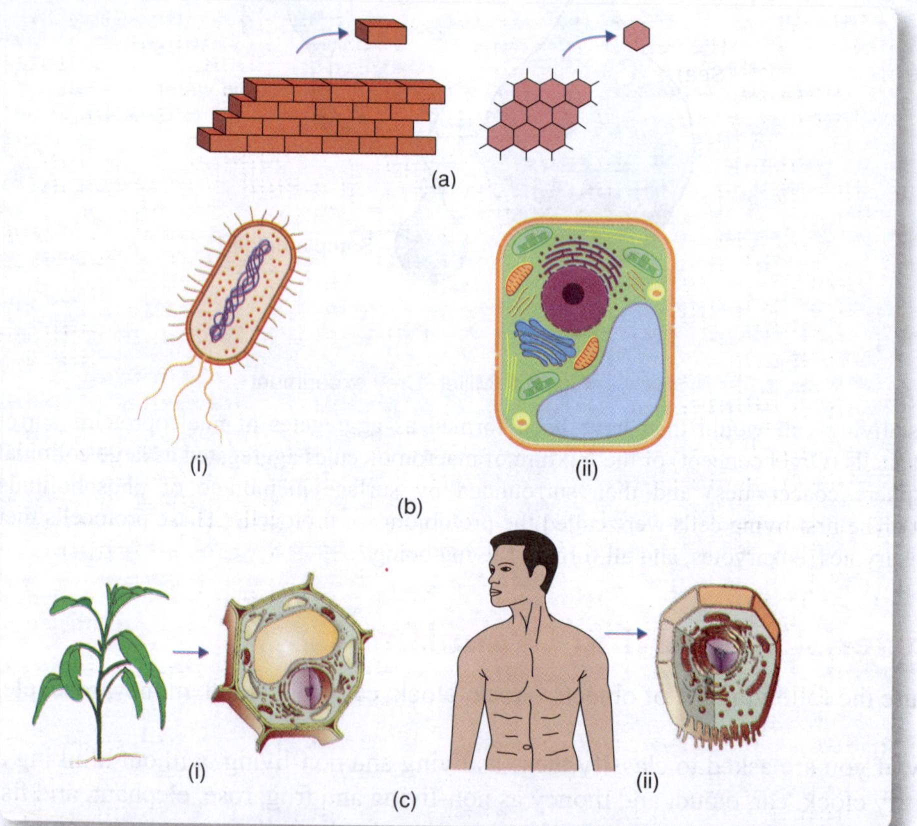

Fig. 2.5 (a) Cell as a basic unit of life; (b) Prokaryotic cell and eukaryotic cell; (c) Plant cell and animal cell

2. **Metabolic activities** A multitude of diverse chemicals inside living beings and their complex chemistry indicates that the ability to manipulate matter and energy is an important characteristic of living beings. Metabolism is the sum of the chemical reactions that help the cell to process nutrients/food in order to provide the diverse range of molecules and to the energy for vital activities to sustain life.

The term *metabolism* is coined from the Greek word *metaballein*, which means 'to turn about, change, or alter'. Cellular metabolic processes build (anabolism) and break down (catabolism) complex molecules in a step-by-step manner. Synthesis and degradation are accompanied by storage and release of energy (Figure 2.6).

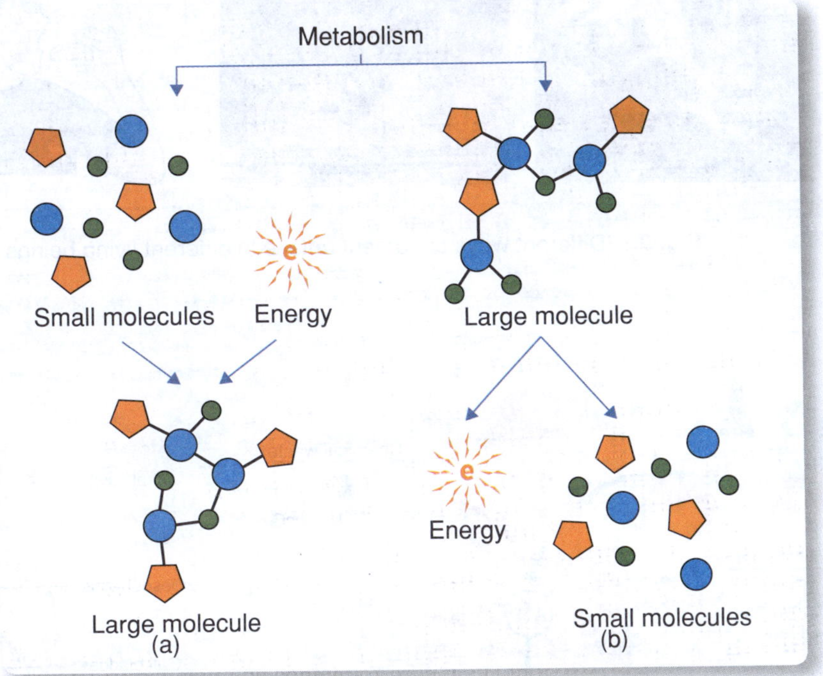

Fig. 2.6 Types of metabolic pathways: (a) Anabolism; (b) Catabolism

Metabolism has three essential facets—nutrient uptake, nutrient processing, and waste elimination. Living organisms take in nutrients from the environment by expending energy by processes such as eating or swallowing (animals), absorption, and ingestion (plants and microbes) (Figure 2.7).

The energy stored in the chemical bonds of these nutrients (raw materials) is processed and manipulated by a network of chemical reactions to yield the molecules and energy requirements for manufacturing new parts, repairs, growth, reproduction, etc. The key players in catalyzing the chemical reactions are enzymes. However, not all materials are valuable to living systems, and portions of nutrients that are not useful or harmful are eliminated as waste. Excess heat generated is also dissipated as a waste product (Figure 2.8).

Fig. 2.7 Different ways of nutrient uptake in different living beings

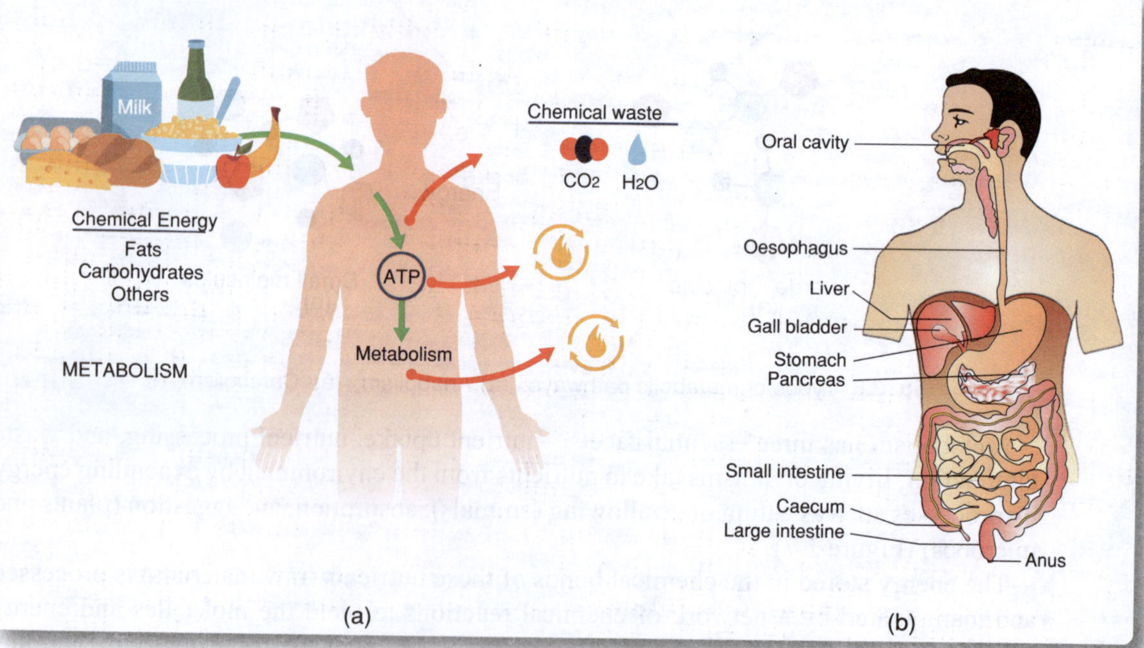

Fig. 2.8 Nutrient processing and waste elimination

3. **Generative processes** Another distinct characteristic of living systems is their ability to grow and reproduce. Growth and reproduction are generative processes. Growth is an increase in the size of an individual. Usually, there is an increase in the number of cells, which adds to the structure. Reproduction is a biological process from which offsprings are produced by one parent or both parents (Figure 2.9). Reproduction is essential to maintain the continuity of species. Growth and reproduction cannot occur without metabolism.

Fig. 2.9 Growth and reproduction

Reproduction can be of two types: asexual reproduction and sexual reproduction. In asexual reproduction, there is involvement of only one parent from whom offsprings are produced, while in sexual reproduction fusion of gametes from two parents of different sexes creates new organism(s). Asexual reproduction is usually seen in single-celled organisms and occurs by different methods such as binary fission (bacteria), budding (yeast), fragmentation (sea star), regeneration (hydra), etc. In the majority of plants, animals, and humans, the natural way of reproduction is sexual reproduction, where the male and female gametes fuse to form a genetically distinct offspring (Figure 2.10).

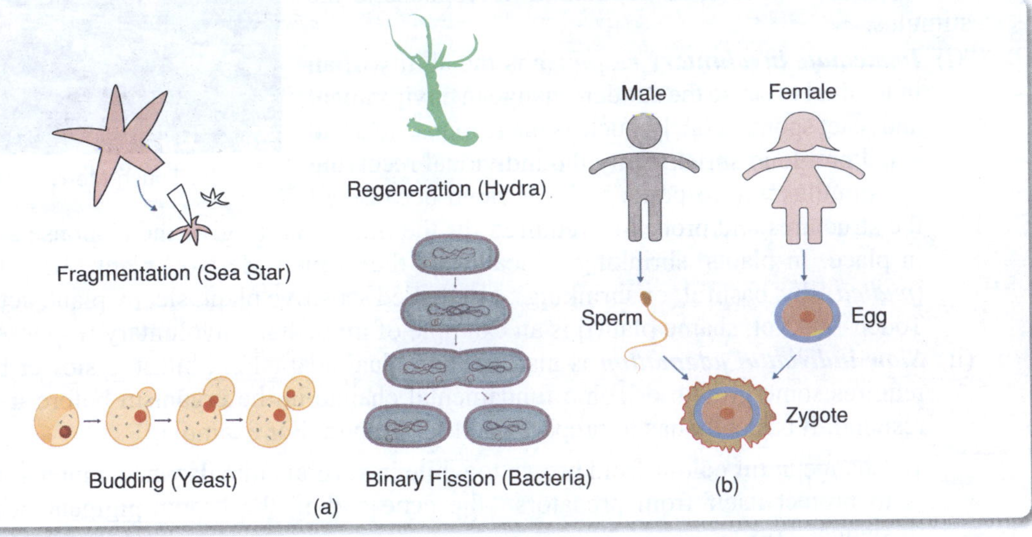

Fig. 2.10 Two forms of reproduction: (a) Asexual; (b) Sexual

Interesting Fact 2.1

While growth and reproduction are distinct in most organisms, in microbes like bacteria, the terms growth and reproduction can be used synonymously. Bacterial growth is the growth of the population and not the size of 'individuals', as it is in other organisms. When we say bacteria are growing, it signifies that these unicellular entities are multiplying, essentially reproducing by division. An individual bacteria is not even visible to the naked eye; what we see is a colony (population) of a huge number of individual bacteria. Interestingly, a biology textbook is the only place where multiplication and division mean the same!

4. **Responsive processes** Reacting to stimulus has been the hallmark of living beings. The immediate environment has an influence on the living being, and an organism reacts to changes in the environment. Chemotactic responses in microbes, phototactic responses in plants, and stings by animals in response to threat are a few obvious examples.

   The responses from different living beings for the same stimulus may vary. For example, if you push a bird, it will fly; if you push a frog, it may jump, if you push a dog, it might growl or attack, and so on. Similarly, we see different human beings reacting differently to the same stimulus—if you tap on someone's shoulder, someone might like it, finding it friendly; another might get offended, etc. A morning alarm also triggers different responses in different individuals—one wakes up, the other ignores, another snoozes the alarm, and so on.

   Response by living beings is in itself a complex enigma that cannot be curtailed to universal, predictable mechanistic laws. Observing the adaptation of living beings in nature and analysing the same to understand the fundamental tenets is one of the interesting facets of biology.

Living organisms can respond in three different ways—(i) a spontaneous response, which is an immediate, involuntary response by an individual, (ii) a delayed meaningful change at the individual level, or (iii) a gradual shift or change in population in response to the stimulus.

(a) Touch-Me-Not plant

(i) ***Immediate involuntary response*** is the ability of an individual to sense the sudden change in environment and react spontaneously, such as the response to a bad smell or a loud shriek. Only the individual receiving the stimulus will respond, and it is rapid because all the structures and processes required for the individual to give the response are already in place. In plants, shrinking of leaves of the 'Touch-Me-Not' plant [*Mimosa pudica* (*pudica* 'shy, bashful, or shrinking'; also called sensitive plant, sleepy plant, action plant, Touch-Me-Not, shame plant)] is an example of immediate involuntary response.

(ii) ***Slow individual adaptation*** is also an individual adaptation, but it is slower because it requires some growth or some fundamental change in the organism before it is able to respond. It basically has to prepare itself to respond. For example:

1. Change in fur colour from brown to white in several animals when winter approaches, to protect itself from predators—the gene coding the brown pigment needs to be switched off.
2. Body's response to low oxygen levels—by rapid and deeper breathing and increase in RBC levels.
3. Body's immune response to microbial infection—exposure to pathogen leads to immune changes and hence production of attacking cells, which neutralize the pathogen.

(b)

(iii) ***Gradual shift in population*** is when there is an even slower change in the genetic make-up of a population over generations. The more suitable individuals to the stimulus are selected by natural selection, and the process is called *evolution*. *Evolution is a population adaptation process that enables species to adapt to long-term changes in the environment*. Only this type of response has a change in the genetic make-up and leads to a directional change in characteristics of species or populations over several generations. For example, drug-resistant microbes, insecticide-resistant moths, evolution of human brain, etc. (Figure 2.11).

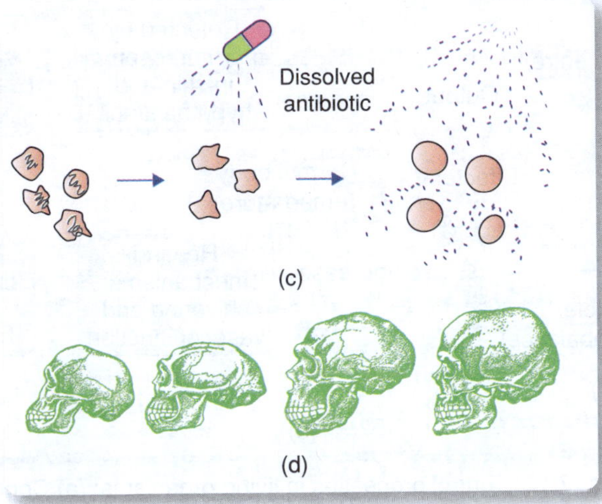

Fig. 2.11 Responsive processes in living organisms: (a) *Mimosa pudica* (pudica 'shy, bashful, or shrinking'; also called sensitive plant, sleepy plant, action plant, Touch-Me-Not, shameplant); (b) Change in fur colour in rabbit; (c) Evolution of population antibiotic resistance; (d) Human skull evolution—brain size

5. **Control processes** Orderliness in structure, function, processes, and activities is another important feature that contributes to the complex design of living beings. Living organisms have mechanisms called control processes that ensure that all life activities happen in the right sequence (coordination) and at the right speed/rate (regulation).

The metabolic reactions are coordinated and linked together into specific pathways to ensure an efficient stepwise handling of nutrients. This is achieved with the help of enzymes that catalyze the reactions in the right sequence and rate and, thus, control the amount of nutrients processed. Coordination occurs at the organism level too. In a six-legged insect, during walking, the muscles of the legs are coordinated (contract and expand), and similarly in plants, regulatory chemicals ensure a proper sequence of events resulting in an orderly sequence of flowering and fruit development (Figure 2.12).

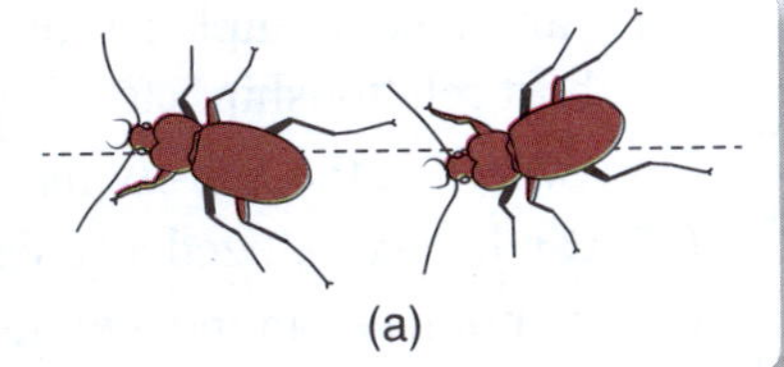

Regulation involves controlling the rate of processes. Many activities inside living organisms are interrelated and regulated. Such regulation is important for the body to maintain a constant internal environment called *homeostasis*. Homeostasis refers to an organism's ability to regulate various

physiological processes to keep its internal state steady and balanced. For example, in case of intense physical activity, O_2 is used up rapidly, and its level falls and CO_2 level rises. To maintain a constant level of these gases, rapid breathing ensues with forceful pumping of the heart—Breathing Regulation. The two activities occur simultaneously at the right time at the correct rate to get adequate blood to the lungs for O_2–CO_2 exchange, thus ensuring the required levels in the blood. Similarly, to regulate the body temperature, one sweats when one is too hot and shivers when one is too cold—Temperature Regulation. These are examples of homeostasis.

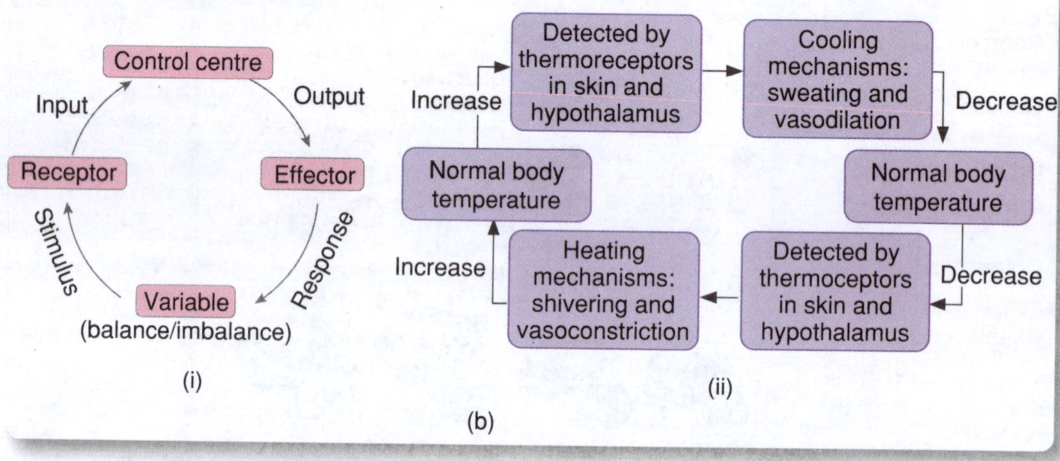

Fig. 2.12 Control processes in living organisms: (a) Coordination of muscles during walk; (b) Homeostasis

Thus, living beings show the earlier-mentioned five characteristics that non-living things do not possess. Although these characteristics are fundamental to all living organisms, they need not be present in every living organism at all points of time. For example, individuals grow and reproduce only at certain times, but the ability to do so, even for some part of life, is important! This knowledge goes beyond our basic understanding of the differences between the living and the non-living—to eat, breathe, live, grow, reproduce, or die.

Virus: Living or Non-Living?

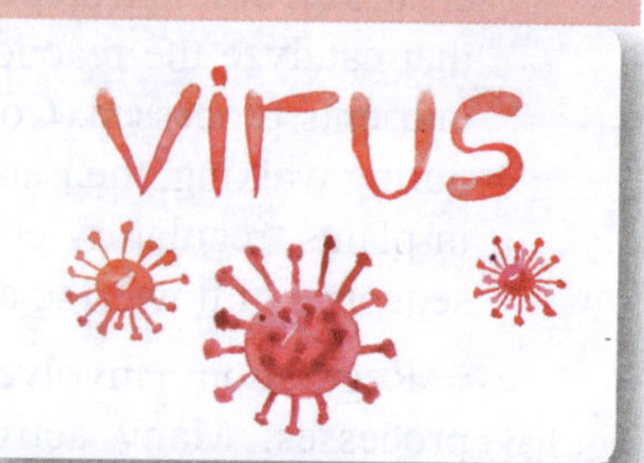

Viruses are at the borderline of living and non-living! . They show some characteristics of living, such as, they:

(a) multiply, but only when inside a living cell
(b) have genetic material but do not have a cellular structure
(c) can cause complex infections and show a pathogen–host relationship, etc.

However, like the non-living, they:

(a) can be crystallized as chemicals,
(b) do not have an independent existence
(c) are not cells, have no growth, and do not show metabolism of their own.

Table 2.2 Virus Versus Cell

Characteristic	Virus	Cell
Structure	Simple DNA or RNA core, capsid	Complex internal structure with cell membrane, cytoplasm, mitochondrion, nucleus, and organelles
Size	Size grows small, 20–300 nm	much larger 1–100 nm
Reproduction	Only within a host cell	Independent cell division, either asexually or sexually
Genetic code	DNA or RNA	DNA
Growth and development	No	Yes, in mutlicellular organisms, cells increase in number and differentiate
Obtain and use energy	No	Yes
Change over time	Yes	Yes

Hence, whether viruses are living or non-living is debatable; and they are, therefore, called 'acellular infectious particles.' A comparison of acellular infectitious particle virus with cell is given in Table 2.2.

Interesting Fact 2.2

Are these living?? Let's analyse .... Are the characteristics good enough?

Interestingly, the automobile or car takes in nutrients in the form of fuel, processes, and gives out waste products. Thus it carries out one of the important characteristics of a living organism. It also has automation and control and is responsive to some extent. But it is not living! It fails in other features such as generative processes, and ability to grow and divide, which are missing.

Objects like money and cloud have the ability to grow and reproduce (generative) to some extent; but they would still not qualify to be called living as they fail in metabolic, responsive, and control processes.

Some automated objects like thermostats and automatic washing machines are responsive to temperature changes and programmed to respond to different activity cycles, respectively, maybe with great precision like living but again fail in generative processes.

All these objects fail and do not have the basic unique structural organization—Cell. The most important fact is that non-living things do not have the same basic structural and functional unit called cell, which is true in all living organisms. Elephant, frog, rose plant, eagle—all at the fundamental level are made up of cells. But non-living obects like steel, metal, wood, plastic, glass, etc., do not have the same fundamental unit—cell.

Experimental Insight 2.2 Cell as the Basic Unit of Life

Cell is the basic, fundamental unit of life. How did they know this?? Because of their small size, they cannot be seen with naked eye. The advent of the field of microscopy opened up possibilities to observe all organisms at the cellular level and realize that fundamentally all organisms have the basic unit as a cell.

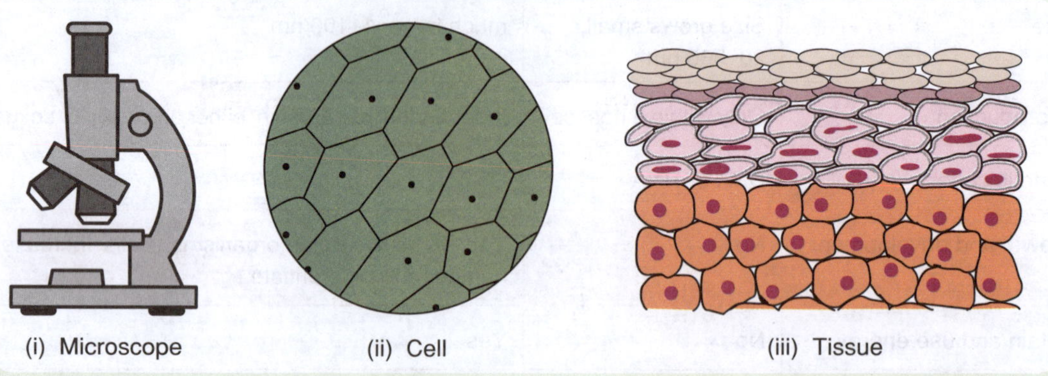

(i) Microscope (ii) Cell (iii) Tissue

Antonie van Leeuwenhoek observed the first living cells and called them *animalcules*. Robert Hooke coined the term 'cell' while observing the thin slice of dead cork. The cell theory was first proposed by Schleiden and Schwann. Matthias Schleiden was a German botanist who first proposed, in 1838, the idea that all plants consist of cells. Theodor Schwann was a German zoologist who independently asserted that all animals and plants are made up of cells. Rudolph Virchow stated that cells arise from previously existing cells.

Leeuwenhoek

Robert Hooke

Schleiden

Schwann

Virchow

SUMMARY

- Three theories about the origin and evolution of life—Panspermia Theory, Theory of Spontaneous Generation, and Biochemical Evolution Theory.
- Urey–Miller experiment was done to prove the Theory of Biochemical Evolution.
- Historic developments of life on Earth with timelines
- Characteristics of life are common structural organization and cells and types of cells.
- Metabolism involves nutrient uptake, processing, and waste elimination for energy and matter changes.
- Generative processes involve growth and reproduction.
- Types of reproduction—sexual and asexual reproduction

- Responsive processes could be spontaneous individual response, individual adaptation, and population adaptation or evolution.
- Control processes are coordination and regulation—homeostasis.
- Homeostasis is an internal process to stabilize the body against sudden changes by regulating physiological processes.

EXERCISES

A. Objective-Type Questions

1. Panspermia theory about the origin of life proposes that
   (a) Life was created by God and implanted onto Earth.
   (b) Life came from outer space.
   (c) Life emerged spontaneously from inanimate objects.
   (d) Life began from inorganic to organic molecules in the primordial soup.
2. Planet Earth was formed approximately ---------.
   (a) 4.6 billion years ago (b) 0.4 billion years ago
   (c) 4.6 trillion years ago (d) 4.6 thousand years ago
3. The generic term used to refer to the first living cells on Earth is
   (a) Eukaryotes. (b) Fungi.
   (c) Protobionts or protocells. (d) Maggots.
4. Identify the right sequence of events in the evolution of life on Earth.
   (a) Self-replicating RNA – bacteria – eukaryotes – multicellular organisms – plants – human beings
   (b) Eukaryotes – bacteria – self-replicating RNA – multicellular organisms – plants – humans
   (c) Humans – eukaryotes – bacteria – self-replicating RNA – multicellular organisms – plants
   (d) None of the above
5. All biochemical reactions required for a cell to grow, reproduce, and undergo repairs are collectively termed as ___________.
   (a) Anabolism (b) Catabolism (c) Metabolism (d) Reproduction
6. The two major components of generative processes are
   (a) Growth and respiration. (b) Growth and reproduction.
   (c) Reproduction and respiration. (d) Reproduction and metabolism.
7. It is anticipated that the polar ice caps will start melting if global warming continues, slowly turning the Earth into a huge body of water with no land. In such a situation, over some time, organisms with swimming/aquatic skills will survive while others will become extinct. Such a phenomenon can be termed as ___________.
   (a) Ecosystem (b) Evolution
   (c) Bioremediation (d) Population explosion
8. Two scientists who have contributed to cell theory are
   (a) Watson and Crick. (b) Needham and Redi.
   (c) Pasteur and Spallanzani. (d) Schleiden and Schwann.

B. Short-Answer Questions

1. For entities as follows, identify one characteristic of life that each one of them fails to fulfil to be classified as living.
   (a) A humanoid (b) COVID-19 virus
   (c) Cloud (d) Mercedes Benz car

2. Use one word each to describe the five hallmarks of life.
3. 'Life gives rise to life!' Do you agree with this statement? Justify.
4. Discuss the different theories proposed for the origin and evolution of life on Earth.

C. Self-Study Questions

1. Read about the creation and functioning of humanoids such as Optimus, and Ameca, and their features. Evaluate with your knowledge gained.

FURTHER READINGS

1. Dill, K.A. and L. Agozzino, Driving forces in the origins of life, *Open Biology*, Vol 11, Issue 2, Feb 2021.

USEFUL WEBLINKS

1. https://assets.cambridge.org/97805216/80547/excerpt/9780521680547_excerpt.pdf
2. https://qsstudy.com/mention-the-living-and-non-living-characteristics-of-virus/

VIDEO LINKS

1. https://www.generationgenius.com/videolessons/living-vs-non-living-things-video-for-kids/

Diversity of Living Organisms—Taxonomy and Phylogeny

LEARNING OBJECTIVES

After reading this chapter, you will be able to:

- Understand biodiversity and the levels of organization of living things
- Know what taxonomy and the scientific way of naming organisms is
- See how the 3-domain, 6-kingdom classification was arrived at and what it is
- Learn taxonomic categories along with their examples
- Understand phylogeny and evidences that help in phylogenetic classification
- Know the specific characteristics of the domains Bacteria, Archaea, and Eukarya
- Study the acellular pathogens not conforming within the classification system
- Grasp the benefits of model organisms in understanding life

CHAPTER TAKEAWAY

How organising, naming, and classification helps to catalogue and study large diverse groups of organisms on Earth

This chapter presents an insight into organising, naming, and classification that helps to catalogue and study large diverse groups of organisms on the Earth, as biodiversity of living beings on our planet is immense.

PRELUDE

All around, we see a great variety of living organisms; from plants, to microbes, to insects, to birds, and a variety of animals that inhabit, both, land and water. To know how many species are there on the Earth is one of the most important questions in science. Latest estimates indicate that there are 8.7 million eukaryotic species on this planet, of which only 1.2 million, i.e., less than 15% have been identified and classified. To identify, regroup, and recognize various organisms, a systematic study is required. It is also important to have a clear scientific nomenclature for the identified species, as common names used locally vary from culture to culture and a unique identification may be hampered. Study of characteristics of living organisms can be done using the hierarchical levels of organization of living organisms–from atoms to the entire biosphere. After classifying organisms based on their phenetic and phylogenetic features, the understanding of each group of organisms with their specific characteristics is critical. Model organisms identified in each category help to delineate important aspects of their functioning. Outliers such as unique infectious agents that are acellular and do not conform to the groups of classification systems add new dimension and intrigue to the entire field of biology. Thus, taxonomy and phylogeny have emerged as two important aspects of biology. These studies help us to understand and maintain the biodiversity of the ecosystem, on which humanity depends!

Diversity of Living Organisms

When one looks around, even within a small stretch of nature—a nature park, ocean, forest, etc.—one can clearly see a wide variety of life on Earth. Scientists have estimated that the approximate number of different species on the Earth today could be as high as 30 million. Of which, a total of about 2 million living species have already been identified and new species are being discovered all the time. Biodiversity is the term used to refer to the enormous diversity of living beings on the Earth. This diversity is further exemplified by the enormous genetic variation within individual species and their diverse interactions with each other and with their environment (Figure 3.1).

Fig. 3.1 Diversity of living beings on Earth

Levels of Organization of Living Things

Study of living organisms follows a hierarchy that helps to examine life on a scale from small to large. The natural flow and structure of organization of life scales from an atom to the entire biosphere. The complexity at higher levels is composed of things at lower levels. Biological systems and their impacts are interlinked to some extent. The levels of organization for living

things start with atom → molecule → organelle → cell → tissue → organ → organ system → individual organism → population → community → ecosystem → biosphere.

The definition of each and their interrelationships is given in Table 3.1, as follows.

Table 3.1 Hierarchical Levels of Biological Organization

Level of Organization	Characteristic	Example	Remark on a Larger Context
Biosphere	Worldwide ecosystem	All regions of Earth—crust, water, and atmosphere—that sustain life	All ecosystems of the universe
Ecosystem ↑	Community and its physical environment	Forest, ocean, valley	Interplay of both biotic and abiotic factors
Community ↑	Populations of different organisms occupying the same area	Populations of trees, insects, birds, fungi, bacteria that interact at any location	Individual populations interact to become communities
Population ↑	Group of individuals of the same kind occupying the same area	Human population, tiger population	Global tiger population is about 5500 in 2023[1]
Individual organism ↑	Multiple organ systems functioning together to form an independent living unit	Human, mushroom, plant, yeast, bacterium, amoeba	Some organisms are uncellular and some are multicellular
Organ system ↑	Two or more organs interacting for action	Circulatory system, nervous system	Circulatory system contains blood, blood vessels, and heart for transport of blood
Organ ↑	Collection of multiple types of tissues for a specialized function	Brain, heart, liver, eye	An organ may have nervous tissue, connective tissue, blood vessels
Tissue ↑	Organized aggregation of cells with a particular function	Nerve tissue, muscle tissue, layers of skin	Function together in a specalized activity
Cell ↑	Smallest unit with characteristics of life	Prokaryotic cell, nerve cell, muscle cell, heart cell	Smallest unit—independent existence (unicellular) or part of a larger structure (multicellular)
Organelle ↑	Parts of cells with specific function	Ribosome, mitochondria, nucleus	More the organelle, more complex the structure and function of cells
Molecules ↑	Unit of two or more atoms bonded together	CH_3, CO_2 Specific arrangement of atoms	Living things have bio-molecules like proteins, carbohydrate, DNA
Atom ↑	Smallest unit of an element	Hydrogen, oxygen	Retains the characteristic of that element

Taxonomy and Naming of Organisms

Taxonomy is the science of describing, naming, and classifying organisms. The term 'taxonomy' comes from the Greek word 'Taxis', which means arrangement. Taxonomy forms the

[1]WWF Tigers Alive Initiative

basis to catalogue and understand biodiversity at all levels and in all niches, which,in itself, remains one of the biggest challenges of the 21st century. To understand the overwhelmingly diverse set of living organisms and to describe them or talk about them, it became imperative to name each one of them. During the Middle Ages, Latin was used as a scientific language of choice, and any new organism or species discovered was named in Latin. This system ensured that irrespective of native language, if descriptions were made in Latin, people across the world understood that it would be a species name. However, there were a few challenges that still persisted:

(a) Latin word descriptions for some organisms were long (15 words or so) and difficult to comprehend.
(b) As communication was slow, there could be multiple Latin names for the same species.
(c) Common colloquial names remained, making the situation even more difficult.

Therefore, a proper code for nomenclature, which was simpler yet robust and scientific to describe biological species, was necessary. A binomial (two-name) system of nomenclature was introduced by Carolus Linnaeus in the 10th edition of *Systema Naturae*. Every organism was then named with two epithets or words—one for genus and the other for species. Conventionally, the unique scientific name given to each organism was written with the first letter of the generic name in capital letters and all the other letters in small. Scientific names are to be underlined or italicized.

Some examples depicted in the form of genus and specific epithet are as follows:

- *Panthera leo*—lion
- *Felis catus*—house cat
- *Canis familiaris*—domestic dog
- *Canis lupus*—wolf
- *Homo sapiens*—humans

Classification of Living Organism

Classification of living organisms can be done on the basis of many characteristics, such as cellular nature (e.g., unicellular and multicellular), morphology/structure (e.g., prokaryotes and eukaryotes with their organelle differences), functional diversity (e.g., autotrophs and heterotrophs), biochemical variation (e.g., ability to metabolize proteins and nucleic acids with nitrogenous waste products, like ammonia—ammonotelic, urea—ureotelic, and uric acid—uricotelic), ecological niches (e.g., terrestrial, aquatic), etc.

Modern systematics uses the evolutionary relationships (phylogeny) with findings from molecular biology, cytology, genetics, and biochemistry to arrive at the systematic classification of living organisms. Modern systematics evolved from the initial two-kingdom classification by Linnaeus (i.e., plants and animals) to three-kingdom classification by Haeckel (i.e.,Animalia, Plantae, and microbes in Protista) and subsequently to four- and five-kingdom classifications by Copelan (i.e., Animalia, Plantae, Protista, and prokaryotic Monera) and Whittaker (i.e., Monera, Protista, Fungi, Plantae, and Animalia), respectively (Figure 3.2).

In 1990, Carl Woese proposed the six-kingdom classification system under three domains of Bacteria, Archaea, and Eukarya. This classification was based on the differences at molecular level in their 16S rRNA, cell membrane structure, and sensitivity to antibiotics. The major difference in this classification method was the splitting of Eubacteria and Archaebacteria into separate domains.

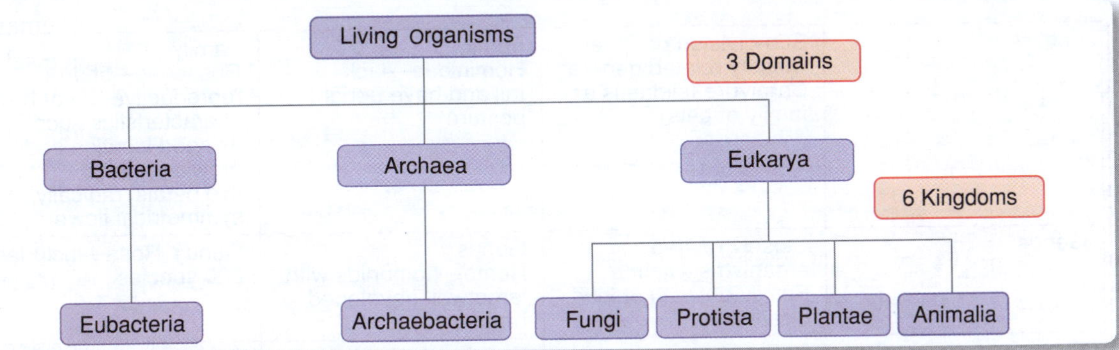

Fig. 3.2 Six-kingdom classification with three domains, Carl Woese (1990)

Carolus Linnaeus, in addition to the binomial nomenclature, also developed the hierarchical classification system to organize and place organisms into logical groups (Figure 3.3; Table 3.2) and his contributions credited him to be the 'Father of Taxonomy'. In the hierarchical classification system, there are seven basic categories of hierarchy within each domain, which are as follows:

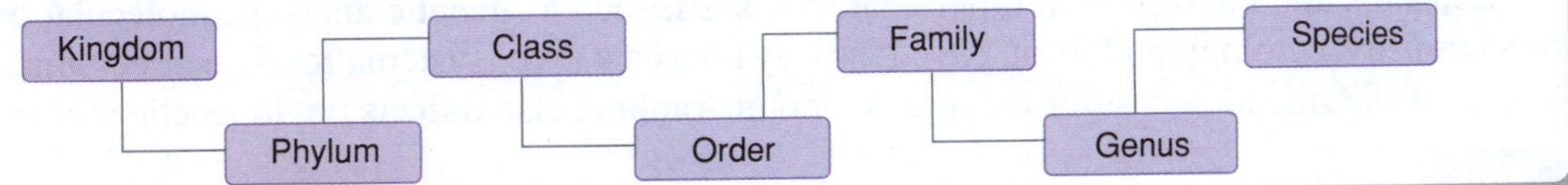

Fig. 3.3 Hierarchical classification system

Interesting Fact 3.1

Table 3.2 Taxonomic Categories and Their Definitions with Examples

Taxonomic Category	Taxon Feature	Example 1: Classification of Humans	Example 2: Classification of Rose Plant
Domain	Highest hierarchical classification	Domain Eukarya—eukaryotic cells with nucleus and organelles	Domain Eukarya—eukaryotic cells with nucleus and organelles
Kingdom	Sub-division of Domain Eukarya has four kingdoms while bacteria and archaea have one each	Kingdom Animalia—heterotroph, motile with specialized tissue	Kingdom Plantae—non- motile, multicellular, contain chlorophyll, and autotrophs
Phylum or division	Sub-division of Kingdom for plants, instead of phylum, division is used All kingdoms have more than one phyla	Phylum Chordata—stiffening rod vertebrae in the back	Division Magnoliophyte—broad-leaf flowering plants
Class	Sub-division of Phyla Birds, reptiles, mammals, fishes are classes in phylum Chordata of Animalia	Class Mammalia—hair and mammary glands	Class Magnoliopsida—dicots with net-like veins
Order	Sub-division of Class, Carnivora—meat eating within class Mammalia	Order Primates—large brain and opposable thumbs	Order Rosales—with thorns or stiff hairs and the fruits are dry fruits, drupes, or aggregates

[*Contd*]

Family	Sub-division of Order Closely related genera Carnivore felidae is a family of cats	Family Hominidae—lack tail and have upright posture	Family Rosaceae—shared reproductive characteristics such as seed location and growing habits five petals, radically symmetrical flowers
Genus	Closely related organisms within a Family but with unique distinctions	Genus Homo—homonids with large well-developed brain	Genus Rosa—includes 200 species
Species	Different groups within a Genus, each differing in one or few traits	Species sapiens—all existing humans on this planet	Species rubiginosa, alba, etc.

The evolution of the classification system from two to six kingdoms progressed from phenetic classification (based on determinative identification—obvious external similarities) to phylogenetic (reflecting genetic similarities and evolutionary relatedness). Phylogeny—study of evolutionary relationships and coupled with other characteristics, such as genetic analysis, molecular biology, biochemistry, cytology, and so on, paved the way for a new era of systematics. The three-domain, six-kingdom classification ensures that the taxonomic ranking also reflects phylogenetic relationships.

Tips

Best learning outcomes in biology! Bacteria are the oldest forms of life on Earth. The engulfing of one bacterial species by another led to the rise of eukaryotes containing cell organelles such as mitochondria and chloroplasts.[2]

To achieve this, evolutionary biologists use several lines of evidence available, such as the following (Figure 3.4):

(a) ***Fossils*** are remnants of history as physical evidence preserved in nature. The discovery and study of fossils help scientists to carbon-date them and study organisms in a time sequence by linking the intermediate evolutionary changes with the extremes to understand the descendants. However, fossils are limited by their availability and may not be able to define all species.

(b) ***Comparative anatomy*** Similar structures indicate relatedness among organisms and vice-versa.

(c) ***Life cycle information*** Similarities in the complex stages of life cycle could be used as another line of evidence.

(d) ***Biochemical studies*** Biochemical similarities such as the presence of unique molecules between two species indicates relatedness.

(e) ***Genetic and molecular analysis*** Ribosome analysis, variations in chromosome numbers, structure, behaviour during meiosis, and variations in genes and proteins, all of which adds to the phylogenetic classification.

The current view of the tree of life encompassing the total diversity of sequenced genomes is a continuing endeavour.

[2] Ricklefs, Ecology: The Economy of Nature, 8e, 2018 © W.H. Freeman and Company.

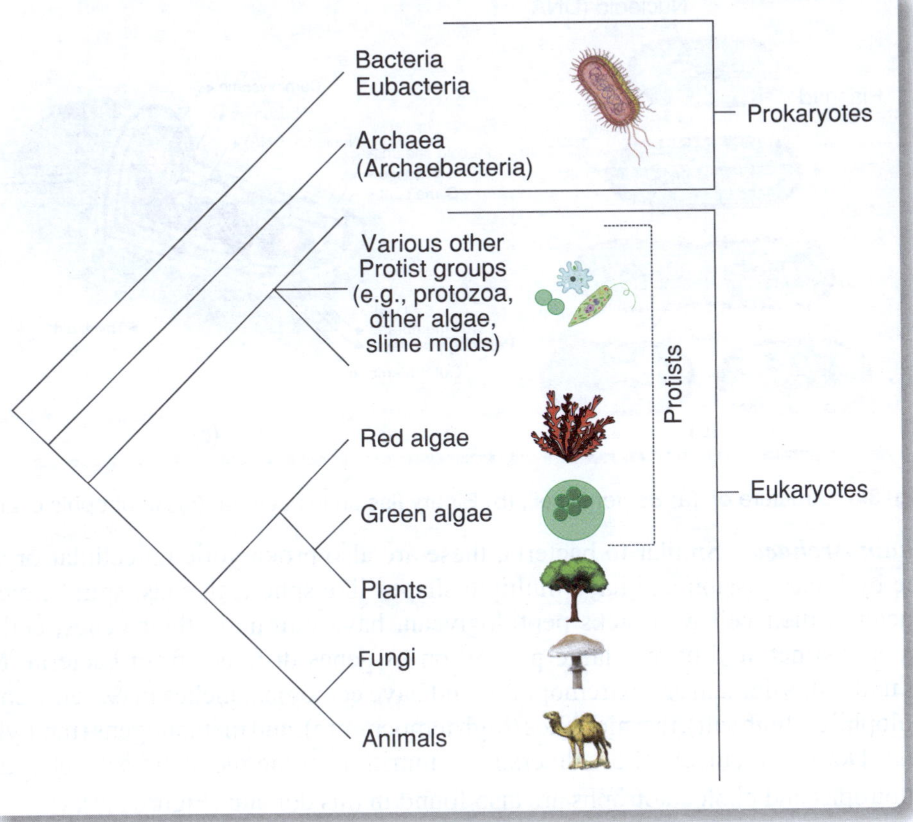

Fig. 3.4 The evolution of life on Earth

Source: Adapted from Fig. 1.6, Ricklefs, *Ecology*: *The Economy of Nature*, 8e, 2018 © W.H. Freeman and Company.

Characteristics of Three Domains: Bacteria, Archaea, and Eukarya

A broad understanding and characterization of three domains—two prokaryotic in nature and one eukaryotic in terms of cellularity—is as follows (Figure 3.5):

Domain Bacteria This domain includes single-celled, small, prokaryotic organisms (no well-defined nucleus), existing in different shapes, such as cocci (spherical), bacilli (rod shaped), and spirals, having peptidoglycan in their cell wall, reproduce asexually by binary fission, and have locomotory organelle as flagella. Based on oxygen requirement, the domain includes members that are aerobic, anaerobic, or facultative. Most in this domain are heterotrophs, the exception being cyanobacteria (autotrophs). Bacteria have a broad niche and are observed to survive in all habitats and environments.

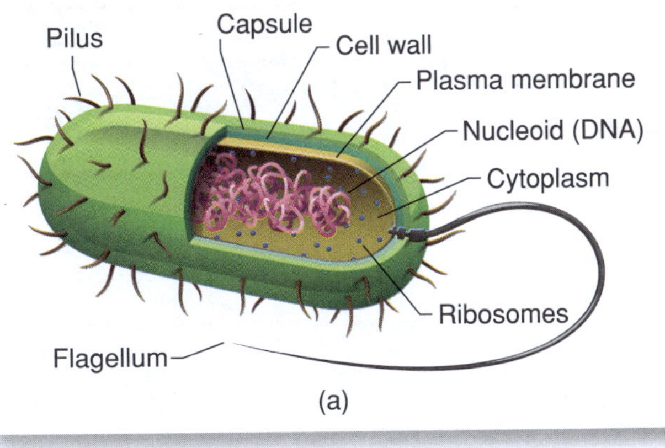

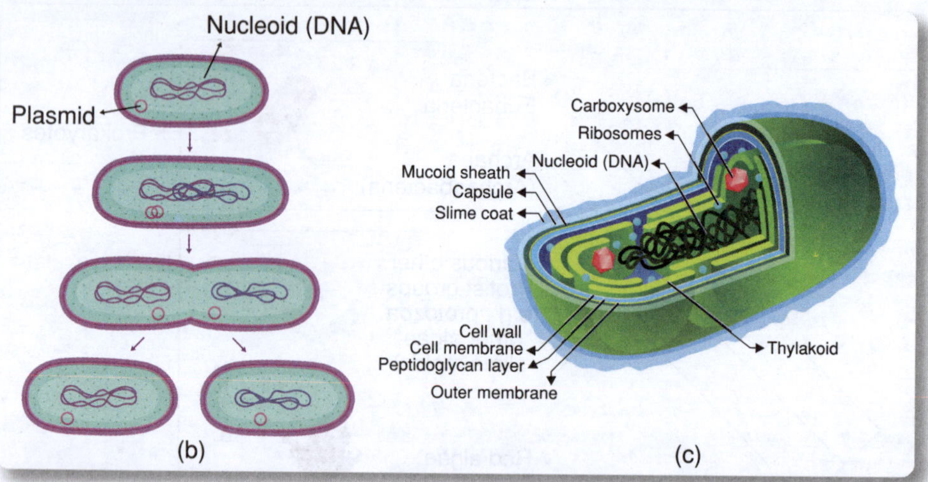

Fig. 3.5 Structure of: (a) Bacterial cell; (b) Binary fission in bacteria; (c) Autotrophic cyanobacteria

Domain Archaea Similar to bacteria, these are also prokaryotic unicellular organisms that reproduce by binary fission and have multiple shapes like spherical, rods, spirals, etc. However, unlike bacteria, their cell wall lacks peptidoglycan, have introns in their genes, cell membrane chemistry is distinct, and have a large proportion of genes different from bacteria. Most of the members under this domain are extremophiles and have ecological niches in extreme environments such as halophiles (high salt), thermophiles (high temperature), and methanogens (methyl-producing anaerobes). Domain members show diversity in nutrition; some members are heterotrophs while chemoautotrophs and photoautotrophs are also found in this domain (Figure 3.6).

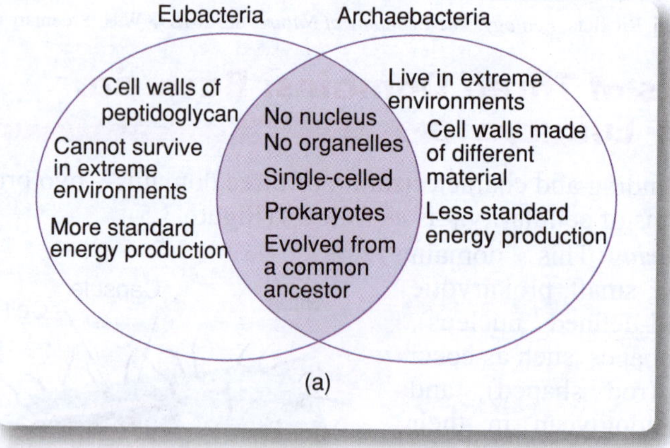

Archaebacteria

- Ancient bacteria
- Rigid cell wall
- Habitat
  - High salt concentration
  - High temperatures
  - Oxygen-free marshy areas

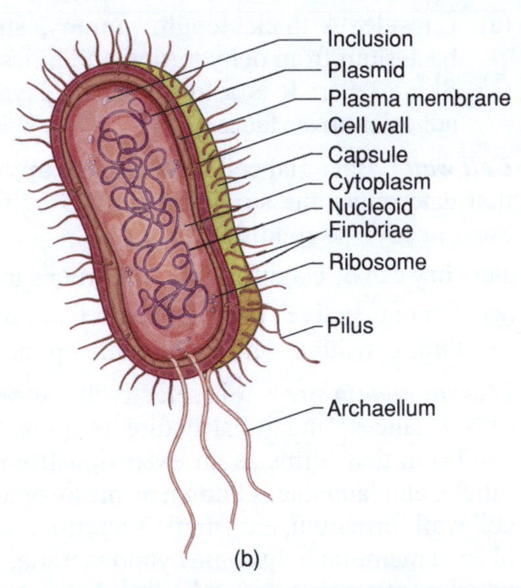

Fig. 3.6 Archaeabacteria: (a) Differences with eubacteria; (b) Extremophiles

STRUCTURE OF BACTERIA

Bacteria are unicellular organisms lacking a true nucleus and cell organelles. They are associated with the prokaryotic family having a cell wall surrounding the cell membrane, a fluid-filled matrix i.e. cytoplasm, no well-defined nucleus, and genetic material is not surrounded by a nuclear membrane, rather it is present in nucleoid region in cytoplasm. Besides that, some small circular DNA are also present except genomic DNA (single chromosome/circular DNA), which provides a particular morphological feature like resistance to antibiotics. Bacteria can be classified on the basis of shapes like bacilli (rod-shaped), spirilla (spiral), coccus (sphere), and vibrio (comma-shaped).

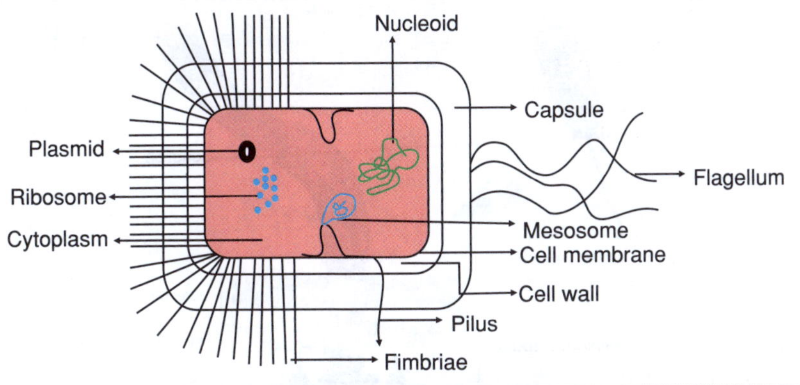

Fig. 3.7 A generalised structure of bacteria

Bacterial cells are enclosed by a chemically complicated cell envelope that consists of three layers:

1. ***Outermost layer*** Glycocalyx is the outermost layer made up of polysaccharides and proteins that help in cell adhesion and its protection from surrounding. It can be of two different types:

(a) Capsule: A thick, tough, gummy, sticky, and well-defined layer that protects the bacterium from dehydration and hides from host immune responses.
(b) Slime layer: It is a loose sheath layer with a lower organisation that helps in cell adhesion to surfaces.

2. ***Cell wall*** Bacterial cell wall is made up of peptidoglycan (murein or mucopeptide) layer that determines the structural rigidity of the cell, preventing the cell from collapsing or burst in adverse conditions.

Further, they can be classified into two groups in response to the staining procedure of gram stain.
(a) Gram negative: Those bacteria that do not take up the gram stain are gram negative.
(b) Gram positive: Those who take up the gram stain are gram positive.

3. ***Plasma membrane*** It is selectively permeable in nature, which regulates the movement of substances, and its structure is quite similar to eukaryotes. A unique membranous extension that forms as an extension from the plasma membrane in the form of vesicle, tubule, and lamellae is known as mesosome, which acts as a site for respiration, helpful for cell wall formation, enzymatic secretion, replication, and increasing the surface area of the plasma membrane. In some cyanobacteria, some other types of extensions are also present, which contain pigments called as chromatophores imparting their blue-green colour due to a pigment called phycobilin, which helps in the generation of ATP and NADPH and plays an important role in capturing light energy.

SPECIAL SURFACE STRUCTURE OF BACTERIA

Bacterial cells are very peculiar in nature. Some of them show motility (mobile) while others are non-motile (immobile).

1. ***Flagella*** It is a filamentous, hip-like structure extending from its cell wall that is composed of flagellin protein. It is composed of three parts called as filament (longest portion extending from cell surface to outside), hook (connect filament to base), and basal body (rod-like structure consisting of rings).

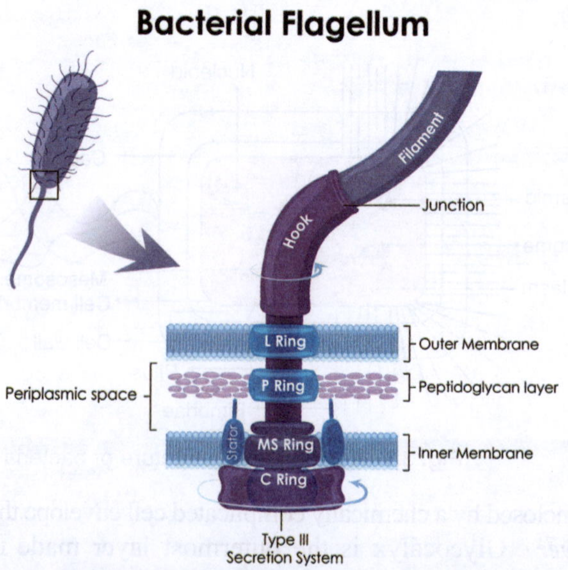

Fig. 3.8 Structure of bacterial flagella

The flagella also act as a rotatory motor for the easy movement of bacteria in the forward direction. It moves using a proton motive force, which acts as an electrochemical gradient instead of using high-energy sources like ATP for different ions like H^+ and Na^+ to pass across the cell membrane pushing the rotor. Later, the rotor spins, turning the flagella into a propeller (an apparatus having blades to push water forward for the bacterial movement). If it rotates in a counterclockwise direction, the bacteria move in a straight line; otherwise, in an anticlockwise direction, the bacteria move in a random way by spinning. The basal body is made up of four different rings, i.e., MS-at base region, PL-distal region, for gram-negative stain and two rings (S and M) for gram-positive stain with a rod in between them. The LP ring is responsible for the rod rotation; the C ring is composed of FliM and FliN protein, which helps in changing direction, i.e., clockwise or counterclockwise, and the MS ring acts as a foundation upon which other rings rely and further helps in the rotation of motor. Additionally, another ring called the T ring (MotX, MotY) is also responsible for different functions of the motor.

Other surface structures of bacteria like pili and fimbriae are also present, which are not responsible for cell mobility. The pili are tubular structures responsible for the attachment of bacterial cells to a surface made up of a special protein called pilin. True pili or sex pili play a role in the transfer of DNA (conjugation process) from donor to recipient cell. The fimbriae are bristle-like fibres of 3–10 nm in diameter composed of helically arranged protein subunits emerging out of the cell that help in the attachment of bacteria to rocks in streams and also to host tissues.

Ribosomes They are 15 nm × 20 nm in size and are associated with the plasma membrane of the cell and formed of two different subunits (50S and 30S), which together form 70S prokaryotic ribosome. Its main function is protein synthesis, and several ribosomes may attach to a single mRNA, forming a chain called polyribosome or polysome, which further translates mRNA into proteins.

Inclusion bodies Non-membrane bound system, which acts as a reserve material in prokaryotic cells lying freely in the cytoplasm, e.g., phosphate granules, cyanophycean granules, and glycogen granules. Additionally, gas vacuoles are present in blue-green, purple, and green photosynthetic bacteria.

Domain Eukarya The common thread in domain Eukarya is the eukaryotic cell with a well-defined nucleus in all organisms. According to the chemical theory of evolution, it is believed that eukaryotic cells emerged from prokaryotes through endosymbiosis. Organelles such as mitochondria and chloroplasts were independent cells that were engulfed by larger cells (endo-) and became part of their existence due to a mutually beneficial interaction (symbiosis). This cell created therein became the forerunner of eukaryotes. This is evident as organelles such as mitochondria and chloroplasts have their own DNA, 70S ribosome, analogous to prokaryotic cells, and are double-membraned structures—one from the original symbiont and the other is the host vacuole membrane that engulfed it. Due to the integration of these and other membranous organelles, eukaryotic cells are much larger in size than prokaryotic cells.

Domain Eukarya consists of four distinct kingdoms; they are as follows.

(i) ***Protista*** These are eukaryotic, single-celled organisms, a very diverse group of organisms. This kingdom includes plant-like autotrophs (algae), animal-like heterotrophs (protozoans), and fungi-like heterotrophs (slime molds). Some organisms reproduce asexually, but sexual reproduction is also seen in some members of this group. All groups within protists respire aerobically. These are a broad and diverse group of eukaryotes distinct from the other three kingdoms of eukarya and, thus, considered as a valid taxonomic category.

(ii) ***Fungi*** Non-motile, multicellular eukaryote (only member yeast is a unicellular fungi), with a cell wall made of chitin, heterotroph, and important decomposers of the ecosystem. Most have saprophytic nutrition, few are parasitic, and some are mutualistic. They show

both sexual and asexual modes of reproduction, and the structurally multicellular fungi have a network of hyphae with spores. Unicellular fungi—yeast is a facultative aerobe, important for fermentation reactions, and divides asexually by budding. Fungi are further classified into different phyla (Figure 3.9).

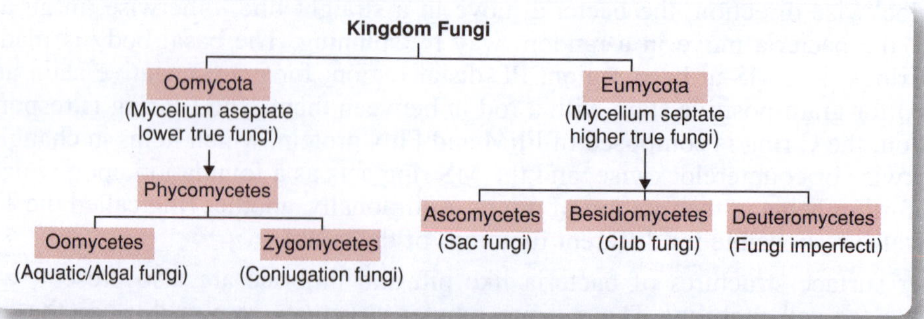

Fig. 3.9 Classification of Fungi

(iii) ***Plantae*** Non-motile eukaryotes, multicellular, contain chlorophyll, synthesize their own organic matter by photosynthesis, and have cell wall made of cellulose. Two lines of plants evolved from green algae—vascular plants (development of vascular—water-transporting tissue) and non-vascular plants. Vascular plants further evolved into seed-producing plants, under which the flowering plants and conifers are distinct lines (Figure 3.10). More than 300,000 species of plants have been identified, out of which about 85% are flowering plants, 14% are mosses and ferns, and the remaining 1% are conifers.

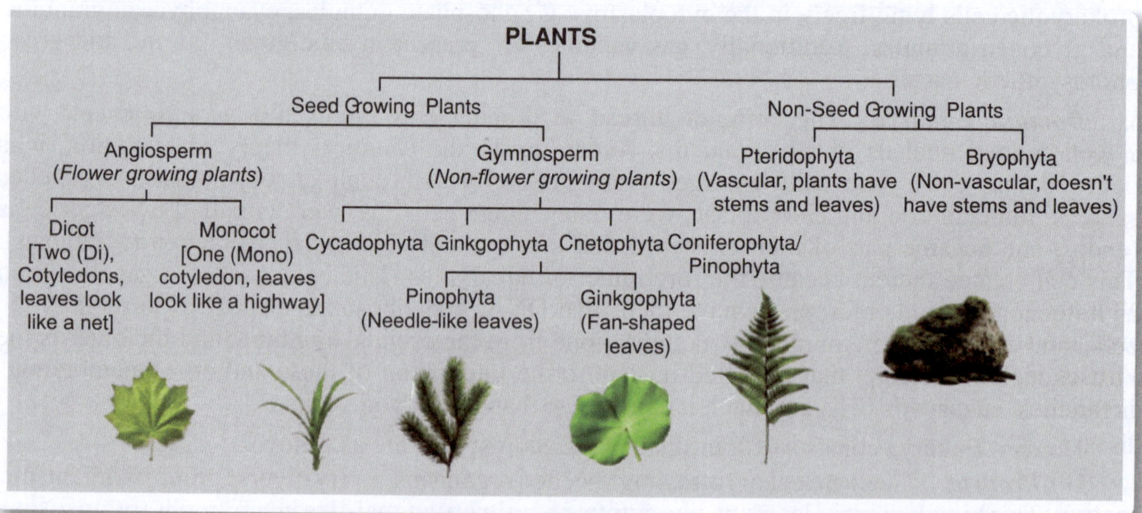

Fig. 3.10 Evolution of plants and different categories of plants

(iv) ***Animalia*** The largest taxonomic group—over a million species! Huge diversity in size, from microscopic animals to gigantic elephants and whales. Eukaryotic, heterotroph, multicellular, and motile (exceptions like sponges and corals are sessile as adults). Most reproduce sexually, though some lower forms may have asexual reproduction too. Lower

forms of animals lacked specialized tissues, but many have evolved to develop specialized tissues, organs, and organ systems, and this has contributed to the diversity in animals that we see (Figure 3.11).

Fig. 3.11 Diversity of Animalia

The six kingdoms discussed so far include cellular forms of life. However, there are some particles that are acellular but are capable of causing diseases, making them important to study and understand. Viruses, prions, and viroids fall under this group (Figure 3.12).

Acellular Infectious Agents

Viruses are obligate intracellular parasites made up of a nucleic acid core surrounded by a protein coat (capsid). They are considered to be non-living because they do not reproduce on their own and show some characteristics of life only when they are inside a host (other living beings). Viruses infect and use as hosts almost all species of organisms found on Earth. Viruses are of different types and classified on the basis of parameters such as genetic material, host specificity, disease pathogenicity, size, and shape, etc.

Viroids are small, infectious strands of RNA that are mostly plant pathogens. *Prions*, on the contrary, are infectious proteins that can be passed on from one organism to another and cause diseases such as the transmitted spongiform encephalitis. Viruses and viroids can be replicated within the cells that they infect, but prions cause deformation of existing proteins to cause the disease.

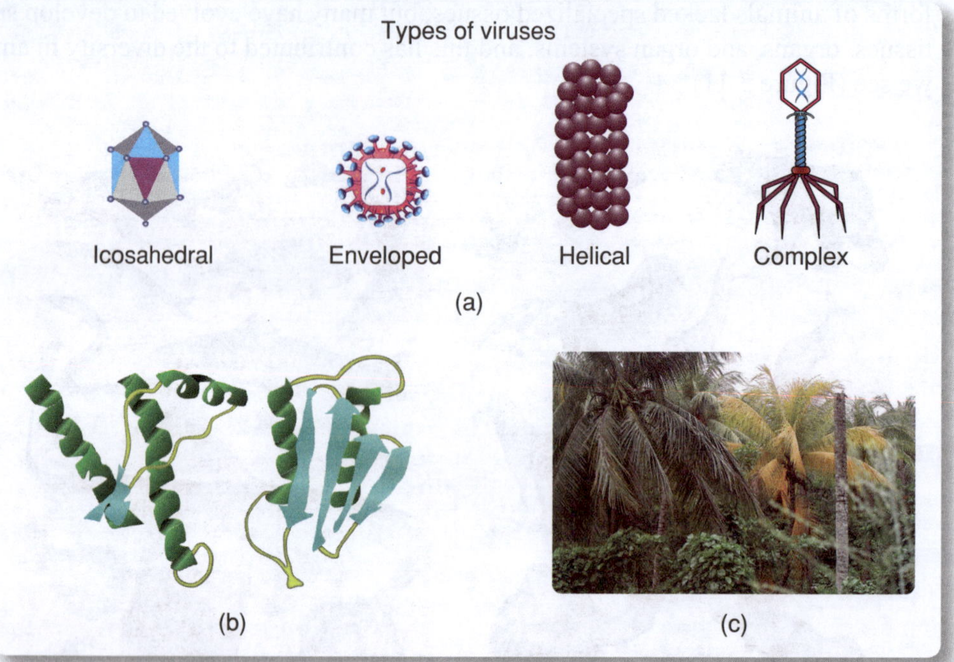

Fig. 3.12 Acellular infectious particles: (a) Viruses; (b) Prion proteins; (c) Plants infected with viriods

Model Organisms

Among the living organisms studied so far, there are a few that are identified and listed as 'model organisms'. Model organisms are those that are widely used for research as their biology is well known, are easy to maintain in laboratory settings, have short generation time, and have particular experimental advantages. There are three categories of model organisms commonly used for various studies; they are as follows:

Mammalian models

- Mouse (*Mus musculus*) and Rat (*Rattus norvegicus*) are major models for genetic research because of the ease with which the genome could be manipulated and their genetic and physiological similarities to humans.

Non-mammalian models

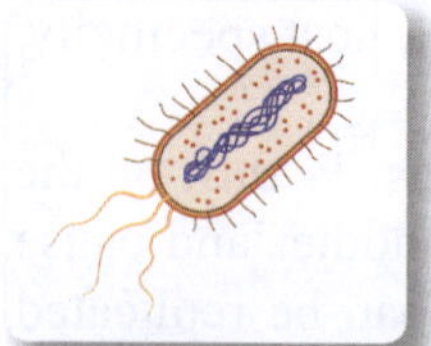

- Bacterium (*Escherichia coli*) Single-celled organism used for studies on basic biochemistry and molecular biology. An understanding of DNA replication, protein synthesis, genetic code, etc., have emerged from studies on *E.coli*, easily amenable to genetic manipulation, and, hence, sought after for recombinant DNA studies.

- Yeast (*Saccharomyces cerevisiae*) The simplest single-celled eukaryote used for human genetics and disease research.

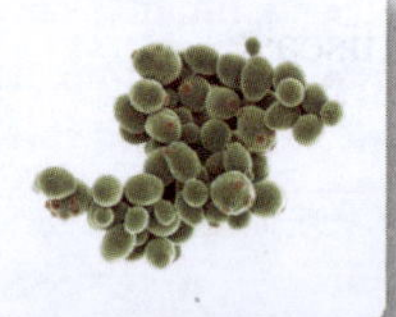

- Nematode (*Caenorhabditis elegans*) A small transparent worm that shares genes and pathways with humans, with genome size and number of genes close to humans, and ideal for studies at the molecular level.

- Fruit fly (*Drosophila melanogaster*) Its genome is 60% homologous to humans, produces a large number of offspring and new lines could be generated quickly, and is used to study genetics and developmental biology.

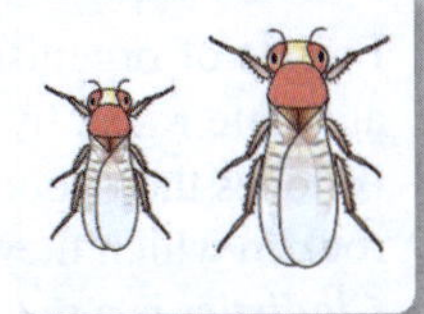

- Zebra fish (*Danio rerio*) A small tropical freshwater fish that is easy to breed. Many human genes have counterparts in zebrafish, which is used for studying developmental biology and human diseases.

- Frog (*Xenopus laevis*) Amphibian embryology model, vertebrate, ease of gaining access to eggs and embryos, and used for biochemistry and developmental biology studies, chromatin studies, inhibitor studies, and microscopy.

Plant models

- *Arabidopsis thaliana* For plant biology research, this small flowering plant is widely used as a model for genetics and molecular biology research, primarily due to its small genome size and ease of producing mutations.

Interesting Fact 3.2

Extinction is a major part of the history of our Earth.

- At least 680 vertebrate species have gone extinct since the 16th century.
- Nearly 33% of reef-forming corals face the threat of extinction.
- There has been a 68% average decline in the population sizes of mammals, birds, amphibians, reptiles, and fishes between 1970 and 2016.
- By 2016, more than 9% of all domesticated breeds of mammals used for food and agriculture had gone extinct, and at least 1000 more spices remain threatened.
- The current observed extinction rate is 2.6 vertebrate species per 10,000 yearly.
- The distributions of 47% of land-based flightless mammals and almost a quarter of threatened birds may already have been negatively affected by climate change.
- Climate change poses a threat to over 4161 species.
- Current negative trends in biodiversity and ecosystem show that we will not meet 80% of the assessed targets of the Sustainable Development Goals concerning poverty, hunger, health, etc.

Experimental Insight 3.1 Phylogeny

Conservation biology is an important application of taxonomy and phylogeny. It entails a comprehensive understanding of the existing biodiversity, its ecological and environmental origins, and systematic efforts to maintain the biodiversity for the benefit of humans.

Fossil evidence of phylogeny

Fossils of organisms are found in sedimentary rocks and sediments of water bodies. It is possible to age-date rocks by comparing the radioactive isotopes they contain. Older rocks have less radioactive isotopes than younger rocks. Fossil age could be ascertained, as fossils are usually of similar age as the rock in which they are embedded.

Cladistics is a method of phylogenesis used to assess the degree of relatedness among organisms within a group based on their genetic similarity. The name emerges from branches of a family tree 'clades'. This tool assumes that a common group of organisms emerges from a common ancestor; changes in characteristics are common over time; and the differences that arise indicate the branching points.

Steps include: list the characteristics of the group being studied, determine the characteristics that are ancestral (by comparing with a close out-group) and those that are derived, and, then, create a branching diagram (the cladogram).

Cladistics bunches creatures by determined characteristics; hence, selection of characteristics is a critical component, and it should compare and reflect from the right evolutionary group and from the same genetic background. For example, emergence of wings of butterflies may not be a valid comparison with wings of birds. Only homologous similarity is evidence that two species are evolutionarily related. An example of a cladogram is as follows:

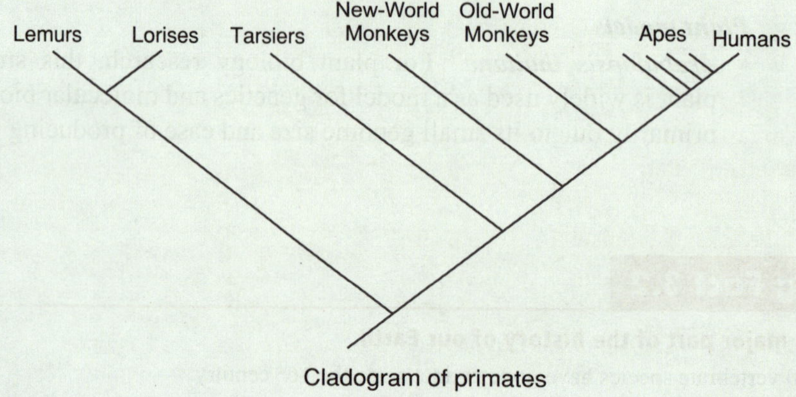

Cladogram of primates

SUMMARY

- Study of living organisms follows a hierarchy: The levels of organization of living things start with atoms → molecules → organelle → cell → tissue → organ → organ systems → individual organisms → population → community → ecosystem → biosphere.
- Taxonomy is a science of describing, naming, and classifying organisms.
- The convention of naming organisms: Every organism was named with two epithets or words—one for genus and the other for species. The unique scientific name given to each organism was written with the first letter of the generic name in capital letters and all the other letters in small, with the full scientific name underlined or italicized.
- Six-kingdom classification with three domains was proposed by Carl Woese (1990).
- In the hierarchical classification system, there are seven basic categories/hierarchies within each domain, from kingdom to species.

- Phylogeny is the study of evolutionary relationships, and coupled with other characteristics, such as genetic analysis, molecular biology, biochemistry, cytology, etc., they paved the way for a new era of systematics.
- Each kingdom has unique characteristics that are used for classification. A detail of each kingdom and its characteristics is explained.
- Details of acellular infectious particles such as viruses, viroids, and prions (not part of the 6-kingdom classification) are provided.
- Model organisms used for biological research are described.

EXERCISES

A. Objective-Type Questions

1. Which of these is not an acellular infectious particle?
   (a) Prion (b) Viroid
   (c) Virus (d) Algae
2. Thermophiles and halophiles belong to
   (a) Plantae (b) Archaebacteria
   (c) Eubacteria (d) Fungi
3. Which of these is not a 'model organism' for research?
   (a) *Escherichia coli* (b) *Mus musculus*
   (c) *Panthera leo* (d) *Arabidopsis thalliana*
4. Humans are classified under phylum Chordata because they
   (a) have hair and mammary glands. (b) have a stiffening rod called vertebrae.
   (c) contain chlorophyll. (d) are prokaryotic.
5. Which is the right sequence in hierarchical classification?
   (a) Kingdom – phylum – class – order – family (b) Phylum – order – class – kingdom – family
   (c) Kingdom – phylum – order – class – family (d) Family – class – order – phylum – kingdom
6. Which of these is not a unicellular organism?
   (a) *Saccharomyces cerevisiae* (b) *Escherichia coli*
   (c) Amoeba (d) *Mus musculus*
7. Which of the following is not used in phylogenesis as evidence for classification?
   (a) Fossils (b) Comparative anatomy
   (c) Molecular and genetic data (d) All of the above
8. Archaebacteria and eubacteria, both
   (a) are prokaryotic. (b) have peptidoglycan in cell wall.
   (c) are extremophiles. (d) show absence of ribosomes.
9. Scrapie and Kuru are diseases caused by
   (a) HIV (b) RNA
   (c) Prions (d) Fungi
10. Under which kingdom will you classify a multicellular, non-motile, autotrophic eukaryote with chloroplast?
    (a) Protists (b) Fungi
    (c) Animalia (d) Plantae

B. Short-Answer Questions

1. Identify one example of mutual support and cooperation among different species that led to the evolution of a larger group.
2. There is a lot of media focus on the endangerment and extinction of some species such as tiger, house sparrow, etc. Is this bias valid, and should it be true for organisms from other kingdoms as well?
3. Viruses are not made up of cells! Identify the theories on the likely origin of viruses.
   Hint: Refer to the 4th paper in Further Readings.
4. Viruses replicate whereas other organisms reproduce! Explain the distinction.

FURTHER READINGS

1. Hug, L., Baker, B., Anantharaman, K. et al. A new view of the tree of life. Nature Microbiology, 1, 16048, Apr 2016.
2. Lipscomb, D. Basics of cladistic analysis. George Washington University, 1998.
3. Sweetlove, L. Number of species on Earth tagged at 8.7 million. *Nature*, 2011.
4. Agol, V.I. (2008). The Origin and Evolution of Viruses. In: Skjeltorp, A.T., Belushkin, A.V. (eds) Evolution from Cellular to Social Scales. NATO Science for Peace and Security Series B: Physics and Biophysics. Springer, Dordrecht. https://doi.org/10.1007/978-1-4020-8761-5_7.

USEFUL WEBLINKS

1. https://www.trvst.world/biodiversity/biodiversity-facts-statistics/
2. https://www.britannica.com/science/phylogeny/Taxonomic-systems

CHAPTER 4

Biochemistry of Life

LEARNING OBJECTIVES

After reading this chapter, you will be able to:

- Understand the biomolecules of life
- Learn how biomolecules are synthesized and broken down
- Present the structural details and functions of four major macromolecules of life—carbohydrates, lipids, proteins, and nucleic acids
- Get acquainted with the significance and purpose of these macromolecules to life
- Realise the critical role of water and its special properties for life

CHAPTER TAKEAWAY The chemistry behind what makes life tick

This chapter presents the biomolecules of life, i.e., the elements that all matter is made up of. Living systems are composed of various types of molecules formed from these elements. The branch of biology that deals with the study of chemistry of living things is called 'Biochemistry'.

PRELUDE

The hierarchical organization of life and the characteristics of living organisms have clearly indicated that we, like all plants, animals, and other organisms, are collections of atoms–molecules interlinked by chemical bonds. Life is a chemical process, and the chemical nature of life makes it important to gain the knowledge of basic chemistry and chemical interactions for understanding the depth of biological existence. The chemistry that deals with life is called Biochemistry. The study of life involves learning about the structure and function of organisms, what they are made up of, how their energy requirements are met, and the answer to all these questions lie in the interface of chemistry and the associated changes. From the oxygen we breathe, the food we eat, the waste we secrete, these are all in some way outcomes of chemical changes. Organic compounds with hydrogen and other elements covalently bonded to carbon are the primary molecules of life. In nature, living cells synthesize biomolecules such as carbohydrates, lipids, proteins, and nucleic acids, and these molecules serve to uphold the characteristics of life. Living systems assemble, rearrange, and degrade these organic macromolecules through the process of metabolism to meet their needs. Understanding the biochemistry of living systems and the biological molecules is a key aspect to study.

Biomolecules of Life

Living systems are composed of various types of molecules formed from these elements. The branch of biology that deals with the study of chemistry of living things is called 'Biochemistry'. The main chemical elements in living systems that make these biomolecules are Carbon (C), Hydrogen (H), Oxygen (O), Nitrogen (N), Phosphorus (P), Sulphur (S), and Calcium (Ca). Out of

these, C, H, O, and N account for 96% of biomass in all living organisms. These elements make up inorganic compounds such as water, inorganic salts, as well as organic macromolecules like carbohydrates, lipids, proteins, and nucleic acids. [Table 4.1]

At the core of organic molecules is a carbon skeleton composed of rings or chains of carbon along with specific functional groups attached to the carbon skeleton. The nature of the carbon skeleton and the functional groups account for the chemical properties of the molecule. *Macromolecules* (macro → large) are very large organic molecules made up of smaller sub-units. Some of these large macromolecules exist as *polymers*, which are combinations of smaller building blocks, called *monomers*, bonded together. Examples of polymers include carbohydrates, proteins, and nucleic acids. Lipids, on the contrary, are macromolecules but not polymers. Some of these macromolecules are synthesized by the body, and some of them are obtained from dietary sources.

Table 4.1 Average Composition of a Cell

Component	Percentage of the Total Cellular Mass
Water	70–90
Proteins	10–15
Nucleic acids	5–7
Carbohydrates	3
Lipids	2
Ions	1

In macromolecules, the individual sub-units are joined together by *dehydration synthesis* reactions with the removal of water molecules (de → remove; hydro → water; synthesis → joining). This reaction occurs when two smaller sub-units come together and an —OH group is removed from one and H is removed from another, which then combine to form a new water molecule. The remaining two segments then combine to form a larger molecule. Dehydration synthesis reactions can be recognized by removal of water molecules and presence of multiple small molecules on the reactant side and large molecules on the product side [Figure 4.1(a)]. These reactions are also called *polymerization* and *condensation* reactions.

The reverse reaction is known as hydrolysis, where a large organic molecule is split into two or more component parts with the addition of water molecules (Figure 4.1(b)).

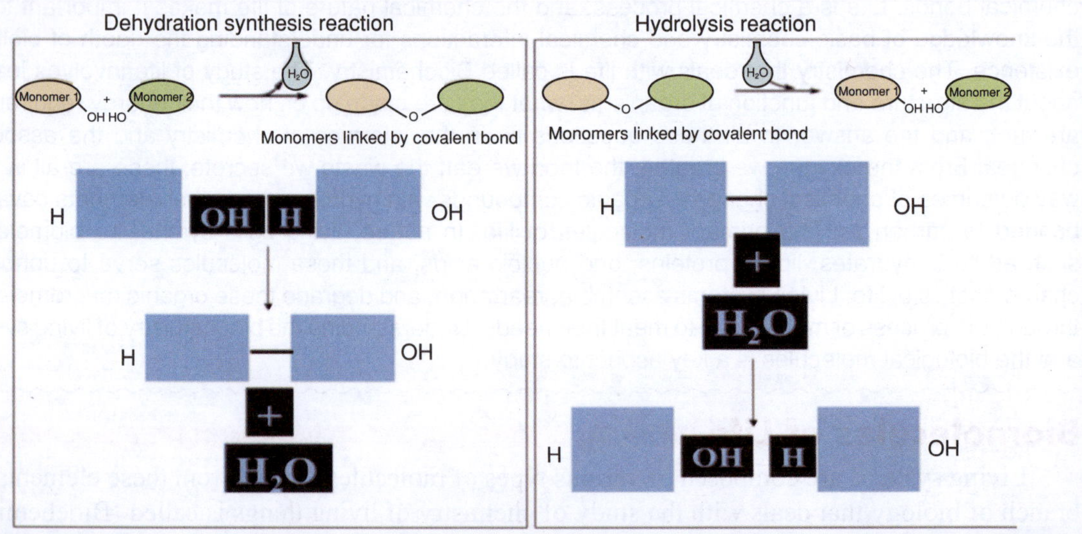

Fig. 4.1 (a) Dehydration synthesis; (b) Hydrolysis

Carbohydrates

Monomers of carbohydrates are the simple sugars composed of carbon, hydrogen, and oxygen atoms linked together in the ratio CH_2O—equal numbers of carbon and oxygen and twice the number of hydrogen. For example, $C_3H_6O_3$, $C_6H_{12}O_6$, etc. Depending on the number of carbon atoms present in simple sugar, sugars are called triose, tetrose, pentose, hexose, and so on. Individual simple sugar is called a *monosaccharide* (e.g., glucose, fructose, galactose), two sugars joined together form a *disaccharide* (e.g., lactose, maltose, sucrose), and multiple sugar units linked together form *polysaccharide* (e.g., starch, cellulose, glycogen).

Individual monomers are joined together by dehydration synthesis reaction. The covalent bonds that join sugars together to form carbohydrates are called the *glycosidic* bonds. Some monosaccharides can undergo modifications to form derived monosaccharides such as deoxyribose, gluconic acid, glucuronic acid, glucosamine, and so on, which also are useful to organisms (Figure 4.2).

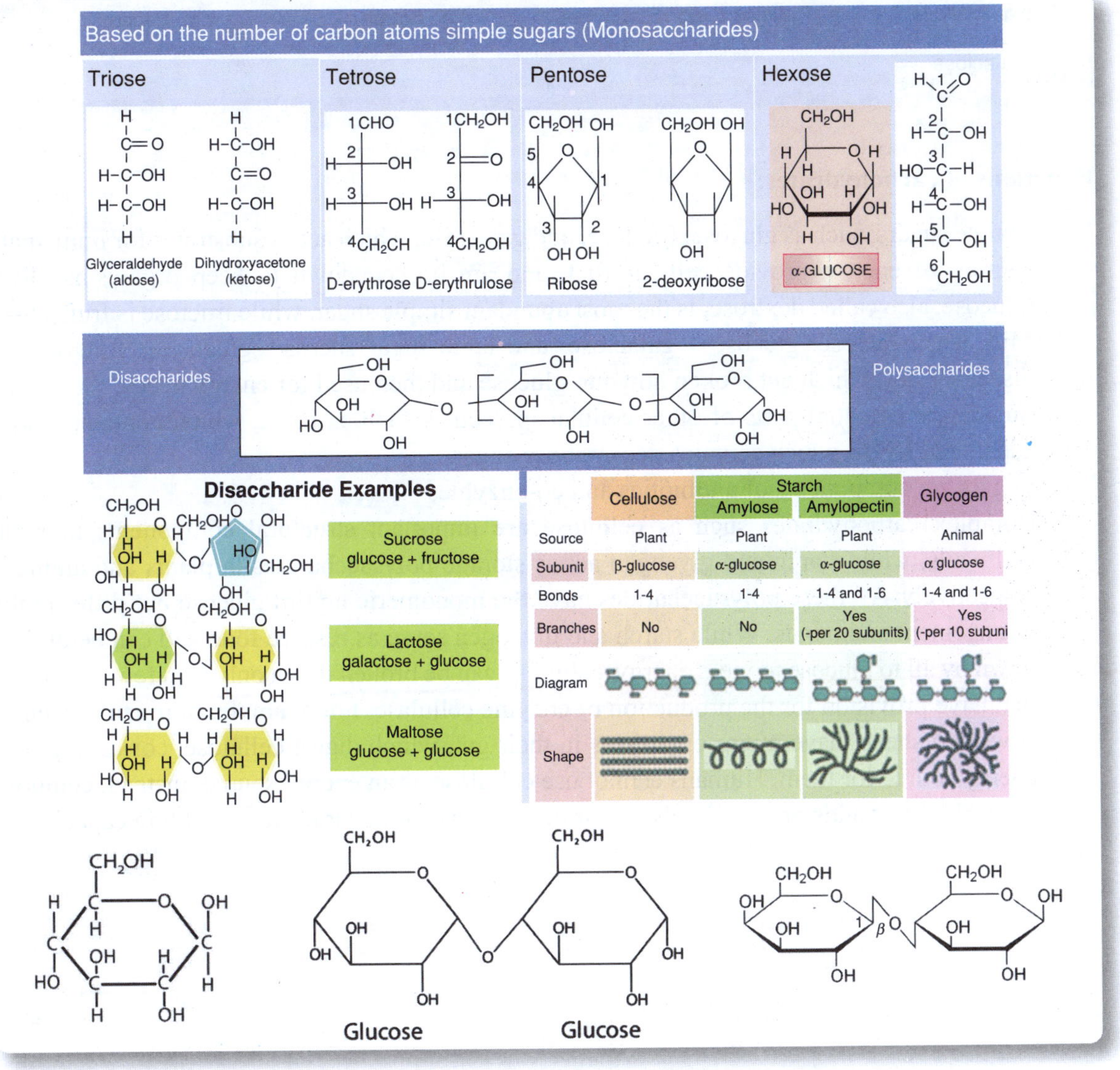

Based on the number of carbon atoms simple sugars (Monosaccharides)

Triose: Glyceraldehyde (aldose), Dihydroxyacetone (ketose)

Tetrose: D-erythrose, D-erythrulose

Pentose: Ribose, 2-deoxyribose

Hexose: α-GLUCOSE

Disaccharides — Polysaccharides

Disaccharide Examples

- Sucrose: glucose + fructose
- Lactose: galactose + glucose
- Maltose: glucose + glucose

	Cellulose	Starch: Amylose	Starch: Amylopectin	Glycogen
Source	Plant	Plant	Plant	Animal
Subunit	β-glucose	α-glucose	α-glucose	α glucose
Bonds	1-4	1-4	1-4 and 1-6	1-4 and 1-6
Branches	No	No	Yes (-per 20 subunits)	Yes (-per 10 subuni
Diagram				
Shape				

Glucose — Glucose

Fig. 4.2 Types of carbohydrates, with examples

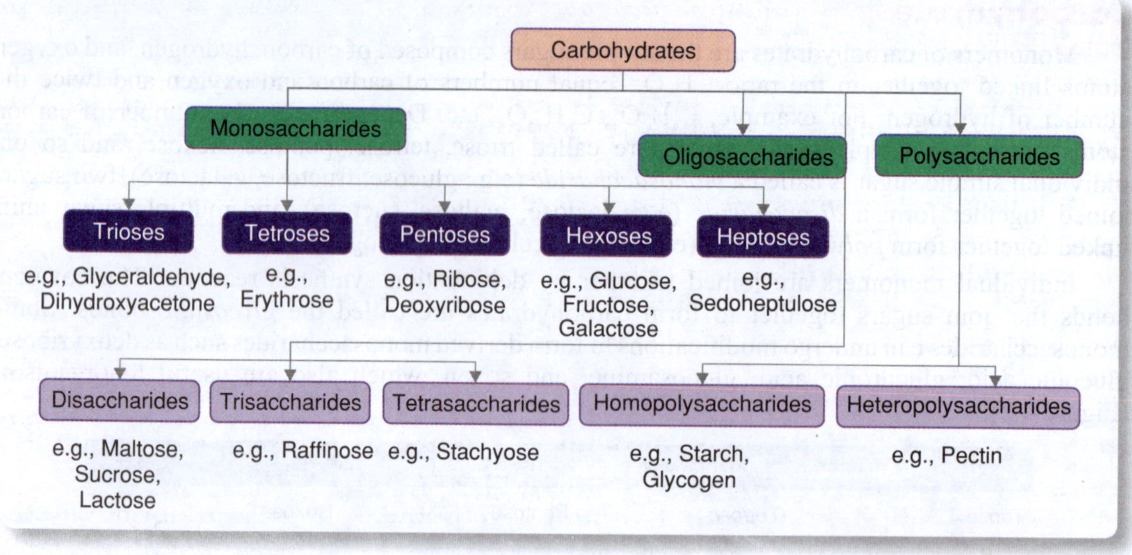

Functions of carbohydrates:

1. Simple sugars, such as glucose, fructose, and galactose, which act as substrates for immediate sources of energy, provide cellular fuel, and are broken down to keep organisms alive. Glucose, also called dextrose, is the most abundant simple sugar, while fructose is fruit sugar.
2. Milk sugar lactose: glucose + galactose; and table sugar sucrose: glucose + fructose are disaccharides, which get broken down to glucose and then used for energy needs.
3. Sugars are essential parts of other cellular molecules such as RNA, which contains ribose sugar, and DNA, which contains deoxyribose sugar.
4. Sugars are components of antibiotics and co-enzymes.
5. Complex carbohydrates such as cellulose are important structural components in plant cell walls, while starch and glycogen act as storage polysaccharides in plants and animals, respectively. All these polysaccharides have the monomeric unit of glucose but differ in the arrangement and bonds. While starch and glycogen serve as reserve food and can be broken down by all to glucose to yield energy, cellulose can be broken down only by a few organisms that have provision for the production of enzyme cellulase. For example, ruminants, such as cows and sheep, have micro-organisms in their gut, which digest cellulose, converting it to energy source for them. Humans cannot use cellulose as an energy source; instead, cellulose from plant diet adds fibre and bulk to our diet because of its large water-holding capacity.

Interesting Fact 4.1

Cis Fats Versus Trans Fats

(i)

(ii)

S. No.	Cis Fat	Trans Fat
1.	Chains of hydrogen atoms are on the same side of double bond giving a kink to the fatty acid	Hydrogen atoms are on the opposite side of the double bonds of the carbon chain, making the fat molecule straight
2.	Unsaturated cis fats increase the levels of good cholesterol (HDL) and reduce the bad cholesterol (LDL). Reasonably good health impact unless consumed in unusually high quantities	Trans fats are detrimental to health. Lowers good cholesterol and increases the level of bad cholesterol in the body. Harmful to heart health and could cause cardiac death
3.	Cis fatty acids occur naturally	While some natural trans fats occur in meat and dairy products, the majority of trans fats come from processed foods (i.e., hydrogenated oils)

Source: In 2002, the National Academy of Sciences concluded that trans fatty acids, whether of plant or animal origin, are non-essential and prove no benefit to human health.

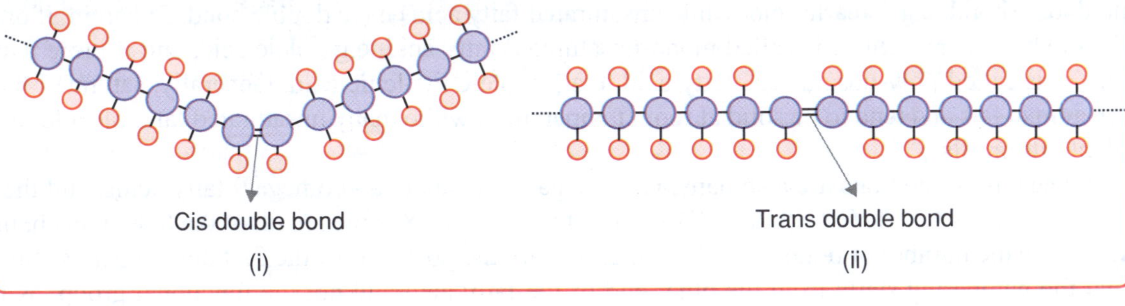

(i) (ii)

Lipids

Lipids are hydrophobic organic macromolecules that do not dissolve in water but are soluble in non-polar solvents such as ether, acetone, chloroform, benzene, etc. Lipids are not polymers, such as carbohydrates, proteins, or nucleic acids, and are, in general, called *fats*.

Lipids are of three types: (a) Simple lipids are esters of fatty acids with various alcohols, e.g., fats and waxes; (b) complex or compound lipids, which usually consist of a fatty acid, an alcohol, and one or more other groups such as phosphorus or nitrogen, e.g., phospholipids; and (c) steroids, which although do not resemble lipids structurally but are included because of their hydrophobicity and insoluble nature in water, e.g., cholesterol.

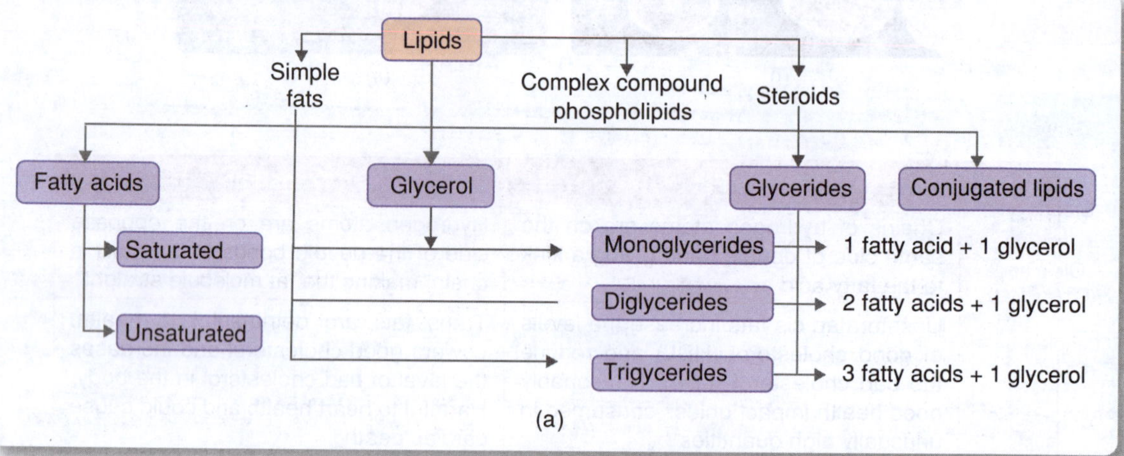

(a)

Simple lipids The building blocks of true fats are glycerol and fatty acids. Glycerol is $C_3H_5(OH)_3$—a carbon skeleton with three alcohol groups attached to it. A fatty acid is a long-chain carbon skeleton that has a carboxyl functional group.

Fatty acids, components of simple lipids, can be either saturated or unsaturated depending on whether all the C atoms are saturated with H atoms. Saturated fatty acids have all single bonds and no double bond, e.g., stearic acid, while unsaturated fatty acids have double bonds, wherein, if one double bond is present, it is called mono-unsaturated fatty acid, e.g., oleic acid, and if more than one, it is called poly-unsaturated fatty acid, e.g., linoleic/linolenic acid. Generally, saturated fats are animal fats and tend to be solid at room temperature, while many unsaturated fat molecules are plant fats or oils.

The unsaturated fats are also named as omega 3, omega 6, and omega 9 fatty acids, and they are indicated as C18:Iω9; C18:2ω6; C18:3ω3, etc., where C18 is the length of the fatty acid chain, I, 2, 3 are the number of double bonds, and 3, 6, 9 are the positions of the first double bonds. Note that the numbering starts from the other end of the fatty acid and not the functional group, as is usually the practice in IUPAC nomenclature. Plant oils are, hence, rich sources of omega-3 and omega-6 fatty acids.

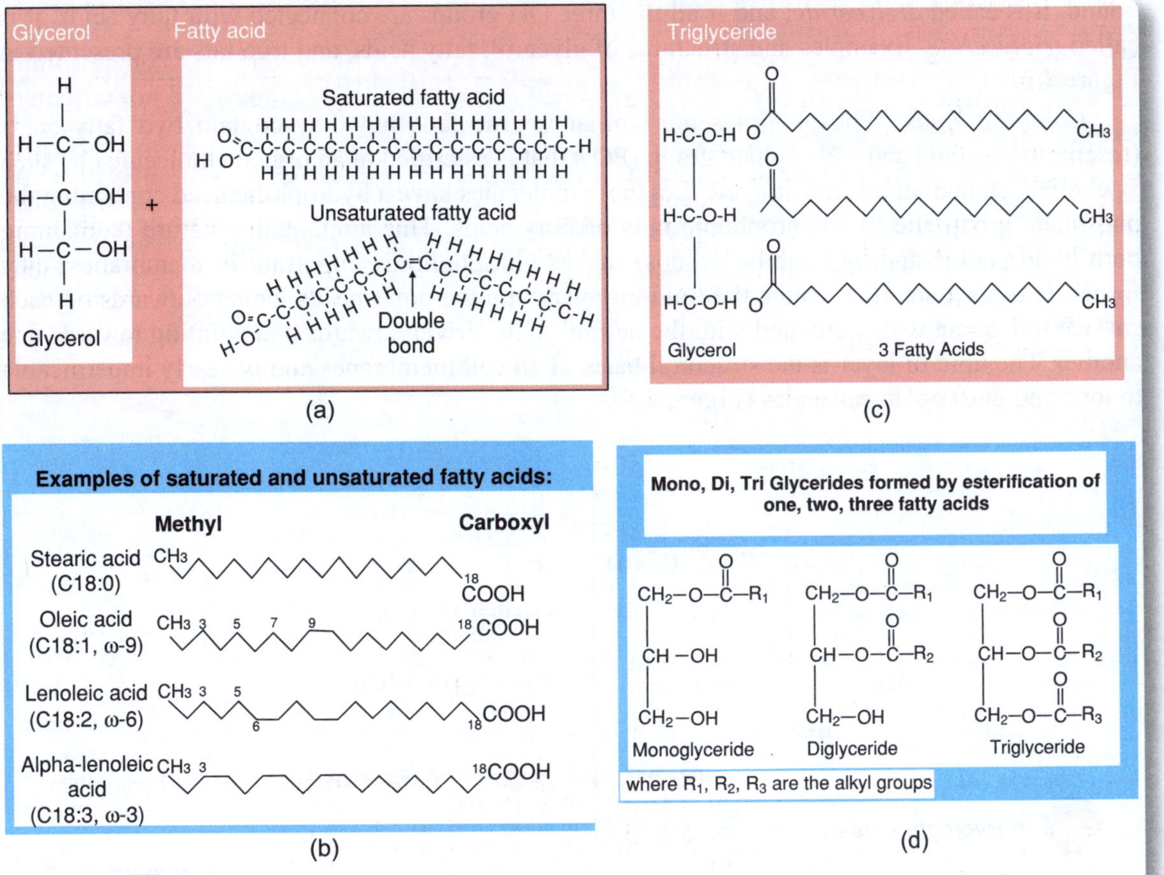

Glycerol

Fatty acid

Saturated fatty acid

Unsaturated fatty acid

Double bond

Glycerol

(a)

Triglyceride

Glycerol

3 Fatty Acids

(c)

Examples of saturated and unsaturated fatty acids:

Methyl **Carboxyl**

Stearic acid (C18:0)

Oleic acid (C18:1, ω-9)

Lenoleic acid (C18:2, ω-6)

Alpha-lenoleic acid (C18:3, ω-3)

(b)

Mono, Di, Tri Glycerides formed by esterification of one, two, three fatty acids

Monoglyceride

Diglyceride

Triglyceride

where R_1, R_2, R_3 are the alkyl groups

(d)

Fig. 4.3 Types and structures of fatty acids and true fats: (a) Glycerol + fatty acid = Lipid; Nature of fatty acids - saturated or unsaturated fatty acids; (b) Examples of saturated and unsaturated fatty acids with omega (ω) nomenclature; (c) Structure of triglyceride; (d) Difference among mono, di, and triglyceride

Table 4.2 Difference Between Saturated and Unsaturated Fats

S. No.	Saturated Fatty Acids	Unsaturated Fatty Acids
1.	No double bond between two carbon atoms	One or more carbon–carbon double bonds
2.	Melting point is higher	Melting point is lower
3.	Solid at room temperature	Liquid at room temperature
4.	Converted into unsaturated fatty acids by dehydrogenation	Converted into saturated fatty acids by hydrogenation
5.	Examples: Palmitic acid, stearic acid	Examples: Linolenic acid, arachidonic acid

Dehydration synthesis reactions connect fatty acids to glycerol through ester bonds to form fats. If one fatty acid is attached to glycerol, it forms a *monoglyceride*; if two fatty acids are

bound, it is called *diglyceride*; and if all the three OH groups are connected with fatty acids, it is called *triglyceride*. Examples and structures of glycerol, fatty acids, and true fats are presented in Figure 4.3.

Complex lipids Phospholipids are similar to true fats, but they contain two fatty acids (resemble true fats) and a phosphate group (PO_4) that is esterified to an organic molecule (X) such as choline, ethanolamine, inositol, etc. So, these molecules have a hydrophilic head containing the phosphate group and two hydrophobic tails of fatty acids. This amphipathic nature (containing both hydrophobic and hydrophilic groups) makes phospholipids important in membranes; they form a two-layer structure called the lipid bi-layer, with the polar heads facing outwards on each surface to interact with water and with the neutral 'tails' driven inwards and pointing towards one another. The lipid bi-layer is the structural basis of all cell membranes and is nearly impermeable to ions and most polar molecules (Figure 4.4).

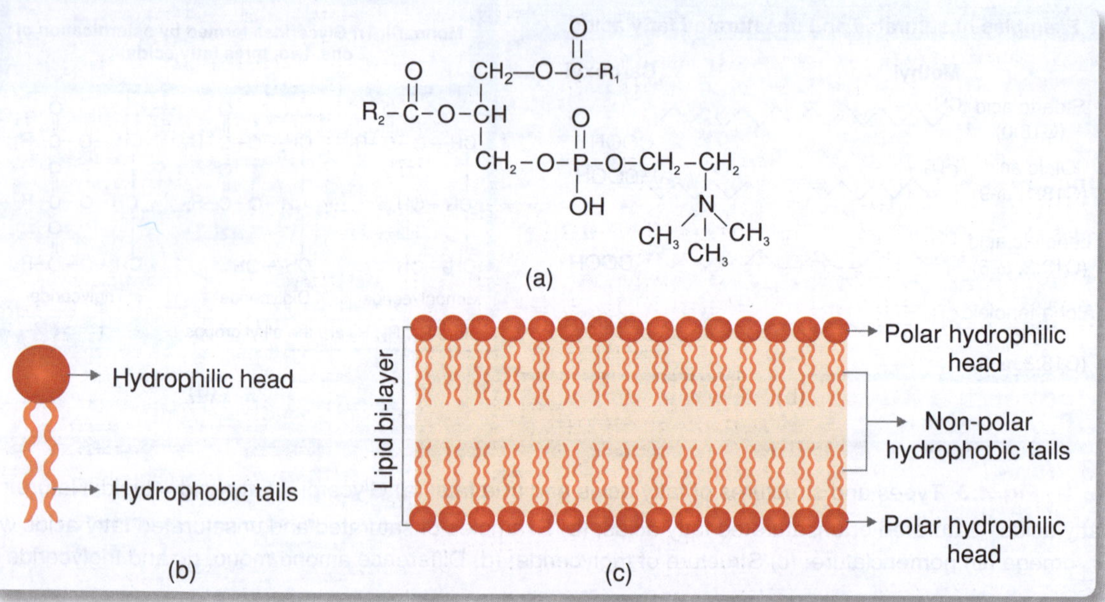

Fig. 4.4 Phospholipid (Lecithin): (a) Types of phospholipid; (b) Its amphipathic nature; and (c) Their organization in membranes

Steroids are characterized by their arrangement of interlocking rings of carbon. The main feature of steroids is the ring system of three cyclohexanes and one cyclopentane in a fused ring system, as shown in Figure 4.5. A variety of functional groups may be attached. The main feature, as in all lipids, is the large number of carbon–hydrogens, which make steroids non-polar. Cholesterol; sex hormones testosterone, progesterone, and oestrogen; and adrenocorticoid hormones aldosterone and cortisol are important steroid molecules involved in various functions in the body.

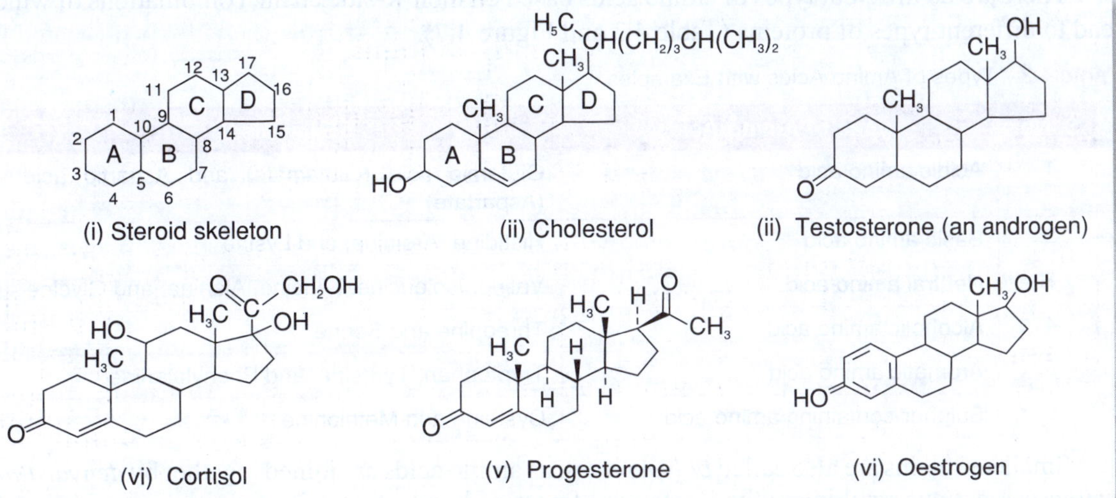

(i) Steroid skeleton (ii) Cholesterol (ii) Testosterone (an androgen)

(vi) Cortisol (v) Progesterone (vi) Oestrogen

Fig. 4.5 Structure of steroids with examples

Functions of lipids:

1. *Storing energy* The primary function of lipids is as an energy reserve. Most of the energy required by the human body is provided by utilizing carbohydrates and lipids (fats). A gram of fat is densely concentrated with energy—it contains more than double the amount of energy than a gram of carbohydrate.
2. *Signalling and regulation* Triglycerides control the body's internal climate, maintaining constant temperature. Triglycerides also help the body to produce and regulate hormones. For example, adipose tissue secretes the hormone leptin, which regulates appetite.
3. *Insulation and protection* Vital organs such as the heart, kidneys, and liver are protected by visceral fat. Fats also help in preventing heat loss from the body and act as an insulating layer in whales, walruses, and seals, protecting them in the cold watery environment.
4. *Cell composition* Phospholipids are the major component of cell membranes of living organisms.
5. *Endocrine* Cholesterol acts as a precursor of Vitamin D and bile acids. Among sex hormones, testosterone is the male hormone involved in the development of secondary sexual characteristics whereas progesterone and oestrogen (or oestradiol) are the female hormones that control the ovulation cycle. Similarly, in adrenocorticoid hormones, aldosterone is involved in regulating the reabsorption of sodium and chloride ions in the kidney tubules and increases the loss of potassium ions whereas cortisol is an important glucocorticoid and has the function of increasing glucose and glycogen concentrations in the body.

Proteins

Polymers of amino acid monomers are called *proteins*. Amino acids are organic monomers made up of carbon, hydrogen, and oxygen; in addition, they also contain nitrogen in the amino group and other side chains (Figure 4.6).

There are 20 different types of amino acids based on their R-side chain, combinations of which lead to different types of proteins (Table 4.3 and Figure 4.7).

Table 4.3 Types of Amino Acids with Examples

S. No.	Types of Amino Acids	Examples
1.	Acidic amino acid	Glutamic acid (Glutamate) and Aspartic acid (Aspartate)
2.	Basic amino acid	Histidine, Arginine, and Lysine
3.	Neutral amino acid	Valine, Isoleucine, Leucine, Alanine, and Glycine
4.	Alcoholic amino acid	Threonine and Serine
5.	Aromatic amino acid	Tryptophan, Tyrosine, and Phenylalanine
6.	Sulphur-containing amino acid	Cysteine and Methionine

Smaller proteins are also called *polypeptides*, as amino acids are joined together by *dehydration synthesis reactions* resulting in the formation of peptide bonds. Because of multiple peptide bonds between amino acids, the term 'polypeptide' is also used to notate smaller proteins (Figure 4.7).

However, proteins are not just about the (primary) sequence of amino acids (which is the primary structure and is crucial), but they also have secondary, tertiary, and quaternary structures, which are critical for the function of the proteins. These depend on the precise folding of the primary sequence; therefore, precisely folded polypeptide chains form *functional proteins* (Table 4.4).

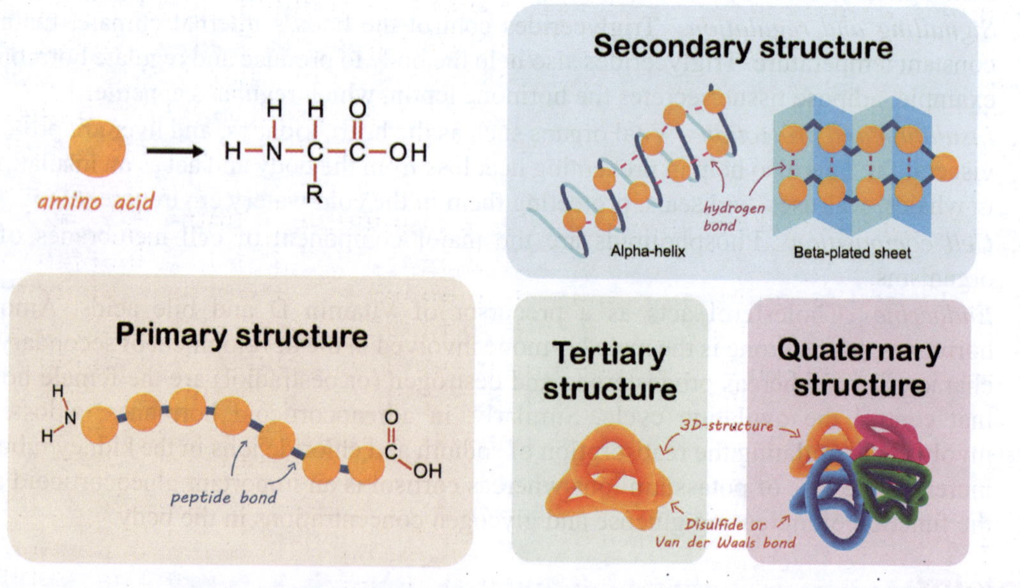

Fig. 4.6 Amino acids

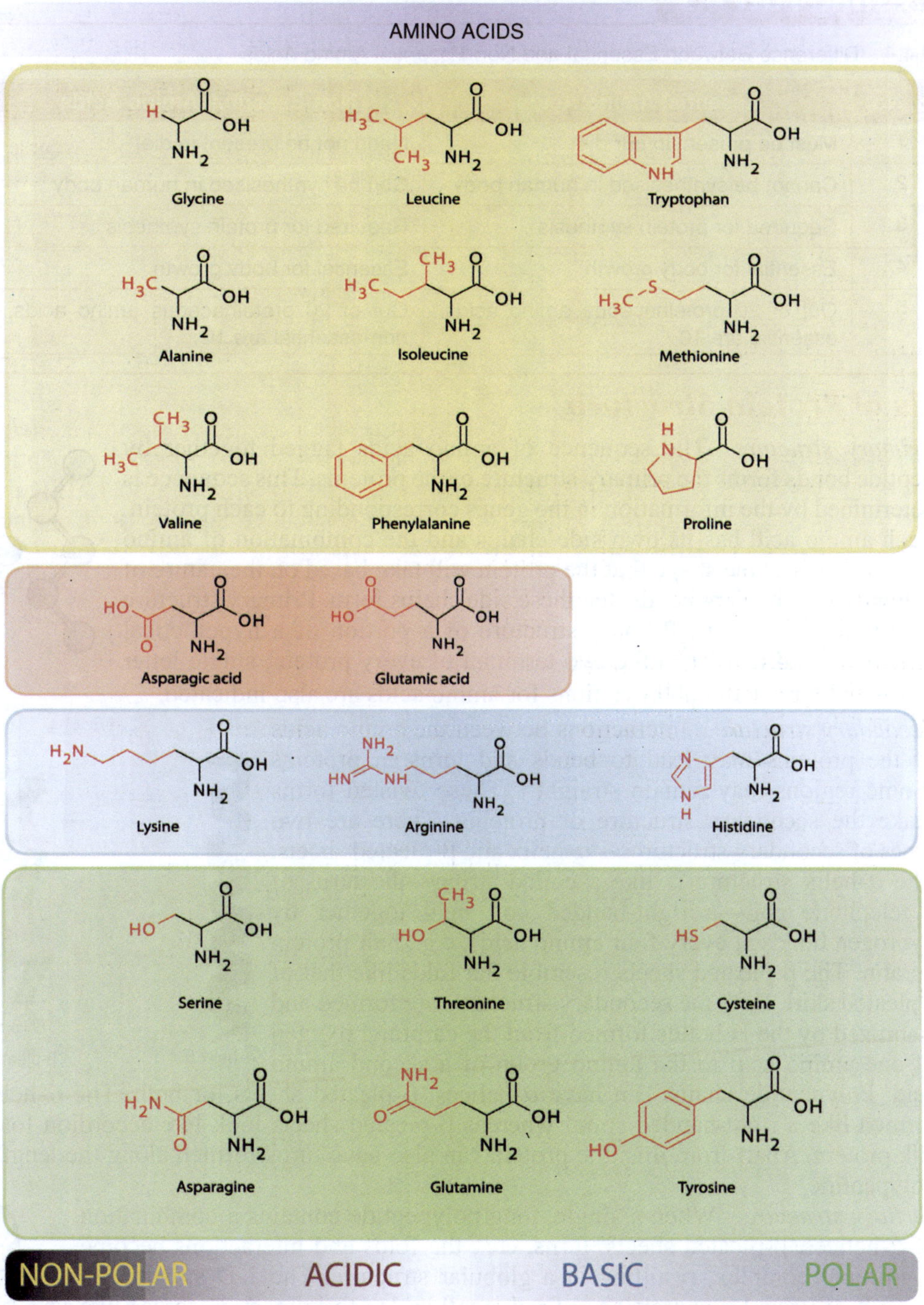

AMINO ACIDS

Glycine, Leucine, Tryptophan

Alanine, Isoleucine, Methionine

Valine, Phenylalanine, Proline

Asparagic acid, Glutamic acid

Lysine, Arginine, Histidine

Serine, Threonine, Cysteine

Asparagine, Glutamine, Tyrosine

NON-POLAR ACIDIC BASIC POLAR

Fig 4.7 Another classification of amino acids

Interesting Fact 4.2

Table 4.4 Difference Between Essential and Non-Essential Amino Acids

S. No.	Essential Amino Acids	Non-Essential Amino Acids
1.	Must be present in our diet	Need not be present in diet
2.	Cannot be synthesised in human body	Can be synthesised in human body
3.	Required for protein synthesis	Required for protein synthesis
4.	Essential for body growth	Essential for body growth
5.	Out of 20 proteinaceous amino acids, essential are 10	Out of 20 proteinaceous amino acids, non-essential are 10

Levels of Protein Structure

(a) *Primary structure* The sequence of amino acids tagged together by peptide bonds forms the primary structure of the proteins. This sequence is determined by the information in the genes corresponding to each protein. Each amino acid has its own side chains and the combination of amino acids influences the shape that the protein will take based on the nature of H bonds and covalent bonds that these side chains form. Primary structure of a protein is *linear*. Primary structure of a portion of a hypothetical protein. N and C refer to the two terminal of every protein; single letter codes and three letter abbreviations for amino acids are also indicated.

(b) *Secondary structure* Interactions between the amino acids of the proteins may lead to bends and turns in proteins (Some regions may remain straight.). These twisted forms make the secondary structure of proteins. There are two types of secondary structures—α-helix and β-pleated sheets. The α-helix structure is like a coiled string—the turns of a telephone coil—a right-handed coil held together by hydrogen bonds at every four amino acids, e.g., hair protein keratin. The β-pleated sheets resemble flat folds like that of a pleated skirt. Both the secondary structures are formed and stabilized by the H bonds formed from the carbonyl oxygen of one amino acid to the amino group of a second amino acid. Polypeptide chains can have α-helices, β-pleated sheets, or both. The α-helices are formed like a right-handed spiral whereas β-pleated sheets look like accordion folds, e.g., silk protein. Apart from this, the protein can also have turns formed along the length of the polypeptide.

α-helix

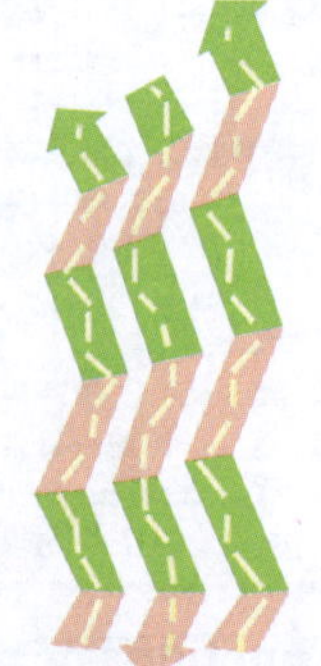

β-pleated sheets

(c) *Tertiary structure* When a single, long polypeptide contains a combination of α-helices, β-pleated sheets, turns, etc., the folds and interactions become even more complex, resulting in a globular structure. The 3-D structure is a result of irregular contortions of polypeptide due to interaction forces like hydrogen bonds, hydrophobic interactions, Van der Waals attraction, and di-sulphide bridges among the R-side chains of the amino acids. These 3-D architectures may take the form of globular protein, fibrous protein, or an intermediate (mix) form, e.g., myoglobin.

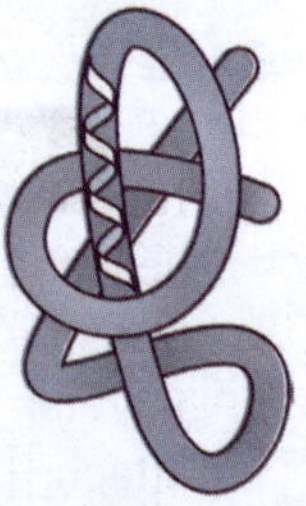

(d) *Quaternary structure* Different polypeptide chains in their secondary or tertiary structures interact with each other to form quaternary structures of proteins. Each interacting polypeptide is held together by di-sulphide bonds. Proteins with only one polypeptide chain do not have quaternary structure, but only proteins with multiple polypeptides have quaternary structure, e.g., haemoglobin, immunoglobulins, or antibodies.

Relationship between Structure and Function of Proteins

The functional properties of proteins are highly dependent on their precise 3-D structures. Altering the structure of protein, i.e., if the 3-D structure of a protein is altered or unfolded by agents such as chemical treatment with mercaptoethanol, increased temperature, or extreme salt or pH conditions, then the function of the protein is lost and the protein is said to be denatured.

Interesting Fact 4.3

Protein structure–function relationship

The nature of hair protein and other regulatory proteins contribute to a functional curly or straight hair. Heat treatments used to straighten hair break the bonds and alter the protein structure to convert curly hair into straight hair. However, this is temporary as the new hair that grows will be as per the genetic disposition of the individual.

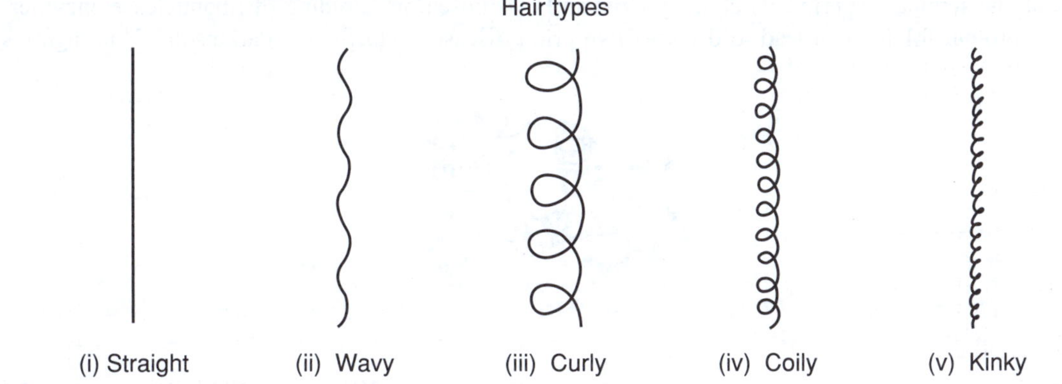

Functions of proteins:

Living systems have thousands of proteins, which can be classified into three major categories based on their functions; they are as follows (Figure 4.8).

1. *Structural proteins* Important for maintaining the shape of cells and organisms. This includes proteins that are part of cell membranes, muscle cells, tendons, blood cells, etc. (e.g., silk fibroin, keratin)
2. *Regulator proteins* Responsible for carrying out and regulating various chemical activities of the cell or organism such as gene expression, immune functions, and other chemical functions. These include enzymes, chaperones, some hormones, etc. (e.g., digestive enzymes, heat shock proteins, glucagon, insulin). Messenger proteins, such as some types of hormones, transmit signals to coordinate biological processes among different cells, tissues, and organs.
3. *Carrier proteins* These proteins help in the transport of various molecules, e.g., proteins attach to cholesterol and help in its distribution by transporting in the form of lipoprotein. Glucose transporter and other proteins help in the transport of materials into the cell.

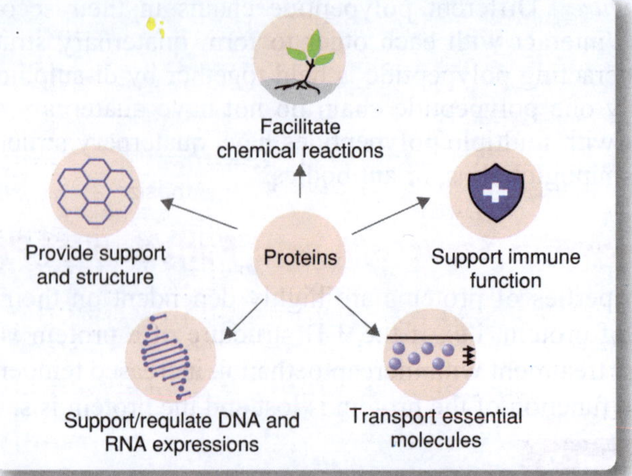

Fig. 4.8 Functions of proteins

Experimental Insight 4.1 Protein Folding

Is a spontaneous and fast process, and all the information required by the protein to fold to its functional conformation/shape is contained in its amino acid sequence. Given the right environment, every polypeptide will fold spontaneously to its native conformation. This is called 'the Anfinsen principle' based on the seminal experiments done by Cristian B. Anfinsen on refolding of ribonuclease enzymes. Error in protein folding can lead to diseases like prion disease, Alzheimer's, Parkinsons, Huntington's disease, etc.

Nucleic Acids

These are complex organic polymers, often considered as one of the most important macromolecules in living beings. They store genetic information and direct protein synthesis; they are of two types: (i) Deoxyribonucleic acid (DNA) and (ii) Ribonucleic acid (RNA). DNA serves as genetic material, and RNA contributes to the process of protein synthesis.

Table 4.5 Difference Between DNA and RNA

S. No.	DNA	RNA
1.	Double-stranded molecule: Two strands held together by hydrogen bonds between purine and pyrimidine residue.	Usually single-stranded molecule but secondary structures may be formed.
2.	Pentose sugar: deoxyribose sugar	Pentose sugar: ribose sugar
3.	Nitrogenous pyrimidine base is: thymine Has four nitrogenous bases: adenine, guanine, cytosine, and thymine	Nitrogenous pyrimidine base is: uracil Has four nitrogenous bases: adenine, guanine, cytosine, and uracil.

4.	Pyrimidine base complementary to Adenine is Thymine	Pyrimidine complementary to Adenine is Uracil
5.	DNA has only one function, i.e., to bear hereditary information	Many species of RNA such as mRNA, tRNA, rRNA have different functions. Genetic material in retroviruses.
6.	Can replicate on its own	Synthesized on a DNA template

Nucleic acids are formed by *dehydration synthesis reactions* like other polymers. The monomeric units of nucleic acids are nucleotides, and the bond that joins nucleotides to form nucleic acid is called the phospho-di-ester linkage (Figure 4.9). Each nucleotide monomer is made up of (i) a nitrogenous base—purine or pyrimidine, (ii) a pentose sugar—ribose or deoxyribose, and (iii) a phosphate group. Purines are adenine and guanine whereas pyrimidines are cytosine, thymine, and uracil (Table 4.5).

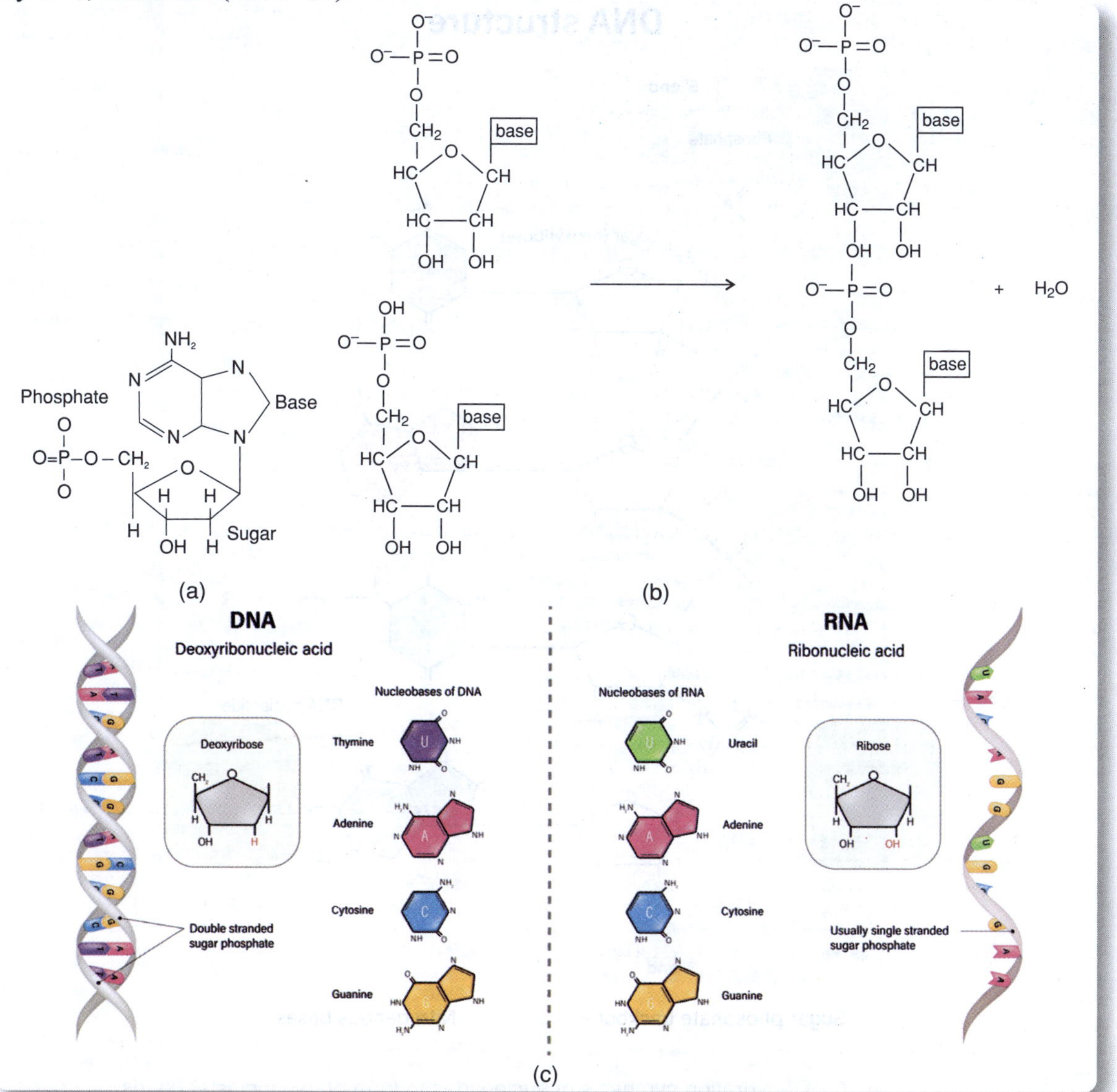

Fig. 4.9 Structural details of DNA and RNA— Nucleotide structure

Table 4.6 Purines and Pyrimidines

S. No.	Characters	Purine	Pyrimidine
1.	Number	A purine is nine-membered.	A pyrimidine is six-membered.
2.	Rings	A purine is double-ringed and larger sized-nitrogen-containing biomolecule.	A pyrimidine is a single-ringed structure and smaller sized-nitrogen-containing biomolecule.
3.	Nitrogen atoms	Purines bear four nitrogen atoms at 1′, 3′,7′, 9′ positions.	Pyrimidines bear two nitrogen atoms at 1′ and 3′ positions.
4.	Examples	Adenine (A) and Guanine (G).	Cytosine (C), Thymine (T), and Uracil (U).

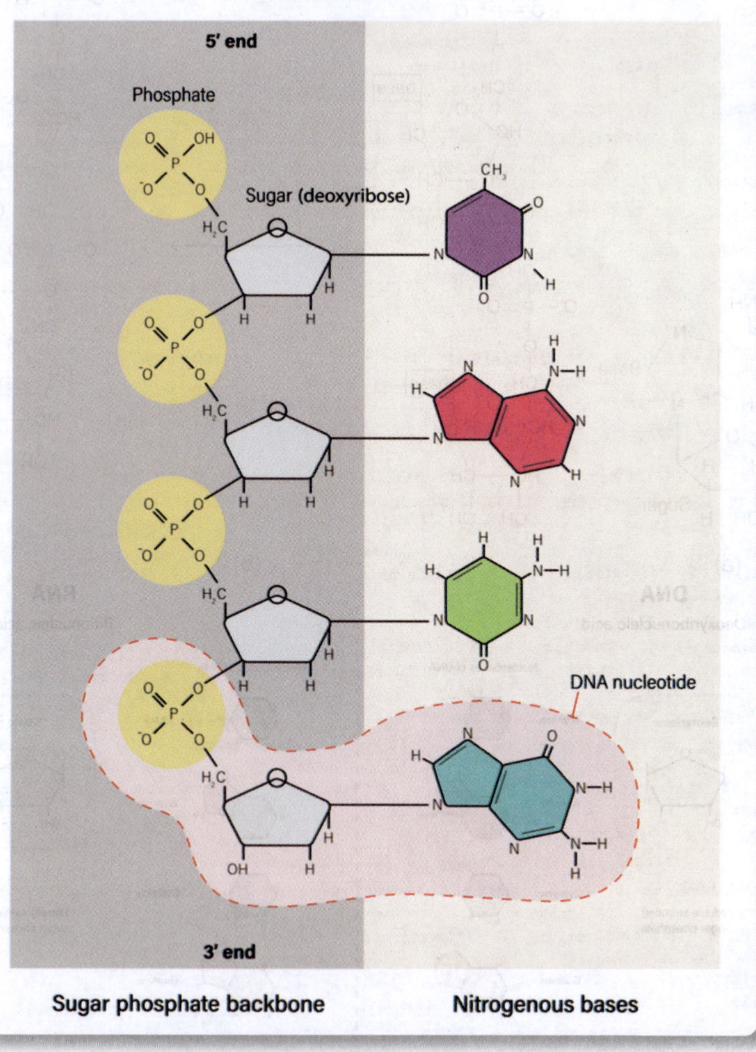

Fig. 4.10 Dehydration synthesis of nucleotides to form phosphodiester bonds: (a) Pentose sugar; (b) Phosphoric acid; (c) Nitrogen base

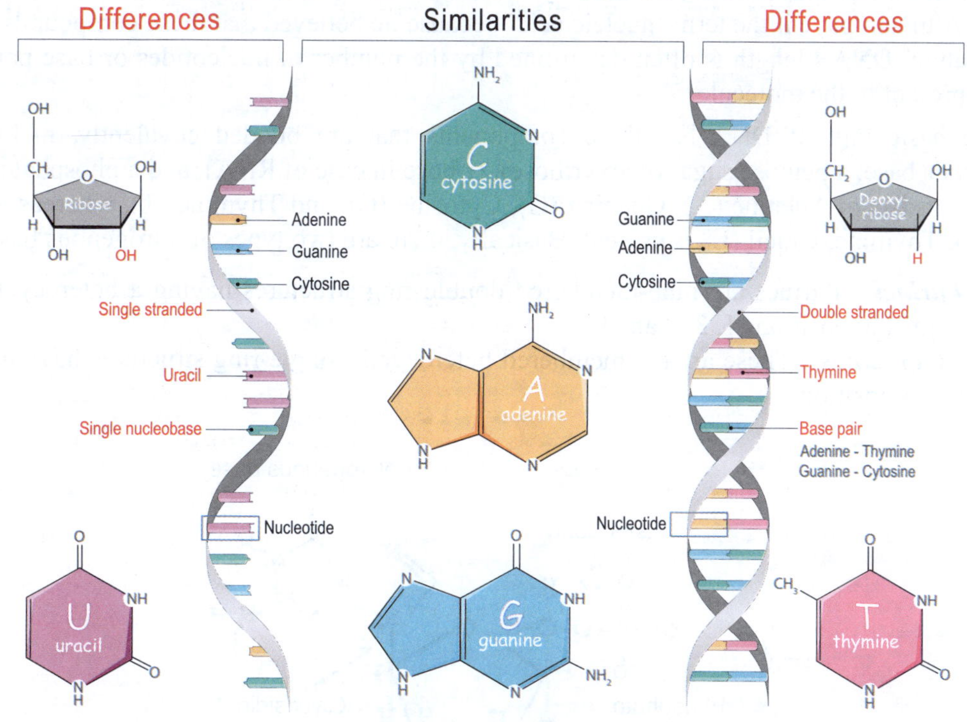

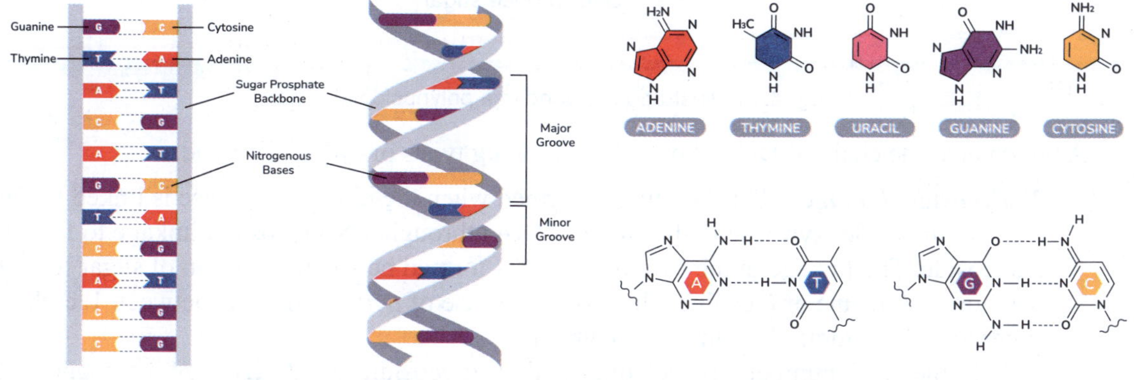

Fig. 4.11 Watson–Crick model: (a) RNA; (b) DNA

DNA

All Types of Bonds in DNA

Nucleotide is a chemical molecule made up of phosphate, pentose sugar, and a nitrogenous base in which DNA and RNA are composed of a series of monomers forming polynucleotides containing various nitrogenous bases. DNA is an acidic and lengthy deoxyribonucleotide polymer present in the nucleus, which was first identified by Friedrich Meischer in 1869. He called it 'Nuclein'. Further, Altmann gave it the term 'nucleic acid' because he believed that these compounds had an acidic nature. DNA's length is often determined by the number of nucleotides or base pairs (bp) that are present in the molecule.

The basic unit of DNA has three components that are bonded covalently, including a nitrogenous base, a pentose sugar (deoxyribose), (ribose in case of RNA), and a phosphate group. DNA is consists of Adenine (A), Guanine (G), Cytosine (C), and Thymine (T), whereas in RNA instead of Thymine, Uracil (U) is present. Basically, there are two types of nitrogenous bases.

1. ***Purines*** Purines are nine-membered double-ring structures having a heterocyclic ring with N at positions 1, 3, 7 and 9.
2. ***Pyrimidines*** These are six-membered heterocyclic single-ring structures having N at 1 and 3 positions.

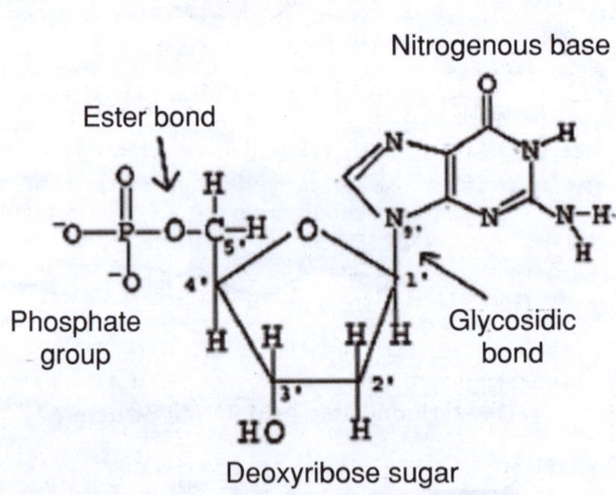

Fig. 4.12 Illustration of bond in a polynucleotide chain

A polynucleotide chain of DNA shows the following types of linkage or bond:

(i) ***N-glycosidic linkage*** This bonding is formed when the nitrogenous base is linked to the pentose sugar (deoxyribose in the case of DNA) through a N-glycosidic linkage to form a nucleoside. The linkage in the case of purine is formed between 1' carbon of sugar to *N*-9 position of Adenine or Guanine whereas in the case of pyrimidine, it is between 1' carbon sugar to *N*-1 position of Thymine or Cytosine.
If N is the total number of nucleotides, then glycosidic bond (G) = N, as sugar and nitrogenous bases are linked by one glycosidic bond in each nucleotide.

(ii) ***Phosphodiester linkage*** It is the main backbone of DNA and is formed between 5′C of sugar of one nucleotide and 3′C of sugar of another nucleotide, which are linked by 3′ – 5′ phosphodiester bond for the formation of a dinucleotide.
For a double-stranded DNA, $P = N - 2$, where N = total number of nucleotides. For a single strand, $N/2 - 1$ phosphodiester bonding is present.

(iii) ***Hydrogen bonds*** Hydrogen bonding is formed between two nitrogenous bases by holding the two strands of DNA with the help of complementary base pairing. This occurs between purine and pyrimidine in which Adenine (A) is bonded to Thymine (T) by two hydrogen bonding whereas Guanine(G) is bonded to Cytosine(C) by 3 hydrogen bonding.
H = ($N/2 \times 2$(A–T fraction pair) + $N/2 \times 3$(G–C fraction pair)) as A–T contributes two bonds while G–C contributes three bonds.

(iv) ***Hydrophobic interactions*** It is present between the nitrogenous bases stacked inside the DNA helix by keeping them inside due to their hydrophobic nature and the sugar-phosphate (hydrophilic) backbone outside. It imparts stability to DNA as the polar groups are kept outside in contact with water.

(v) ***Van der Waal forces*** It is a type of weak force that happens between the aromatic rings of nitrogenous bases. It also gives stability to DNA by a greater number of base stackings.

Sugar in DNA and RNA

The sugar found in DNA is deoxyribose, whereas RNA contains ribose. The major difference between the two sugars lies in their structure in which deoxyribose lacks an OH (hydroxyl) group at the 2′-carbon, whereas ribose contains an OH group at the same position. Their stability and functionality are significantly impacted by this difference. Deoxyribose ($C_5H_{10}O_4$) increases DNA's chemical stability, which makes it less reactive and enables it to store genetic information for a long time. On the other hand, RNA is more reactive due to its catalytic nature and hence prone to degradation due to ribose ($C_5H_{10}O_5$), which is advantageous for its function in protein synthesis. Hence DNA is chemically less reactive and more stable in terms of structure in comparison to RNA. The mutation rate of RNA is more compared to DNA due to its less stability, providing an opportunity for viruses to mutate faster in a shorter lifespan.

A nucleotide is a monomeric unit of DNA—monomers of nucleic acids. They are small but complex molecules and are made up of three moieties: a pentose sugar, a cyclic nitorogenous base, and a phosphoric acid. Thus nucleotides are composed of C, H, O, and N. Each nucleotide is composed of nucleoside and a phosphate group. The nucleoside consists of a pentose sugar and a nitrogenous base.

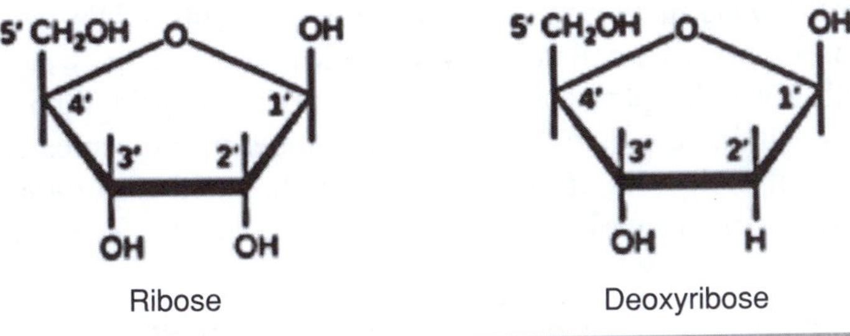

Fig. 4.13 Sugar present in RNA and DNA

A combination of nitrogenous base and pentose sugar is joined at 1'C position by glycosidic bond by one of its nitrogen atoms, usually N3 in pyrimidine and N9 in purine, to form a nucleoside. There are four nucleosides with ribose and deoxyroibose sugars.

Nitrogenous Bases and Chargaff's Rule in DNA Double Helix

Nitrogenous bases are heterocyclic compounds of two types: purines and pyrimidines. The pyrimidines are six-carbon single rings having nitrogen at 1 and 3 positions. There are three types of pyrimidines, namely, cytosine (C), thymine (T), and uracil (U). The purines are nine-carbon double-ring nitrogenous bases with nitrogen at 1, 3, 7, and 9 positions. There are two types of purines, namely, Adenine (A) and Guanine (G).

DNA is double stranded, twisted (right-handed), ladder-like helical structure (proposed by Watson and Crick, 1953), thousands of nucleotides long. The sugar phosphate units of the nucleotides form the backbone, the nitrogenous bases form the cross-rungs of the DNA molecule, and the two strands are joined by hydrogen bonds according to Chargaff's base-pairing rules (Figure 4.11). The base-pairing rule states that adenine (purine) always pairs with thymine (pyrimidine) and guanine (purine) always pairs with cytosine (pyrimidine). The amount of purines and pyrimidines are always equal (A + G = C + T). The amount of adenine is equal to the amount of thymine, and the amount of cytosine is equal to guanine. The ratio A + G/T + C is always a unity. This is Chargaff's rule of base equivalence.

The two strands of the DNA molecule are held together by the hydrogen bonds formed between purine and the pyrimidine bases.

Adenine is paired with thymine by two hydrogen bonds (A = T) while guanine is paired with cytosine with three hydrogen bonds (G ≡ C) (Table 4.6). The sugar and phosphate are on the outer side of the ladder whereas the nitrogenous bases are on the inside of the helix. Nucleotides of a strand are linked to each other through phospho-di-ester bonds. The phospho-di-ester bond between the nucleotides is formed between the phosphate group attached to the fifth pentose carbon of one nucleotide and third pentose carbon with OH of the other nucleotide. Hence, the bond is called the 3′–5′ phospho-di-ester bond (Figure 4.10). The distance between two neighbouring bases is 0.34 nm, and each helical turn accommodates 10 bases (3.4 nm).

The two strands run in opposite directions; one strand ends with 3′ end and is called the 3′ strand whereas the other ends in 5′ end and is called the 5′ strand. The 3′ end has OH group whereas the 5′ end has the phosphate group. Synthesis of nucleic acid happens in 5′–3′ direction.

Chargaff's Rule

Erwin Chargaff (1947) analyzed the base and other contents of DNA of the amount of different organisms. His observations are called Chargaff's rule.

(i) Chargaff observed that purine is equal to the amount of pyrimidine in a DNA molecule (i.e., adenine + guanine = thymine + cytosine); (ii) The number of molecules of adenine (A) is always equal to the number of molecules of thymine (T), and the number of molecules of cytosine (C) is always equal to the number of molecules of guanine (G), i.e., A = T and C = G; (iii) Deoxyribose

sugar and phosphate occur in equimolar proportions; (iv) The ratio of A+ T/G + C is always constant, thus,

$$A + C = G + T \text{ (where } A = T \text{ and } C = G)$$

Interesting Fact 4.4

Chargaff's Rule: In a Nutshell

DNA structure is basically examined by X-ray Crystallography and Chargaff's rule.

(a) X-ray Crystallography technique was used by Maurice Wilkins and Rosalind Franklin in 1953, which suggested that the structure of DNA was a kind of repeating helical type but still a proper model was not figured out by them.

(b) Erwin Chargaff along with his colleagues performed a paper chromatography technique to analyse the base composition of DNA, and his experiments revealed that the relative ratios were not in a random manner. His studies revealed that:

- **(i)** Equal amounts of purines to pyrimidines are present with a ratio of approximately 1.
- **(ii)** Purines found in DNA are adenine and guanine, which are equal to pyrimidines of DNA are thymine and cytosine. $A + G = T + C$.
- **(iii)** The value of $(A + G)/(T + C) = 1$ remain constant for all species.
- **(iv)** This ratio is less than one in prokaryotes, whereas it is more than one in eukaryotes.
- **(v)** Sugar deoxyribose and phosphate residues occur in equal numbers.
- **(vi)** The amount of purine adenine is equal to pyrimidine thymine residues for all DNA samples $(A = T)$.
- **(vii)** The amount of purine guanine is equal to pyrimidine cytosine residue for all DNA samples $(G = C)$.

Further, it is not applicable for single-stranded DNA.

Numerical Questions

1. Calculate the number of phosphodiester and glycosidic bonds in 100 base pairs .
2. In a dsDNA number of phosphodiester bonds are 3250, then find the number of base pairs.
3. A DNA has 30% of T. What could be content of A, G, C, U in it?
4. A DNA has 200 bases; the content of A is 15 %. What is the number of H bonds in total in the DNA?
5. What is the sequence of complementary strand of DNA and RNA if one strand of DNA has a sequence of 5′- AAATTGGCC-3′?

DNA Double Helix

DNA double helix was proposed by Watson and Crick by considering the chemical structure, which is coiled in a right-handed fashion in a clockwise direction. Salient features of this double helical model are:

(i) The two strands of DNA consist of two long strands of polydeoxyribonucleotides coiled in a helical way, where the backbone is formed of sugar–phosphate bonding projecting the bases inside.

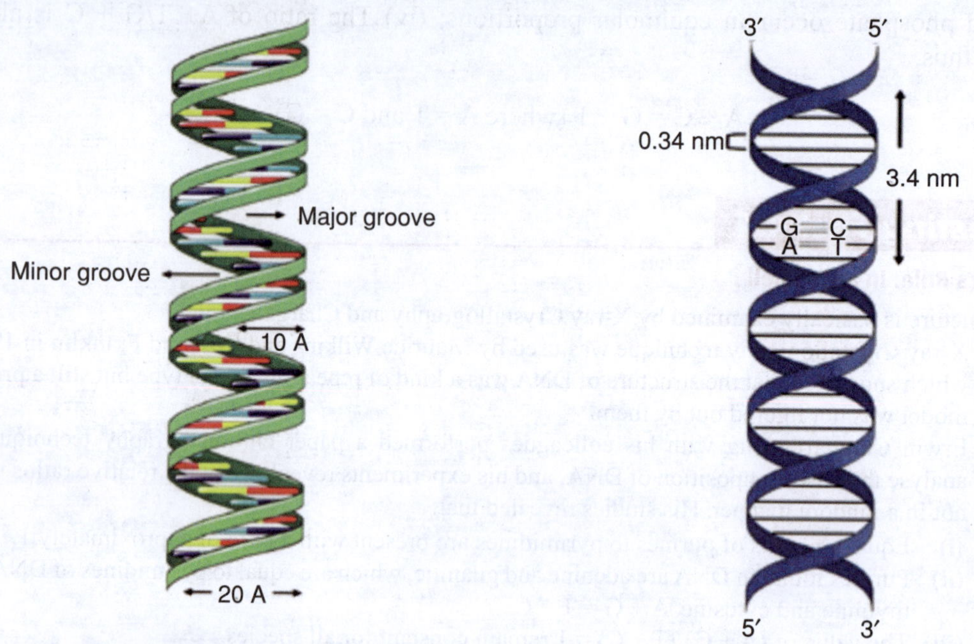

Diagram showing the double helical structure of DNA proposed by Watson and Crick having a diameter of 20 Angstrom having major and minor grooves. The pitch of helix is 3.4 nm having 0.34 nm as adjacent base pair

(ii) The DNA duplex is coiled in a right-handed manner just like a spiral having a diameter of 20 Angstroms, having alternate major and minor grooves. Further, hydrogen bonding is found in the bases of two strands, which form a base pair(bp). Purine always forms a bond with a pyrimidine, which results a uniform distance between the two strands in Adenine, forms 2 hydrogen bonds with Thymine, whereas Guanine forms three hydrogen bonds with Cytosine.

(iii) Each turn of the helix has a distance of 34 Angstroms (3.4 nm) called as pitch of the helix, which contains 10 base pairs in each turn, making the distance between adjacent base pairs 3.4 Angstroms (0.34 nm).

(iv) The two strands of DNA run antiparallel to each other in opposite directions in relation to their sugar molecules. If the direction of one strand is 5′ to 3′, then the direction of the other strand is 3′ to 5′, which is useful in complementary base pairing.

(v) Out of the two strands of the DNA, only one strand carries the genetic information, which is called the template or antisense strand, whereas the other is called the coding or sense strand.

Interesting Fact 4.5

Sugar substitutes Artificial sweeteners and sugar substitutes help to reduce sugar and calories in our diet. A sugar substitute (artificial sweetener) is a food additive that duplicates the effect of sugar in taste but usually has less food energy. This has led to the development of products such as 'sugar free', 'diet coke', etc., which promise to control your sugar levels and waistline. Some sugar substitutes are natural and some are synthetic. Those that are not natural are, in general, referred to as *artificial sweeteners*. Sugar substitutes do not affect the blood sugar level; hence, they are used widely by diabetics.

Product	Common Name/ Brand	Source	Sweetness	Taste Quality	Calories
Sucrose	Sugar	Sugarcane	1*	Traditional standard	1620
Saccharine	Sweet n' Low	Coal derivative	200*	Lingering effect	0
Sucralose	Splenda	Chlorinated sugar	600*	Astringent	0
Neotame	Neotame	Methanol	8000*	Lingering effect	0
Aspartame	Equal, Nutrasweet	Methanol	200*	Approaching sugar	0
Fructose	Corn syrup	Fruit/vegetables	1.2*	Similar to sugar	17
Stevia	Truvia	Stevia plant	300*	Bitter aftertaste	0
Nectresse	Monk fruit	Chinese melon	200*	Similar to sugar	0
Agave	Nectar	Agave cactus	1.5*	Sweeter than sugar	20
Honey	Honey	Bee	1.5*	Flavourful	21

*times

RNA

RNA is found in three forms in living systems. All three forms have significant contributory roles in protein synthesis. As indicated earlier, RNA is usually single stranded, but it can form secondary structures by complementary base pairing.

Messenger RNA, or mRNA, which is a single-stranded copy of a portion of DNA called the gene encoding a specific protein. Synthesis of mRNA is based on the complementary base rules, except that as RNA does not have thymine, 'adenine–uracil' pair is seen instead of the 'adenine–thymine' pairing. Ribosomes are the organelles involved in protein synthesis and contain a type of RNA called ribosomal RNA or r-RNA. Transfer RNA or t-RNA has a clover leaf-like structure and is involved in carrying specific amino acids to ribosomes for protein synthesis. The rRNA and tRNA are also transcribed from the corresponding genes on DNA (Table 4.5).

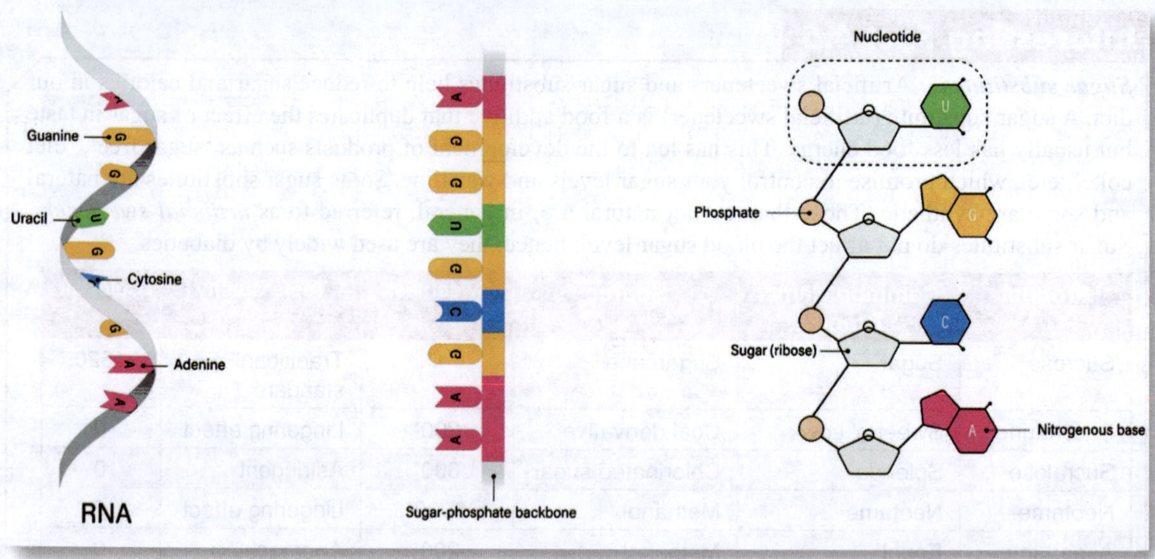

Fig. 4.14 RNA structure

Functions of Nucleic Acids

1. DNA is the genetic material of most organisms. A few viruses have RNA as genetic material.
2. Genetic material stores information that determines the characteristics of cells and organisms. It is also called the 'blueprint of life' and 'master molecule'.
3. DNA (genetic material) replicates and makes copies of itself for the cell to *divide* and also mutates or changes to *adapt* and *evolve*.
4. DNA directs the synthesis of proteins essential for the operation of cells and organisms. RNA helps in protein synthesis by copying the genetic information as mRNA, carries amino acids to the ribosome as tRNA, and is a part of the ribosome as rRNA.

Water

Apart from these organic biomolecules, living cells contain a major constituent, water, along with inorganic ions, minerals, etc. Water is essential for life to exist. Hence, water is used as a criterion in search of life on other planets. Water seems to be a simple molecule consisting of two hydrogen atoms and one oxygen atom joined by covalent bonds but has special features that make it indispensable for life (Figure 4.13). Water has a polar nature due to its electron arrangements and

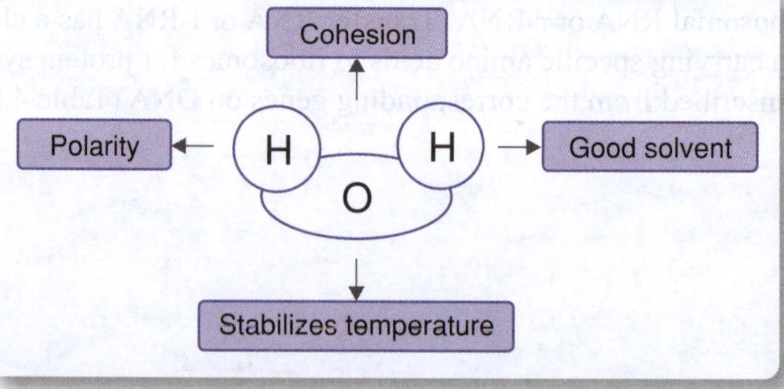

Fig. 4.15 Water molecule and its major properties

bond angles wherein the oxygen end is slightly negative and the hydrogen end is slightly positive. The resulting polarity helps water molecule to form hydrogen bonds with others. Molecules that easily form hydrogen bond with water are said to be hydrophilic (e.g., sugar), and others that are repelled are hydrophobic (e.g., oils).

Water acts as a universal solvent—polar molecules easily dissolve in it—and provides a medium for biochemical reactions to occur inside the cells. Solutes dissolved in water act as reactants, which come in contact with each other, permitting a chemical reaction that results in different products. The *cohesion* property of water also contributes to life. Water molecules co-adhere together through hydrogen bonding, which provides a capacity to resist rupturing under tension. Because of cohesion, nutrient-ridden water moves up through the pipelines of vascular tissues.

As water molecules stick together, large amount of heat energy is required to separate them. Water resists change in temperature! The presence of water helps in buffering against drastic temperature shifts in the body of multicellular organism.

Nearly 70% of the weight of living cells is water and, hence, is one of the most important compounds required by all organisms. The unique properties of water help in the growth and sustenance of life on Earth; without water, life on Earth, as we know it, would not exist.

Interesting Fact 4.6

Facts about water

- Pure water has no colour, taste, or smell.
- About 70% of Earth's crust is covered with water.
- Nearly 97% of water on Earth is salty, and 2% is in glaciers and frozen. Only 1% is available for consumption.
- About 70% of human brain is filled with water.
- A human can live about a month without food but only a week without water.

SUMMARY

- Biochemistry of life was studied, and the critical biomolecules of life were introduced.
- Structural and functional attributes of carbohydrates were discussed.
- Types of lipids and their critical functions were explained.
- Levels of protein structure and the structure–function relationship was explored.
- Nucleic acid types with their structural architecture and biological significance were elaborated.
- Water as an indispensable component of living cells was highlighted.

EXERCISES

A. Objective-Type Questions

1. ____________ is not a polymer.
   (a) Carbohydrate (b) Lipid
   (c) Protein (d) Nucleic acid

2. Which of the following characteristics is not common to biological macromolecules—carbohydrates, proteins, lipids, and nucleic acids?
   (a) All four are organic molecules.
   (b) All four are formed by dehydration synthesis reactions.
   (c) All four contain covalent bonds.
   (d) All four are polymers.
3. Which of the following is not a polysaccharide?
   (a) Cellulose (b) Sucrose
   (c) Glycogen (d) Starch
4. Which one of the following is the odd one out?
   (a) Amino acid (b) Nucleotide
   (c) Monosaccharide (d) Protein
5. The hydrolysis of DNA will yield
   (a) One type of pentose sugar and different nitrogenous bases
   (b) One type of nitrogenous bases and different pentoses
   (c) No nitrogenous base but one type of pentose
   (d) Neither nitrogenous base nor pentoses are found
6. Different proteins are composed of different sequences of __________.
   (a) Sugars (b) Glycerols
   (c) Fatty acids (d) Amino acids
7. Which of the following is true for RNA but not for DNA?
   (a) Has thymine and deoxyribose sugar (b) Has uracil and ribose sugar
   (c) Is double-stranded (d) Shows A=T base pairing
8. __________ is/are the major component of the cell membrane of living organisms.
   (a) Cellulose (b) Progesterone
   (c) Phospholipids (d) Nitrogenous bases
9. Which of the following covalent bonds connects the monomeric units of RNA?
   (a) Glycosidic bond (b) Peptide bond
   (c) Phospho-di-ester bond (d) Di-sulphide bond
10. Lactose is to carbohydrates as __________ is to proteins.
   (a) Insulin (b) Glycerol
   (c) Cholesterol (d) Guanine
11. Molecules that bond with water are __________.
   (a) Hydrophilic (b) Hydrophobic
   (c) Hydrolytic (d) None of the above
12. Water is a good __________.
   (a) Steroid (b) Solvent
   (c) Sugar (d) Polymer

B. Short-Answer Questions

1. Are all the four macromolecules present in your diet?
2. Proteins are broken down into amino acids and used by our body for which purpose?
3. What do you understand by essential amino acid and essential fatty acid? Give examples.
4. Cows can digest the celluloses in grasses but we cannot. Do you know why?
5. What will be the outcome of a dehydration synthesis reaction between two reactants—glucose and galactose?
6. Di-peptide + water = amino acid + amino acid. This is an example of which reaction?

C. Self-Study Questions

1. Connect the dots between biological macromolecules and animal nutrition.

FURTHER READINGS

1. Akella, Somi Devi. Chemistry of Food and Cooking: We are what we eat! Yale-New Haven Teachers Institute, 2020.
2. Pray, L. Discovery of DNA structure and function: Watson and Crick. *Nature Education*, 1(1):100, 2008.

USEFUL WEBLINKS

1. https://www.medicalnewstoday.com/articles/318524

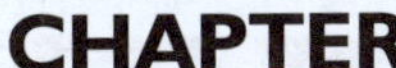

CHAPTER 5

Cell: The Basic Unit of Life

LEARNING OBJECTIVES

After reading this chapter, you will be able to:

- Understand cell and its types
- View cellular envelope—cell wall and plasma membrane
- Attain an understanding of cellular transport through the cell membrane
- Know the structure and function of membranous and non-membranous cell organelles
- Compare the structure and complexity of prokaryotic and eukaryotic cells

CHAPTER TAKEAWAY

Everything about cells and their mechanism of action

In this chapter, various aspects of a cell and the activities performed at the cellular level in living organisms are explained.

PRELUDE

All organisms are made up of cells; some organisms are unicellular and others are multicellular. But what is a cell? When we hear the word 'cell', numerous meanings might pop up in our head, such as a cell battery, a mobile phone, a prison room (jail cell), an intelligence cadre, a compartment or cubicle, and so on. All these have reference to a 'cell', but, for a scientist, and specifically, a biologist, the term cell signifies the basic structural and functional unit of living organisms. It has great significance as the entire existence of any organism depends a lot on the activities and numbers of cells in their bodies. Now, how to study cells? Can we take a 'cellfie'—imaging or photographs of cells! By microscopic examination and by measuring different structural and biochemical parameters, we have been able to understand cells, their types, their structures, different organelles, their specific functions, and so on. So, in essence, the cell is the space delimited by a membrane inside which various organelles are present to perform various functions. To understand further, let us assume a cell to be like a house or a factory inside the body. It has walls or outer membrane, doors that allow entry and exit of molecules and, within, various regions that carry out specific roles such as protein synthesis, storing genetic information, protein packaging, facilitating chemical reactions, carrying out energy reactions, etc.

Cells: Basic Unit of Life

The cell theory states that 'all living things are made up of cells'. The cell is the smallest unit of life that displays characteristics of life. How did the concept of the cell emerge in the very beginning? It is an idea that was developed and modified over several centuries; new insights are being gained even today!

History of Cell—Contributions by Robert Hooke and Anton van Leeuwenhoek

Robert Hooke (1635–1703) was the first one to use the term *cell*. He saw cork slices from the bark of a cork oak tree, under a primitive microscope, and they appeared to him like empty 'cells' or rooms; hence, he called them as cells. He published his observation in 1666 in the journal *Micrographia*. Today, we know that what Hooke saw were only the cell walls surrounding the living portions of plant cells (Figure 5.1).

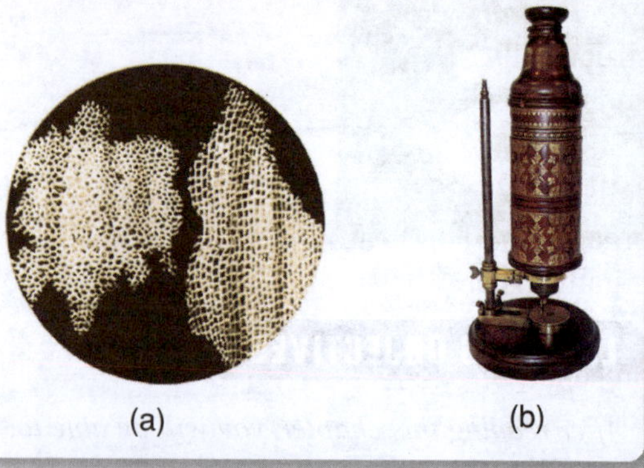

Fig. 5.1 The first microscope by Robert Hooke: (a) Cork tissue; (b) Primitive microscope

Around the same time, Anton van Leeuwenhoek (1632–1723), a Dutch cloth merchant, was inspired by *Micrographia* and Hooke's observations, and he started studying many biological specimens with the magnifying glasses that he had for counting threads. He started improvising on his microscopes and made nearly 400 of them with different magnifications based on his expertise with glass grinding. His microscopes could amplify nearly 270–300 times the object (Figure 5.2). In his curiosity, he started observing multiple things such as pond water, blood, semen, pepper, tartar, etc. He noticed some cells moving and called them as '*animalcules—small moving animals*'. He even communicated with Robert Hooke about his findings and also shared his observations with the *Royal Society of London*.

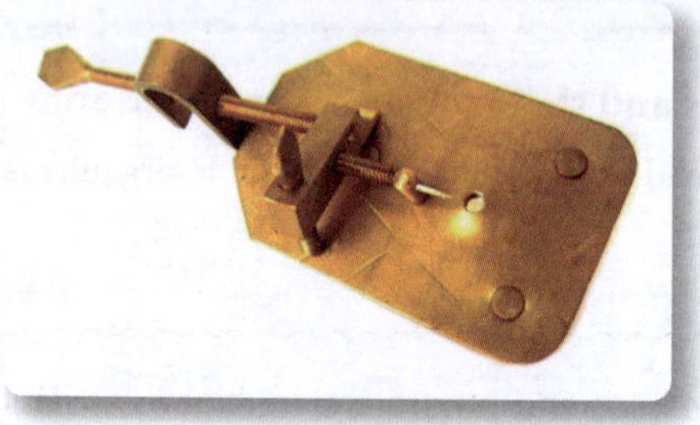

Fig. 5.2 Leeuwenhoek's microscope

However, it took nearly 200 years after these initial observations to finally recognize that 'all living organisms are made up of cells'. In 1838, Matthias Schleiden stated that all plants are made up of cells, and, in 1839, Theodor Schwann published the idea that all animals are made up of cells too.

The term *cell* caught on and different types of cells were identified. The distinction between the non-living cell wall and the living portion of the cell inside was identified, and this living portion was called 'protoplasm', which translates to *first-formed substance*. Better light microscopes, better staining techniques, electron microscopes (Figure 5.3), etc., finally revealed the entire structural morphology of the cell—the central nucleus, fluidic cytoplasm, and all the organelles dispersed throughout the cytoplasm.

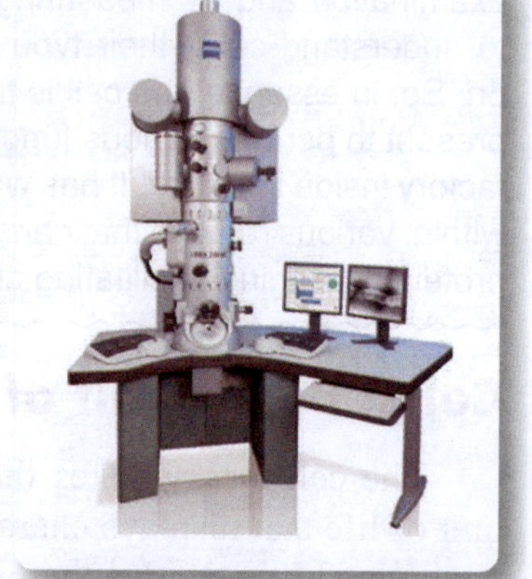

Fig. 5.3 Electron microscope

Types of Cells

There are mainly two different types of cells based on evolutionary organization.

Prokaryotic cells Seen in unicellular organisms such as eubacteria and archaea, these are structurally simple cells, smaller in size (1–2 μm in diameter), and do not have a defined nucleus and other specialized cellular organelles.

Eukaryotic cells Seen in organisms such as fungi, algae, protozoans, plants, and animals, the cells of these organisms are structurally complex with a well-defined nucleus and many specialized organelles. These are typically larger than prokaryotic cells (10–100 times). A few eukaryotic organisms could be unicellular, like yeasts and protozoans, but most are multicellular.

However, what is unifying between these cells, common in all cells of either type, is the presence of outer cell membrane, cytoplasm, ribosome, genetic material like DNA/RNA, and a form of energy currency.

Structure and Organization of Eukaryotic cells

Although eukaryotic cells are larger, there is a limit to cell size. This is because, to stay alive, cells have to efficiently transport nutrients and wastes through its outer membrane. As the size of cell increases, its surface area (area available for material intake) increases by a square (X^2) but its volume (denoting metabolic activity) increases by a cube (X^3). The cell is metabolically active, so this decrease in surface area-to-volume ratio affects the cell's ability to satisfy the metabolic requirements as it is limited by surface area. So, it is beneficial to have a small size (of the cell). Smaller cells with higher surface-to-volume ratios are more efficient in responding to cellular needs. If cells grow large, then they normally divide into two to maintain an appropriate surface area-to-volume ratio.

An egg cell may appear as a large cell (Figure 5.4), but it is important to note that the metabolically active part of the egg cell is only on the surface. Its centre is filled with inactive stored food called yolk; hence, exchange of material is not much impacted.

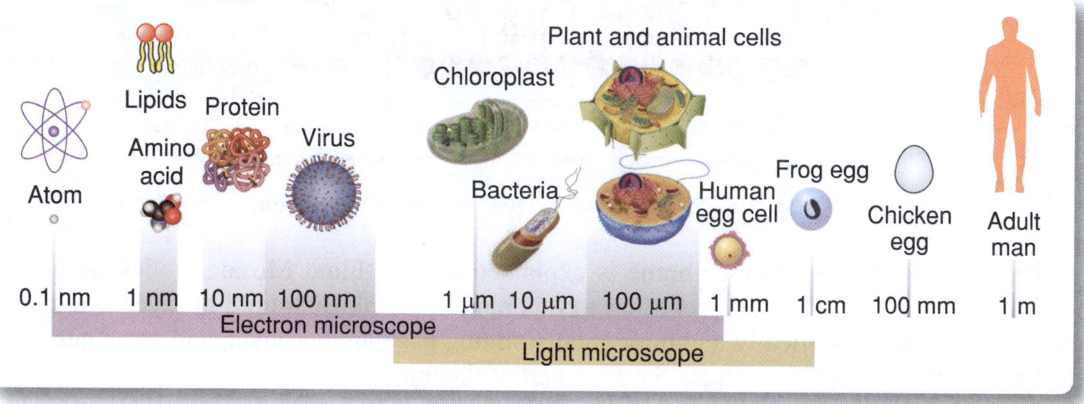

Fig. 5.4 Cell types and their relative sizes vis-a-vis other molecules and structures

Structural Component of Cell

Let us start from the outermost part of the cell and work our way towards the internal core—examine the structure and function of various cellular components.

Plasma Membrane

All cells are enclosed by a plasma membrane, which defines the outer boundary of the cell. This membrane isolates the cell contents from the outer environment and acts as a semi-permeable barrier that selectively allows some substances to pass through while preventing others. This allows critical requirements to be met, requisite wastes to be removed, and, overall, maintain a different internal composition from the surrounding solution.

The plasma membrane, also called cytoplasmic membrane or cell membrane, is a thin structure primarily made of a phospholipid bi-layer. The phospholipid molecules of the membrane have two ends that differ chemically; the phosphate end, which is hydrophilic (water loving—soluble in water), and the other end of fatty acids, which is hydrophobic (fearful of water—insoluble in water). A phospholipid molecule is commonly represented as a balloon with two strings—balloon is the phosphate part and two strings are the fatty acids. When phospholipid molecules are placed in water, they undergo a conformation of double-layered sheet, with the hydrophilic ends facing away from each other, on the outside, and the hydrophobic fatty acids in the centre (Figure 5.5).

Cells inside living organisms contain water inside the cell and are also bathed in a watery solution outside. Hence, the phospholipid forms this kind of bi-layer, which then constitutes the structure of the plasma membrane.

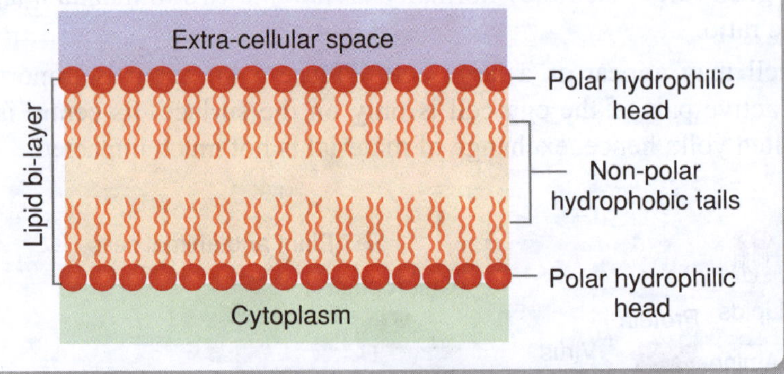

Fig. 5.5 Phospholipid bi-layer in plasma membrane

The structure of plasma membrane is explained by the 'Fluid Mosaic model' proposed by Singer and Nicholson in 1973. According to this model, the phospholipid bi-layer is in constant motion, moves, both, vertically and horizontally sliding past another layer, and has a heavy olive oil-like consistency. This constitutes the *fluid* part of the model.

Other than the phospholipids, the plasma membrane has proteins interspersed across the bi-layer (integral part), which forms the second component of the plasma membrane. These proteins could be seen on the periphery, called peripheral proteins; some on the external side called extrinsic proteins and some on the inner side called intrinsic proteins; and some are transmembrane—across the bi-layer membrane (Figure 5.6).

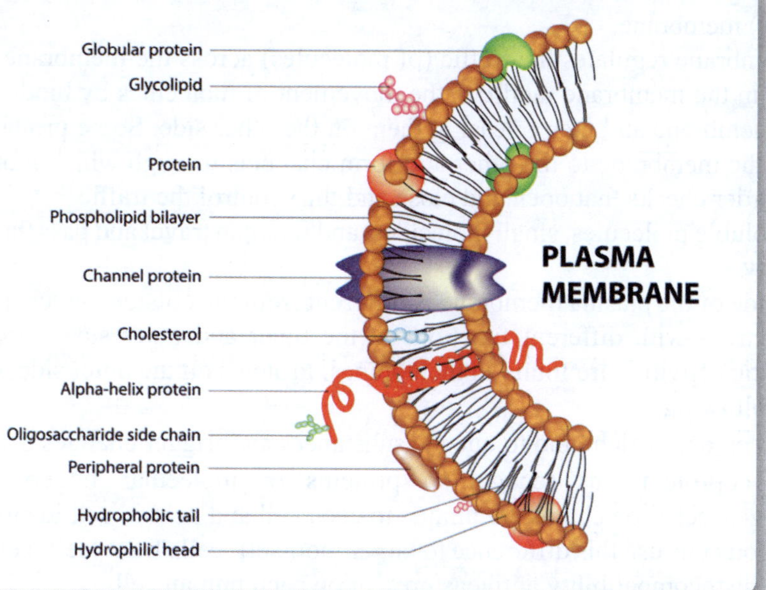

Fig. 5.6 Distribution of different proteins in plasma membrane

Different regions of the membrane contain different proteins and phospholipids, a mixed composition, dispersed in the form of a mosaic, which constitutes the *mosaic* part of the model. Hence, cell membranes are fluid mosaics of phospholipids and proteins. The proteins serve a variety of functions such as transport medium for molecules across the membranes, attachment for other molecules, and identity tags for the cells (Figure 5.7).

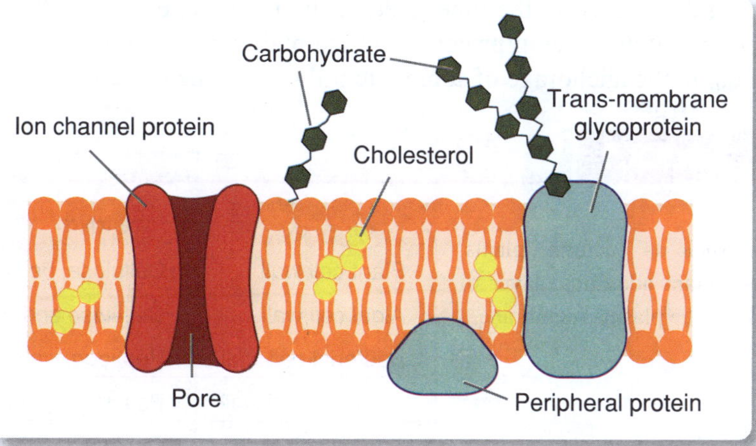

Fig. 5.7 Fluid Mosaic model of plasma membrane

Additionally, cholesterol molecules are also found in the hydrophobic middle part of the membrane, which helps in stabilizing the membrane and keeping it flexible. On the outer side of the membrane, carbohydrates are attached to lipids, and these glycolipids and glycoproteins help in cell recognition, cell–cell interaction, and binding to regulatory molecules.

Functions of plasma membrane:

1. The lipid bi-layer nature of the plasma membrane serves as a barrier between the watery contents of the cell and its organelles and the watery surrounding.

2. The membrane is metabolically active—chemical reactions take place within or on the (surface) membrane.
3. The membrane regulates the traffic (of molecules) across the membrane. Transport proteins present in the membrane facilitate the movement of molecules by binding them on one side of the membrane and then releasing them on the other side. Some proteins extend from one side of the membrane to the other and form channels through which molecules move. They have border checks that open and close and thus control the traffic.
4. Lipid-soluble molecules, small molecules, and ions can travel and pass through the membrane passively.
5. The inside of the plasma membrane is different from the outside, which makes the membrane asymmetric—with different functions at the inner and outer sides. Proteins with specific enzymatic activities are found in membranes. Proteins on the outer side of the membrane act as the following:
   (a) Receptors, which bind to outside substances and trigger changes in cell activity
   (b) Glycoproteins, as recognition proteins or molecular fingerprints. The protein–oligosaccharide chains are unique to each cell and help in cell identification; e.g., white blood cells use this difference to target 'non-self' cells from 'self' cells in the body based on histocompatibility antigens present on each human cell.
   (c) Adhesives—help cells of the same type to locate, stick to one another, and ensure proper tissue architecture.
   (d) Attachment sites for specific microbes such as bacteria, viruses, etc. For example, HIV virus attaches to specific molecules on the surface of certain immune system cells and nerve cells. The knowledge of these attachment sites could be used to design blockers and prevent diseases.
6. Some peripheral proteins on the inner side of the membrane, i.e., on the cytoplasmic side, help in anchoring cytoskeleton proteins and maintaining the structural integrity of the cell. They also help in the anchorage of some integral membrane proteins.

Interesting Fact 5.1

Cell wall

Some organisms such as bacteria, plants, fungi, and algae contain additional layer of cell wall, just outside the plasma membrane.

Type of Cell	Composition
Plant cell wall	Cellulose, pectin, lignin
Algal cell wall	Cellulose and polysaccharides
Bacterial cell wall	Peptidoglycan
Archaeal cell wall	Pseudo-peptidoglycan
Fungal cell wall	Chitin

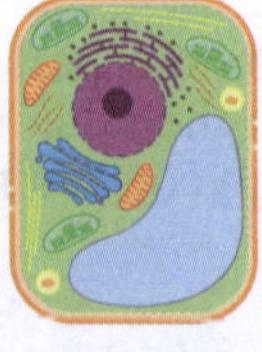
Plant cell

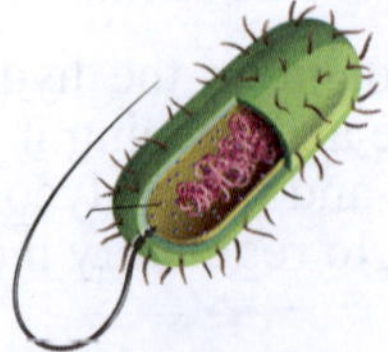
Bacterial cell

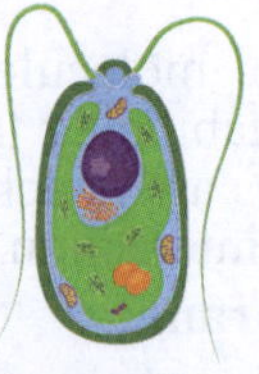
Algal cell

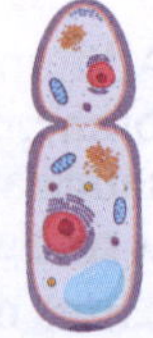
Fungal cell

Experimental Insight 5.1 Fluid Mosaic Model of Membrane

The experimental validation of the Fluid Mosaic model came from experiments using freeze-fracture and deep-etching techniques. Cells were rapidly deep frozen in ultra-low temperature in liquid nitrogen (–196°C), which helps to preserve the 3-D structure of live cells. Thin slices were made with a cold knife to expose the cell membrane. The fracture plane follows the path of least resistance and splits the phospholipid bi-layer, which is then peeled apart to study the protoplasmic and extracellular sides. Deep etching was then done by removing the frozen water and spraying the exposed surface with a mist of platinum. This heavy metal-coated surface was studied under electron microscopy. The carbon film strengthened the platinum, and after digesting the original membrane, it left a platinum–carbon film as a replica of the fractured surface. The observation confirmed bumps and corresponding depression on the two exposed surfaces, which indicated the asymmetrical distribution of proteins in the phospholipid bi-layer. These experiments provided the strongest evidence for the Fluid Mosaic model of biological membranes. It also showed that technical advances help to progressively improve scientific knowledge.

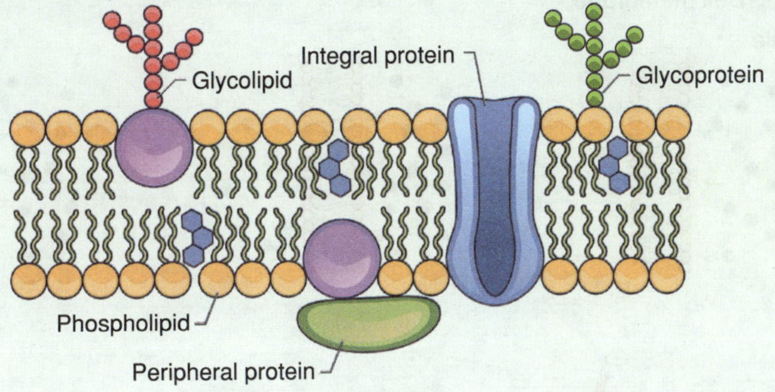

Membrane Transport

For a cell to be alive, it has to metabolize for biological transformation of nutrients, to make raw materials available for synthesis and energy production for sustenance. Metabolism, as we saw, needs nutrient uptake, nutrient processing, and waste elimination. For nutrient uptake and waste elimination, transport of material between the cell and its environment is critical. Useful substances have to be selectively transported inside the cell, and waste materials have to be carefully disposed off. This exchange primarily happens through the cell membrane.

The mechanisms by which this exchange happens can be categorized into five major heads, as follows:

(i) Diffusion
(ii) Osmosis
(iii) Facilitated diffusion
(iv) Active transport
(v) Endocytosis and exocytosis

Diffusion Among the molecules that need to be transported from outside to inside or within the cell are smaller sized simple molecules, lipid-soluble molecules, charged ions, and gases. These molecules are allowed to diffuse passively without any barrier to their free movement. They just diffuse out and in. For example, temporary openings in plasma membrane, due to its fluidic nature, allow small molecules such as oxygen, carbon dioxide, water, and small ions to diffuse through. Similarly, molecules that dissolve in phospholipids, such as vitamins A and D, can also diffuse easily across the membrane.

The cell has no control over the rate or direction of diffusion. What decides this is the relative concentration of specific molecules on either side of the membrane. This process is a passive process because no additional energy is expended, and the energy required is supplied by the kinetic energy of the molecules themselves.

To conclude, diffusion is the unhindered movement of molecules from a region of higher concentration to a region of lower concentration. This difference in concentration across the membrane is called concentration gradient or diffusion gradient. Because of this movement, at some point, if the concentration of the substance becomes equal on both sides of the membrane—there is no concentration gradient, then the system is said to have reached dynamic equilibrium. Why dynamic?—because molecules are still moving in all directions, and why equilibrium?—because there is no net movement or change. This is simply because movement is happening in all directions equally (Figure 5.8).

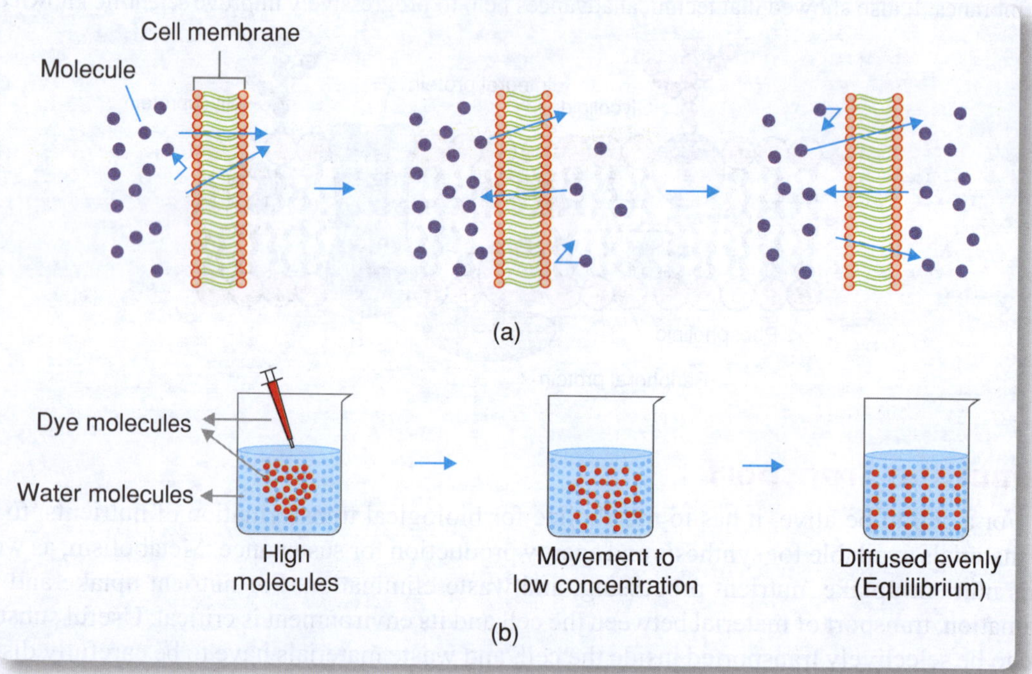

Fig. 5.8 Mechanism of diffusion: (a) Diffusion from high to low concentration and dynamic equilibrium after some time; (b) Dye molecule diffusing in water

Importance of Diffusion in Living Systems

In large multicellular organisms, many cells are buried deep within the body, but their requirements need to be met too. Diffusion can only happen over a short distance as the concentration gradient across the membrane needs to be maintained. So, the circulatory system will bring the molecule near the cell, and from there it will diffuse into the cell. For example, oxygen is essential for life in aerobic organisms. So, when we inhale, oxygen enters the lungs. From there, it diffuses into the blood. This oxygen-rich blood is transported throughout the body, and the cells automatically take up the oxygen by diffusion. This occurs because of higher concentration in the lungs than in the blood and higher concentration in blood than in the cells. Direct diffusion helps

the cells to get oxygen. The reverse is true for carbon dioxide. Cells have a high concentration of CO_2, which diffuses into blood. Blood then takes it to the lungs, from where it is exhaled into the surrounding environment (Figure 5.9).

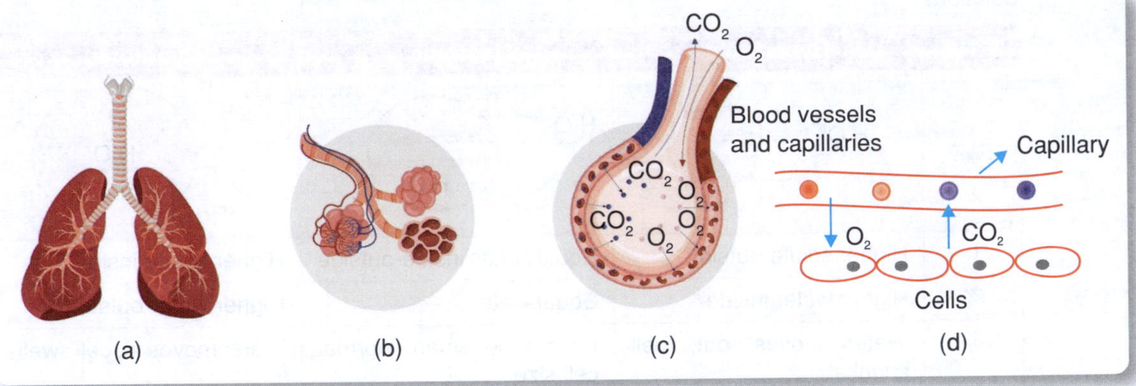

Fig. 5.9 Gas exchange by diffusion in humans: (a) Inhalation and exhalation of gases from lungs; (b) Exchange of gases with blood from lung alveoli; (c) Diffusion of gases from blood into tissue; (d) Gas exchange between cells across the concentration gradient

Osmosis Diffusion of water is called osmosis. A proper amount of water is required for the cell to function efficiently. Too much or too little water will interfere with the cell functioning and metabolism. Osmosis is also a passive process; it also depends on the concentration gradient, and water moves from higher to lower concentration. Hence, cells, as such, cannot control the water movement. In the membrane, there are special protein pores called aquaporins, which also help in water movement. The state of the cell in terms of water balance depends on the environment. There are three possible conditions in which the cells could exist with their environment and that, in turn, will impact their water status, as given below (Table 5.1).

(i) *Isotonic environment* If the concentration of water and dissolved materials of a cell is equal to its surroundings, the cells are said to be in an isotonic state. In this case, there will be no net movement of water in the cells. They neither gain nor lose water as water movement is equal in all directions. For example, organisms that live in oceans, such as jellyfish and sponges, are isotonic to their environment.

(ii) *Hypotonic environment* If the concentration of water in the environment is higher than inside the cell, the cell is said to be in a hypotonic environment, and it will gain water. For example, organisms that live in freshwater ponds and lakes have this situation—less dissolved material and more water in the lakes and ponds. Inside the cells there are many solutes dissolved; therefore, higher concentration of dissolved substances and lower concentration of water! In such a scenario, water from outside will enter the cell (higher to lower concentration), so the cell gains water. To remove this excess water, cells have to expend energy and maintain water balance. Hence, freshwater protozoans have contractile vacuoles that force water out of the cell. Similarly, kidneys of freshwater fishes expel the extra water.

(iii) *Hypertonic environment* If the concentration of water is low in the environment and higher concentration of dissolved solutes than in the cell, the environment is hypertonic, and the cell will lose water and shrink. For example, ocean fishes lose water by osmosis

due to the salty surrounding and need to drink water excessively; but as this water comes with salt, they expend energy to remove excess salt.

Table. 5.1 Impact of Environment on Water Balance in Cells: Hypertonic vs Isotonic vs Hypotonic Solutions

S. No.	Hypertonic	Isotonic	Hypotonic
	H_2O	H_2O / H_2O	H_2O
1.	Higher solute outside	Equal solute inside-outside	Higher solute inside
2.	Higher water inside	Equal water	Higher water outside
3.	Water moves out, cell shrinks	No net movement, normal cell size	Water moves in, cell swells

These impacts of osmosis are true for human cells and plants as well. But the cell wall of plant cells helps them when they gain or lose water. Normally, too much water gain can lead to bursting of cells; but, in plant cells, the turgor pressure increases, and they become rigid but do not burst due to the cell wall. Similarly, when we add salt to salads, the high salt creates a hypertonic environment and the plant cells lose water, but only the protoplasm shrinks leading to *plasmolysis* (in severe cases). So the salad (leaves) become limp and wilt but the cell wall is intact (Figure 5.10).

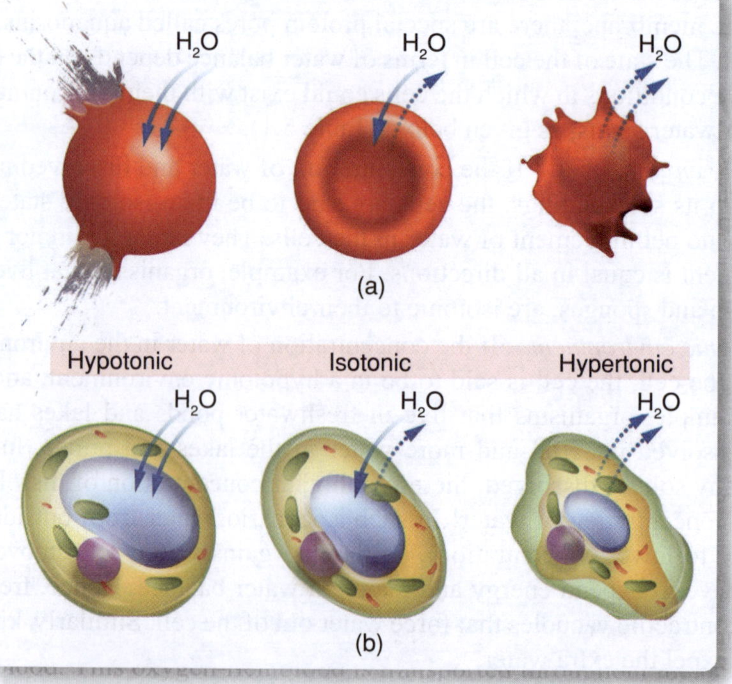

Fig. 5.10 Differential impacts of tonicity: (a) Animal cell (RBC); (b) Plant cell

Interesting Fact 5.2

Examples of osmosis in daily life

(a) Swelling of raisins, almonds, and chickpeas due to osmosis

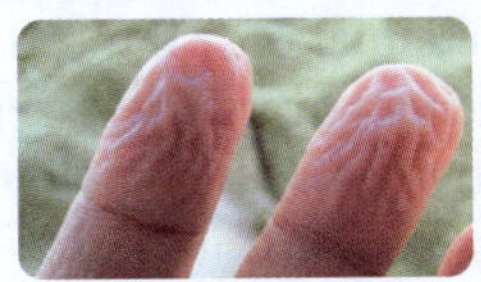

(i)

(ii)

(b) Effect of (i) soaking hand in water; (ii) salt on a slug/earthworm

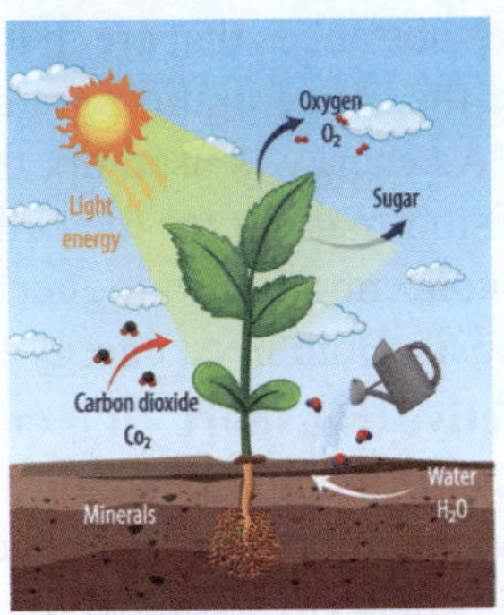

(c) A plant absorbs water and nutrients from soil solution

Facilitated diffusion The hydrophobic core of the plasma membrane impedes the movement of large hydrophilic molecules, polar molecules, and charged ions, even though they are in higher concentration. These molecules diffuse with the help of transport proteins or carrier proteins. The movement of molecules (diffusion) from a region of high concentration to a region of low concentration with the help of carrier proteins is called *facilitated diffusion*. Thus, concentration gradient aids in facilitated diffusion (Figure 5.11).

Facilitated diffusion is specific. Integral membrane proteins specific to a solute or ion act as carrier proteins and help in transporting the molecules from one side of the membrane to the other. For example, glucose and amino acids are transported by facilitated diffusion. In such a transport of solutes, the carrier proteins undergo conformational change and progressively move the solute binding site from one side of the membrane to the other; in the process the solute also gets transported.

The availability of carrier protein controls the rate of transport by facilitated diffusion. As these carrier proteins are specific, the cell can control the movement of specific molecules by regulating the types and numbers of carrier proteins. So different cells can have different distributions of these carrier proteins that characterize the type of molecule transported inside the cell. So, in a sense, in spite of the high concentration of a specific solute, facilitated diffusion is more in control of cell than regular diffusion.

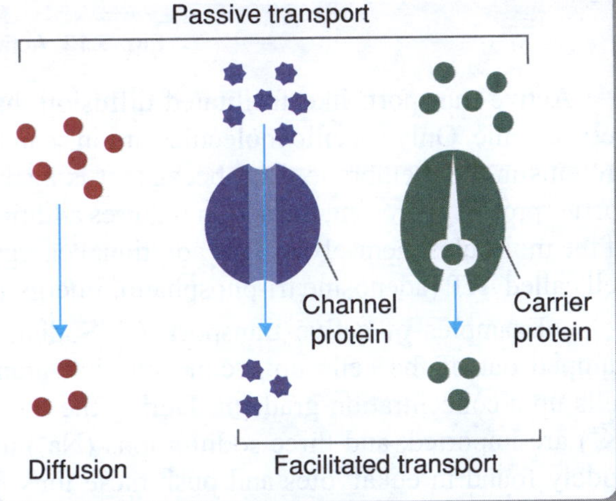

Fig. 5.11 (a) Facilitated diffusion using carrier proteins across the membrane—an example of glucose transport

Diffusion, osmosis, and facilitated diffusion are all passive processes. They work utilizing the concentration gradient and no energy is expended in the process. However, the cell may sometimes require substances that are in lower concentration outside or the cell may need to remove molecules against the concentration gradient. In those situations, the cell has to expend energy and fulfil its needs. As energy is used up, these movements have to be essential and well-coordinated in terms of decision-making for the cell. There are two ways by which the cell can take up molecules from lower concentration to higher concentration to fulfil its needs—active transport and endocytosis/exocytosis. Both need energy to be spent and, hence, are active processes.

Active transport It is an energy-driven mechanism in which membrane pumps make solutes to cross the membrane against the concentration gradient, which means that the net movement is in the direction of higher concentration.

This helps to serve two purposes: (i) to take care of the continuous need of ions and molecules from outside even though their concentration inside the cell is high (keep pumping in from lower to higher concentration), and (ii) to keep the concentration of certain molecules lower than the outside environment (keep pumping out from lower to higher concentration). Such active transport systems help in maintaining the desirable concentration and electric gradients across membranes (Figure 5.12). Various physiological processes in organisms, such as muscle contraction and nerve cell function, depend on these vital gradients.

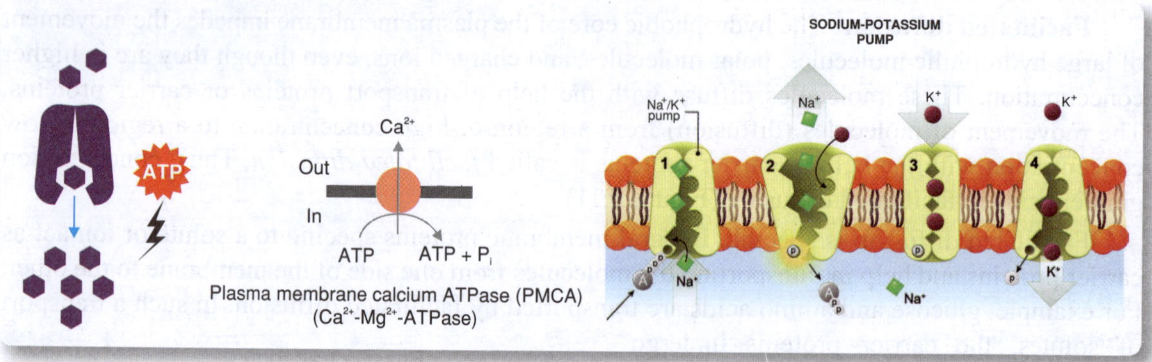

Fig. 5.12 Active transport

Active transport, like facilitated diffusion, happens with the help of carrier proteins and is very specific. Only specific molecules or ions can be moved in this way and are carried by specific proteins in the membrane. But, because it is against the concentration gradient and even though carrier protein is present, the action requires additional input of energy other than the kinetic energy of the molecules themselves. This additional energy is usually supplied by the energy currency of a cell called ATP (adenosine tri-phosphate). Energy in the form of ATP is used up in active transport.

Examples of active transport: (a) Sodium–potassium pumps in which sodium ions are pumped out of the cells up a concentration gradient and potassium ions are pumped into the cells up a concentration gradient. During the transport, one ATP is used up, two potassium ions (K^+) are imported, and three sodium ions (Na^+) are exported. (b) Calcium ion (Ca^{2+}) pumps are widely found in eukaryotes and push these ions from cytoplasm to outside the cell and into the endoplasmic reticulum (ER). So, the ion concentration is typically high outside the cell and inside ER and low in the cytoplasm. This gradient difference has implications for regulatory control of various processes. (c) Proton pumps that facilitate the movement of H^+ ions.

Bulk transport by endocytosis/exocytosis By active transport, the membrane proteins help to move a few ions and molecules at a time. However, if the cell needs larger particles or a collection of materials, it has to transport in bulk. For this, the cell wraps the material around the membrane and transfers it to the other side through vesicles rather than passing the molecules one by one. Such membrane-bound vesicles fuse with membranes and transport them irrespective of whether they are hydrophobic or hydrophilic.

When the contents are moved from outside the cell to the inside, it is called endocytosis, and when the cellular contents are released outside the cell from inside, it is called *exocytosis*. Membranous sacs containing material (exocytic vesicle membrane) from the cell migrate to the plasma membrane and fuse with it, and release the contents to the exterior of the cell (Figure 5.13). For example, cells of the tear glands release tears; brain cells release neurotransmitters, which helps neurons to communicate; and stomach releases digestive enzymes in this way.

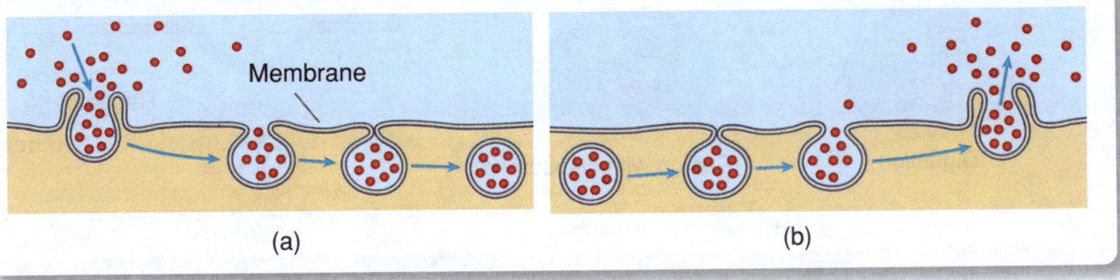

Fig. 5.13 Bulk transport: (a) Endocytosis; (b) Exocytosis

Endocytosis is the reverse mechanism—materials enter the cell. When a patch of plasma membrane balloons inward and pinches off as an endocytic vesicle, it then moves into the cytoplasm.

Depending on the material to be transported, endocytosis can be of three types, as follows (Figure 5.14):

(i) *Phagocytosis*, meaning 'cell-eating'. The material to be engulfed touches the surface of the plasma membrane, which causes the plasma membrane to form an indentation. This small area of plasma membrane sinks inward to form a pocket, which encloses the material to be transported. The pocket then becomes deeper and is pinched off from the membrane into a sac-like vacuole called *phagosome*. This phagosome then fuses with the lysosome. Lysosomes contain lytic enzymes, which then break down the contents of the vacuole and release the material inside the cell.
For example, the white blood cells (WBCs) in our body engulf foreign microbes and, hence, are also known as *phagocytes*. Unicellular protozoans like amoeba engulf and obtain food particles for their need by phagocytosis.

(ii) *Pinocytosis*, meaning 'cell-drinking'. In this process, liquids and materials dissolved in liquids are taken inside the cell in small sacs called vesicles.

(iii) *Receptor-mediated endocytosis* provides a mechanism for the selective uptake of specific macromolecules. Molecules to be internalized bind to specific receptors on the plasma membrane, the concentration of which is called clathrin-coated pits. The membrane

then folds, engulfs, and pitches out to form clathrin-coated vesicles, which then bind to lysosomes for, then, being released into the cells.

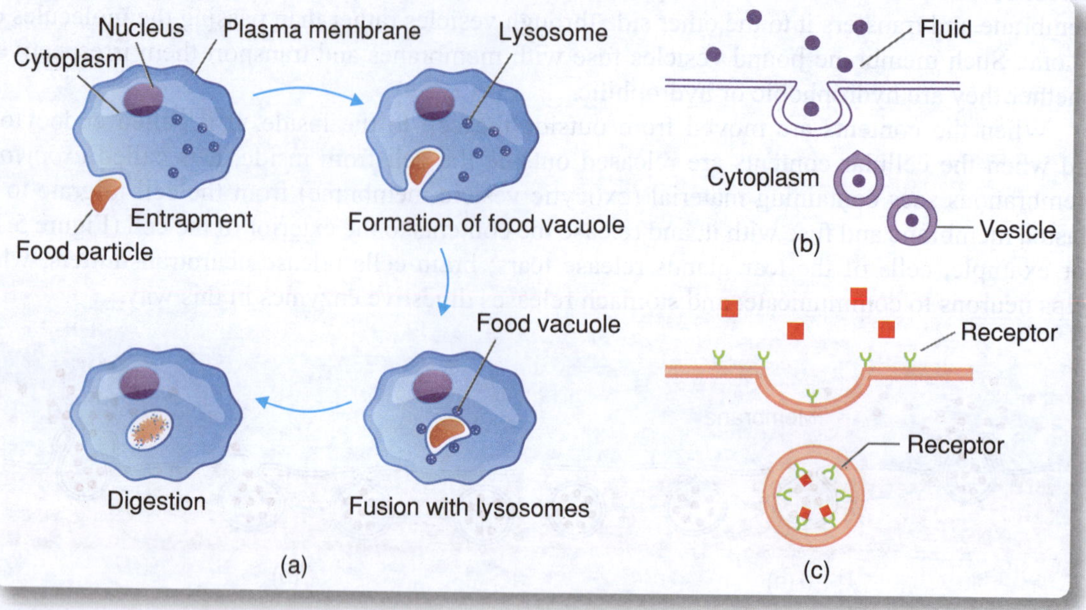

Fig. 5.14 Types of endocytosis: (a) Phagocytosis; (b) Pinocytosis; (c) Receptor-mediated endocytosis

The Endomembrane System—Membranous Organelles

When you study the micrograph of a eukaryotic cell, there are many organelles distributed inside the cell. While some are non-membranous, many of these organelles are made up of membranes and are called *membranous organelles* (Figure 5.15).

In the eukaryotic cell, the advantage of partitioning the cell into intracellular organelles helps in compartmentalization, which allows a large number of activities to proceed simultaneously

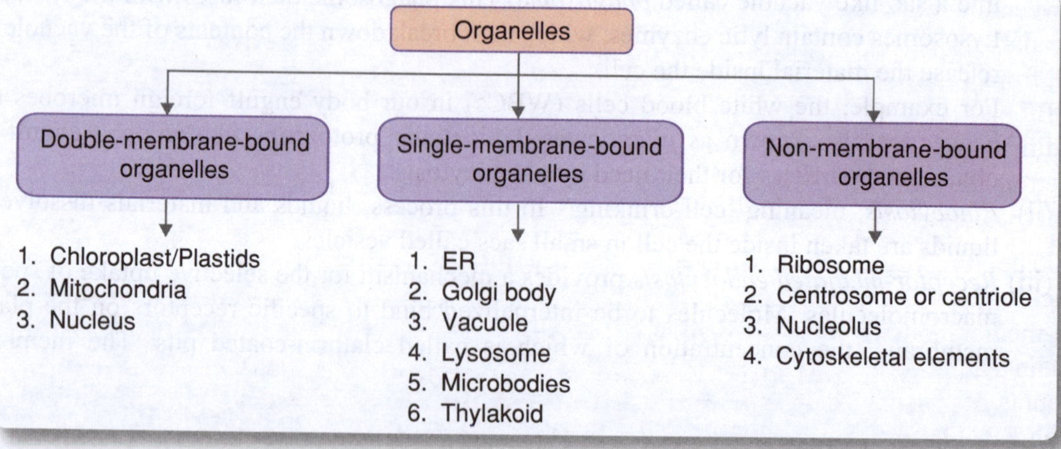

Fig. 5.15 Classification of organelles based on membrane

within a small space (Figure 5.16). Such a distribution allows separation of physically incompatible reactions and, at the same time, ensures that interconnected reactions proceed sequentially and in a coordinated manner at different time periods.

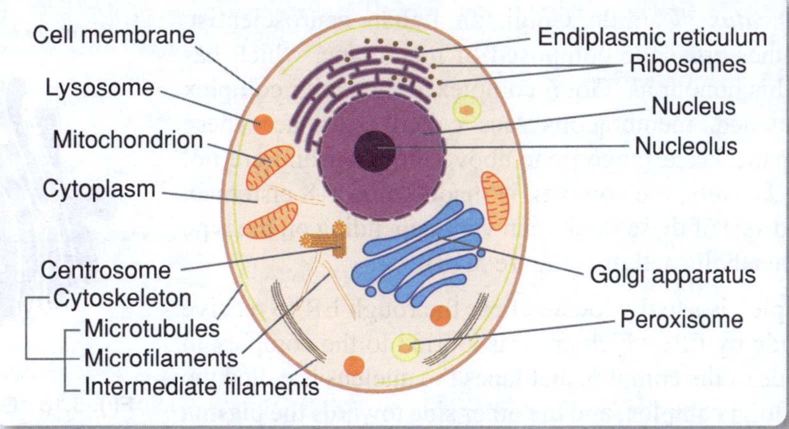

Fig. 5.16 A eukaryotic cell and the organelles

Other than the plasma membrane, the other membranous organelles inside the eukaryotic cell include endoplasmic reticulum (ER), Golgi apparatus, lysosome, and other vacuoles. These constitute the endomembrane system of organelles, which is extensive and dynamic in nature. Each of these membrane organelles has their own unique structure and architecture, which is associated with their specific functions. The details of each membranous organelle are as follows:

Endoplasmic reticulum (ER) is a large network of folded membranes and tubes throughout the cell, which begins at the nuclear membrane and extends into the cytoplasm till the plasma membrane. The folds of the ER provide a large surface area for chemical reactions. The arrangement of proteins on this surface controls the sequence of metabolic activities to occur rapidly and accurately. The spaces between the folded membranes form tubes that serve as canals for the movement of molecules within the cell from one part to another. The main function of ER is concerned with the synthesis, folding, modification, and transport of proteins and lipids. Electron microscopic analysis reveals ER to be of two types: rough ER and smooth ER (Figure 5.17).

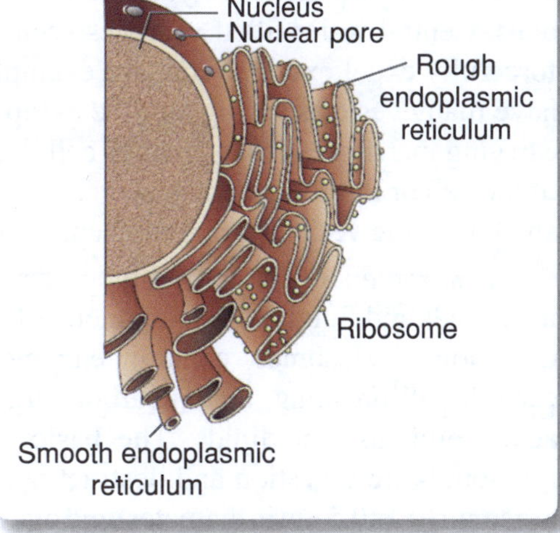

Fig. 5.17 Endoplasmic reticulum—arrangement and types

The rough ER has ribosomes attached to it, which gives it the rough appearance. Ribosomes are non-membranous organelles of the cell involved in protein synthesis. Cells capable of synthesizing large amounts of proteins, such as pancreatic cells and cells of salivary glands, are found to be rich in rough ER. Enzymes of the rough ER are also involved in synthesizing phospholipids, which get inserted into the ER membrane and, thus, help the ER membrane grow. On the contrary, smooth

ER lacks ribosomes but is the site for other activities such as detoxification reactions and fat metabolism. Liver cells detoxify substances such as alcohol, drugs, and poisons; hence, they have higher levels of smooth ER. Similarly, cells in ovaries and testes producing steroid sex hormones are enriched with smooth ER.

Golgi apparatus Camillo Golgi, an Italian neuroscientist, discovered another organelle composed of membrane, which has been named in his honour as 'Golgi complex'. The Golgi complex consists of flattened, membranous sacs called cisternae. These sacs look like pancakes arranged one above the other and are not interconnected. Usually, the complex is made up of 3–8 cisternae. The number and size of these stacks can vary depending on the type of cell and its metabolic activity (Figure 5.18).

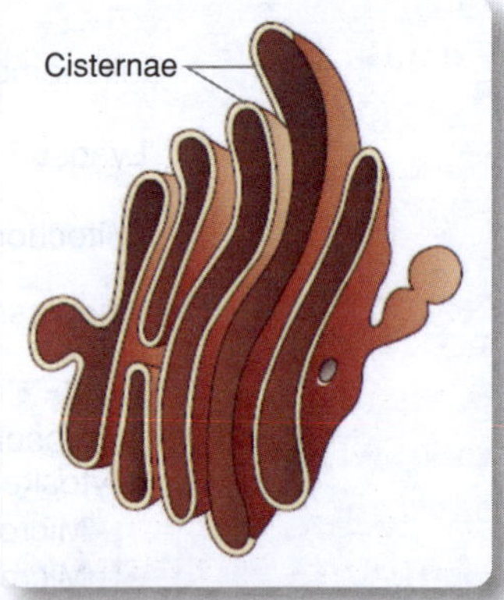

Fig. 5.18 Golgi apparatus—structure and location

Golgi complex is usually located near the rough ER to receive the proteins made by ER, which are transported to the complex in vesicles. The side of the complex that faces the nucleus is called the cis-face of the Golgi complex, and the other side towards the plasma membrane is the trans-face. The vesicles fuse with the cis-face and release their contents. Within the Golgi complex, the proteins are chemically modified and transported to the trans-face, from where the vesicles bud off to be taken to other parts of cells. Chemical modifications within the complex include removing amino acid segments, adding functional groups, and adding lipid and carbohydrate moieties to proteins. Thus, the Golgi apparatus performs three functions of: modifying molecules shipped to it, manufacturing polysaccharides and lipids, and packaging modified proteins within sacs/vesicles for delivery to different locations in the cell. The Golgi apparatus produces different types of vesicles.

Vacuoles and vesicles Many sac-like membrane enclosures called vacuoles and vesicles are found inside the Eukaryotic cells. Vacuoles are larger in nature and vesicles are the smaller ones. Central water-filled vacuoles seen in plant cells and contractile vacuoles in protozoans, to forcefully expel excess water, are examples of vacuoles in living systems. Animal cells typically have many vesicles throughout the cytoplasm; most of these emerge from the Golgi complex and carrying materials such as mucus, cellulose, glycoproteins, insulin, and enzymes to different parts of the cell or the cell exterior. Secretory vesicles release contents to the cell exterior by exocytosis, and endocytic vesicles bring molecules into the cell from the extracellular environment.

Lysosome is a common type of vesicle found in animals and fungi. It is an organelle of intracellular digestion. Lysosomes contain enzymes that are capable of breaking down carbohydrates, nucleic acids, proteins, and lipids. The basic functions of lysosomes are digestion and destruction. Lysosomes are small (0.1–0.5 μm in diameter) and oval or spherical in nature (Figure 5.19). A human cell has about 300 lysosomes. Hydrolytic enzymes are synthesized in the ER and are packaged into lysosomes in the Golgi complex. Cellular components are made up of the four biological macromolecules mentioned earlier;

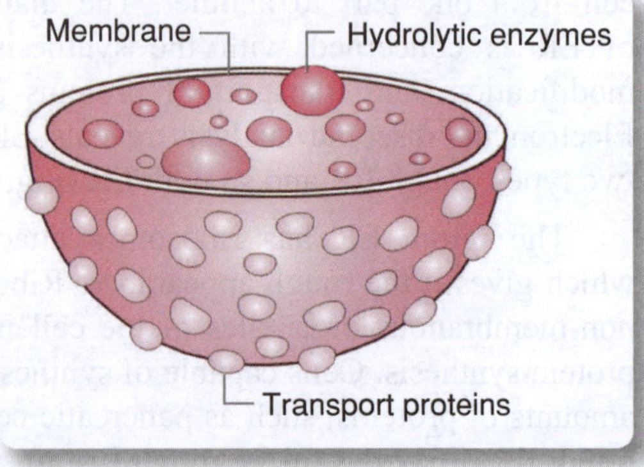

Fig. 5.19 Lysosome

then how do the cellular components remain unaffected? The cell remains protected because of the following reason:

The enclosure of hydrolytic enzymes within an organellar compartment helps in the survival of the cell. Lysosomes act as an acidic compartment of the cell. The hydrolytic enzymes of lysosomes are acid hydrolases and are active in acidic pH but not at neutral pH. Hence, the intralysosomal pH (~5) is maintained. The cytosolic pH on the contrary is between pH 6.5 and 7.2, and this pH difference is effectively maintained by lysosomal membrane. This ensures that hydrolytic enzymes do not function if released into the cell.

To maintain their internal acidic pH, lysosomes must actively concentrate H^+ ions (protons). This is accomplished by a proton pump in the lysosomal membrane, which actively transports protons into the lysosome from the cytosol. This pumping requires expenditure of energy in the form of ATP hydrolysis, since it maintains approximately a 100-fold higher H^+ concentration inside the lysosome.

Functions of lysosomes:

(i) Breakdown of food vacuoles in protozoans and amoeba
(ii) Destruction of disease-causing microbes, such as bacteria, virus, and fungi, by WBCs during phagocytosis, by fusing with phagosome
(iii) Breakdown of worn-out cell organelles and fusing with them and destroying them
(iv) Role in normal development of frog—the cells of the tail of tadpole are destroyed by lysosomes.
(v) In human development, the embryo initially has webbed hands and feet. The skin between the fingers and toes is destroyed by lysosomes. If this does not happen the child is born with webbed fingers or toes and the condition is called 'syndactylism'.

Peroxisomes Another of the organelles that contain enzymes in a sac-like enclosure or vesicle are peroxisomes. These are different from lysosomes as they are not formed by the Golgi apparatus and the enzymes are different (Table 5.2). Peroxisomes are involved in the breakdown of long-chain fatty acids, amino acids, uric acids, etc. and the synthesis of cholesterol and bile acids from cholesterol. The product of oxidative reactions inside peroxisome is hydrogen peroxide, which is toxic, but the enzyme catalase breaks it down to water and oxygen (Figure 5.20).

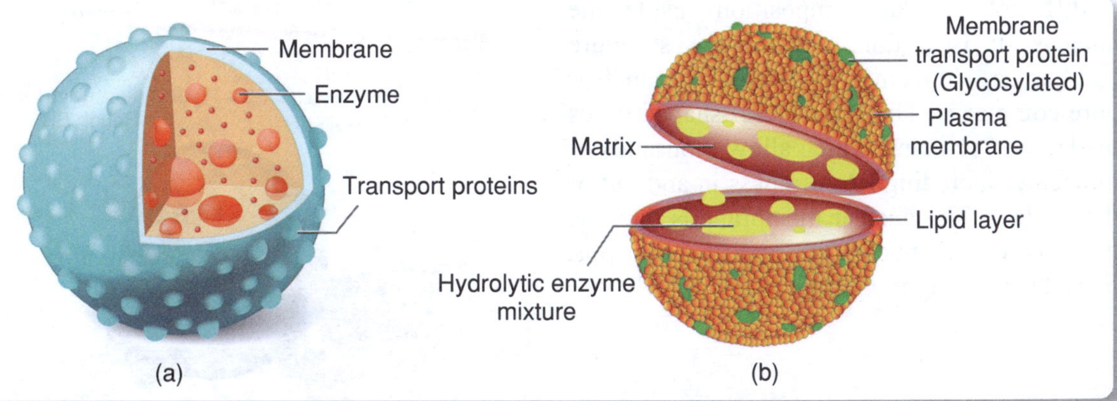

Fig. 5.20 (a) Lysosome; (b) Peroxisome

Table 5.2 Lysosomes Versus Peroxisomes

S. No.	Lysosomes	Peroxisomes
1.	Break down biological polymers like proteins and polysaccharides	Oxidize organic compounds, thus breaking down metabolic hydrogen peroxides
2.	Consist of degradative enzymes	Consist of oxidative enzymes
3.	Responsible for the digestion process in the cell	Responsible for protecting the cell against metabolic hydrogen peroxide
4.	Found in animals	Found in all eukaryotes
5.	Derived from either Golgi apparatus or endoplasmic reticulum	Derived from the ER and are capable of replicating by themselves
6.	Comparatively large in size	Comparatively small in size
7.	Involved in endocytosis, autophagy, and phagocytosis	Involved in biosynthesis of lipids and photorespiration
8.	Degradative reactions do not generate energy	Oxidative reactions in peroxisomes generate ATP energy

Interesting Fact 5.3

Lysosomal storage disease

It is a condition in which one of the hydrolytic enzymes of lysosome is absent, and substrate accumulates and interferes with normal cellular activities.

For example, Tays–Sachs disease, which is a fatal disease of central nervous system, caused by failure to synthesize the enzyme needed for hydrolysis of fatty acid derivatives found in brain and nerve cells. It is a rare, inherited disorder and causes too much of fatty substance build up in tissues and nerve cells of the brain.

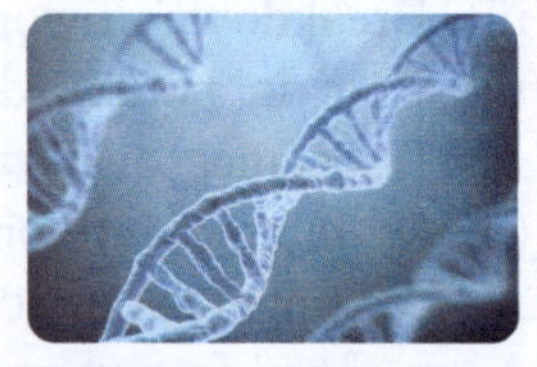

Nuclear membrane Nucleus, in the centre of a cell, is covered by a specialized nuclear membrane in eukaryotes. The nucleus contains the genetic material, and the membrane separates it from cytoplasm, maintaining a different cellular composition inside the nucleus. It is a double-membrane structure and has large openings in it called the nuclear pore complexes. These are barrel-shaped pores made up of proteins, which allow thousands of molecules including RNA to pass in and out of the nucleus. The nuclear membrane is thus the gateway for inter nuclear–cytoplasmic transport of molecules (Figure 5.21).

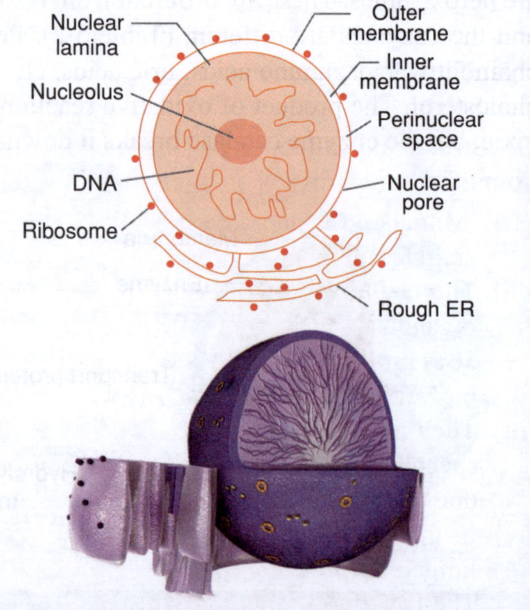

Fig. 5.21 The double-layered structure of the nuclear membrane

Interconversion of Membranes

All the organelles studied so far are membranous in nature, which means that they are made of the phospholipid bi-layer with proteins and other molecules. Interestingly, all these organelles can be converted from one form to another, depending on the need of the cell (Figure 5.22). If we try to understand this interconversion through an analogy, the same cloth material can be used to make different products such as coats, shirts, blouses, kerchiefs, etc., and the wear and tear in these could be repaired with the same base cloth. As an example, plasma membrane is continuous with ER, so if a cell needs more plasma membrane when it becomes bigger, the ER could move and provide the additional plasma membrane. Similarly, the nuclear membrane is also attached to ER, Golgi body could receive membrane sacs from ER, lysosomes are produced by the Golgi complex, and so on.

Fig. 5.22 The interconvertible endomembrane system of eukaryotic cells

Energy Converters—Mitochondria and Chloroplasts

In the cytoplasm of a eukaryotic cell, there are two more extremely important membranous organelles—the mitochondria and the chloroplasts. Both organelles are involved in energy metabolism.

Structurally, although these two organelles are made up of membranes, they are very different from other membranous organelles studied thus far.

(i) Mitochondria and chloroplasts have a double-membrane structure—an outer membrane and an inner membrane.

(ii) The chemical composition of their membranes is very different from other membranous organelles. For example, the outer membrane of mitochondria contains 50% lipid and 50% proteins. Specialized outer membrane proteins called porins help in transport of large molecules and are a critical component of outer membrane.

(iii) They have ribosomes and DNA, which is similar in structure to bacteria; hence there is a speculation that these were prokaryotic organisms that entered Eukaryotic cells and remained due to mutual benefit, as per the endosymbiont theory.

(iv) In this sense, they have a certain level of independence from the rest of the cell such as limited ability to reproduce themselves, with the assistance of nuclear DNA from the cell. The DNA present in these organelles is necessary for their reproduction, without which the cell cannot make these organelles.

Having understood the broad features of these two energy converters, details of their unique structural and functional roles are given below.

Mitochondria are the site of cellular respiration. In cellular respiration, energy-rich molecules such as sugars and other fuels are broken down to yield energy for the cell. Many of the reactions of aerobic cellular respiration reactions take place in mitochondria, thus producing energy in the form of ATP. Hence, mitochondria is also called the 'powerhouse' of the cell.

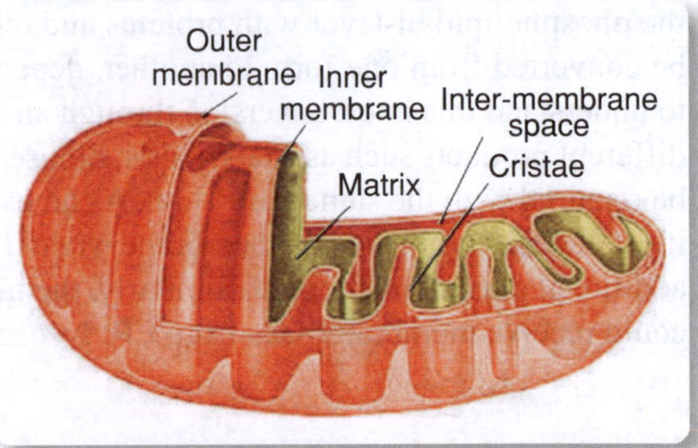

Fig. 5.23 Mitochondrion—structure

Mitochondria are present in virtually all eukaryotic cells, but their numbers and sizes vary depending on the metabolic activity. A typical mitochondria is around 0.5 mm wide and has a length of 2–3 μm.

Structurally, mitochondria is made up of two membranes—the outer membrane that is smooth and covers the organelle and the inner membrane that has multiple folds in it called cristae. The intermembrane space between the outer and inner membranes is 6–8 nm wide. Around the cristae and inside the outer membrane is the fluid-filled compartment of mitochondria called the mitochondrial matrix. Mitochondrial DNA, ribosomes, and related machinery are all present in the matrix. Primarily, the ATP-generating reactions are localized to cristae and matrix as enzymes essential for catalysing sugar metabolism are localized here (Figure 5.23).

Chloroplast Plastids are specialized organelles found in plants. Amyloplasts are plastids typically used to store starch, lipids, and proteins in plants. The other plastid, chloroplasts, are the sites for photosynthesis in plant cells and green algae.

These are double-membrane structures, typically 0.5–2.0 μm wide and 10 μm long. Two regions can be distinctly seen inside the chloroplast: (a) stroma and (b) grana. The double membrane encloses the fluid-filled interior compartment called stroma. In this stromal compartment, there are networks of membrane-enclosed disks and tubes called grana. Each grana is made of flattened closed sacs called thylakoid membranes, stacked one on top of another. Stroma lamellae connect thylakoids of two different grana (Figure 5.24).

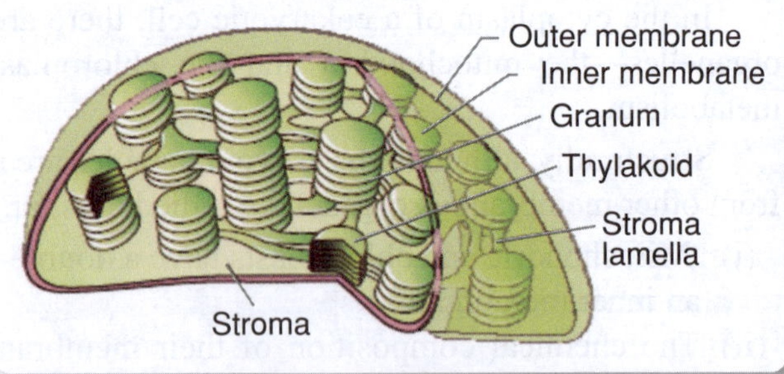

Fig. 5.24 Chloroplast—structure

The main pigment of chloroplast—chlorophyll, responsible for photosynthesis—is present in the thylakoid membranes of grana, which acts as the solar power packs of the cell and helps to trap light energy and convert it to chemical energy by the process of photosynthesis. Stroma is devoid of chlorophyll pigments. The chloroplast stroma contains DNA and ribosomes, similar to the mitochondrial matrix. Interesting to note that there are three internal compartments within the chloroplast—the intermembrane space between the outer and inner membranes, the stromal cavity, and the thylakoid lumen.

Other Non-membrane Organelles

The cell also houses various organelles that are not made up of membranes but have critical functions.

Nuclear components One of the critical components for any cell to function is its genetic material. The genetic material is usually DNA and has instructions for the cell to function. Eukaryotic cells usually house their genetic material housed inside a well-defined nuclear membrane. This creates a central structure called the nucleus, which is the space enclosed by the nuclear membrane and houses the genetic material (Figure 5.25). Prokaryotes do not have a well-defined nucleus and their genetic material is suspended in the cytoplasm. A cell that lacks a nucleus/genetic material dies.

The DNA along with proteins is loosely arranged inside the nucleus in the form of a chromatin (which means coloured material), as it stains heavily compared with the rest of the nucleus. When this chromatin condenses during cell division, it becomes tightly coiled and is then called a chromosome. Chromatin and chromosomes differ in their structural arrangement but are essentially the same. The genetic material of any organism is primarily distributed in the nuclear chromatin material. The number of chromosomes depends on species, e.g., humans have 46 chromosomes in their cells, dog cells have 78, frogs have 26, and so on.

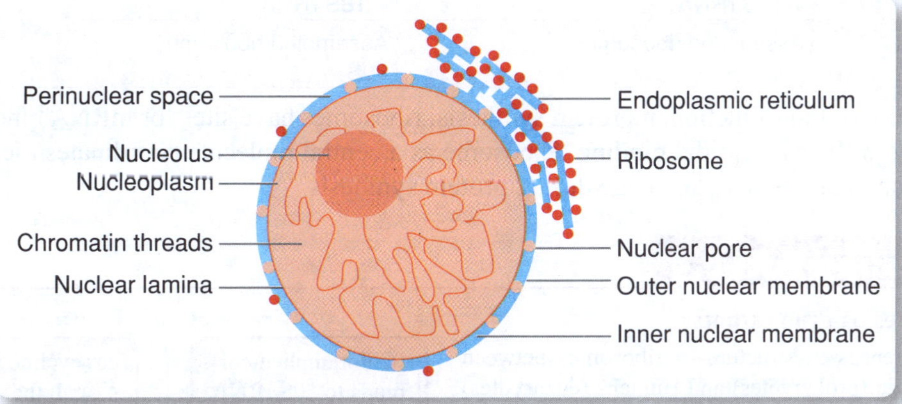

Fig. 5.25 Nuclear components along with a nuclear envelope

Apart from this, the liquid or the semi-liquid substance within the nucleus is called nucleoplasm. Although the majority of the space inside the nucleus is occupied by chromatin, it also contains a spherical dense mass of fibres and granules called nucleolus (plural, nucleoli). Within the nucleolus, the rRNA is transcribed, which combines with proteins to form ribosomal sub-units. These sub-units then reach cytoplasm and form complete ribosomes during the transcription of protein synthesis or associate with ER.

Ribosomes Ribosomes are important for both prokaryotes and eukaryotes as they are responsible for protein synthesis. Ribosomes are composed of RNA and proteins. Each ribosome has two sub-units (Figure 5.26).

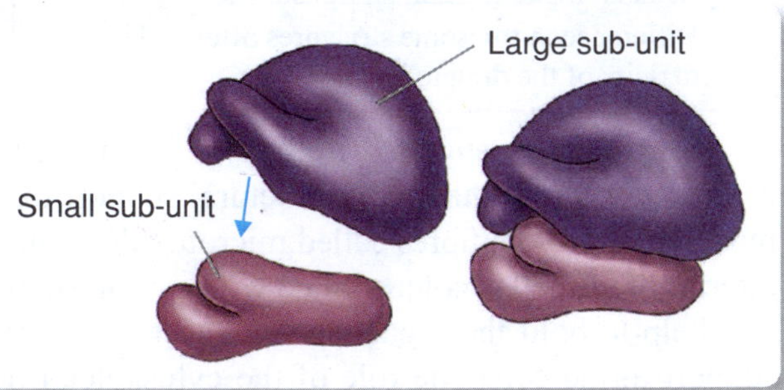

Fig. 5.26 Ribosome

Although all organisms contain ribosomes, there are some distinct differences between the eukaryotic and prokaryotic ribosomes. Eukaryotic ribosomes are larger than bacterial or archaeal ribosomes. Prokaryotic ribosomes are 70S ribosomes, with a smaller sub-unit of 30S and larger sub-unit of 50S (S, Svedberg constant—a measure of rate of sedimentation). On the contrary, eukaryotic ribosomes are 80S ribosomes, with 40S smaller sub-unit and 60S larger sub-unit. The rRNAs form the structural backbone of ribosomes. There are three rRNAs in prokaryotic ribosomes: 16S, 5S, and 23S, whereas the eukaryotic ribosome contains four rRNAs: 18S, 5S, 5.8S, and 28S. In prokaryotes, ribosomes are found free in cytoplasm while in eukaryotes they are found attached to ER and nuclear membrane in addition to their presence in cytoplasm (Table 5.3).

Table 5.3 Prokaryotic Versus Eukaryotic Ribosomes

S. No.	Prokaryotic Ribosomes	Eukaryotic Ribosomes
1.	Large sub-unit (LSU) 50S LSU • 5S rRNA • 23S tRNA	Large sub-unit (LSU) 60S LSU • 26S or 18S rRNA • 5.8S tRNA
2.	Small sub-unit (SSU) 30S SSU • 16S rRNA	Small sub-unit (SSU) 40S SSU • 18S rRNA
3.	Assembled ribosome 70S	Assembled ribosome 80S

Because of their function in protein synthesis, ribosomes have sites for mRNA binding, tRNA binding, as well as a peptidyl binding. Ribosome as a central molecule coordinates the nucleotide information to amino acid language during protein synthesis.

Interesting Fact 5.4

Ribosomes as drug targets

- Differences in structure of ribosomes between bacteria (prokaryotes) and Humans (eukaryotes) is exploited to treat bacterial infections. Ribosome inhibitors are among the most successful antimicrobial drugs.
- Streptomycin, gentamycin, and tetracycline are antibiotics that target the bacterial ribosome, thus interfering with bacterial protein synthesis without impacting human ribosomes.
- Difference in ribosome structures offers selective toxicity of the drug to bacteria.

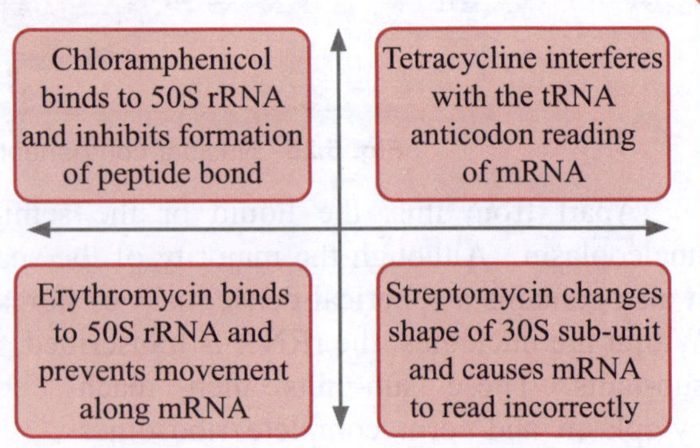

Cytoskeletal network of the cell The characteristic shape of a cell and the internal organization need a framework that gives structural support to the cell. The cytoskeletal network of the cell is made up of protein fibres called microtubules, microfilaments, and intermediate filaments. So, the interior of the cell is not just filled with cytoplasm but has these interconnected cellular components that help to hold the organelles, connect to plasma membrane, and give structural integrity and support to the cell. The role of the cytoskeletal network is especially critical in animal cells as they lack rigid cell walls. The cytoskeleton is dynamic, and the component parts are continuously assembled and disassembled (Figure 5.27).

Microtubules are hollow tube-like structures made of tubulin, whereas microfilaments are made of actin. The intermediate filaments, made of intermediate family proteins that include keratins, lamins, etc., found in different organelles, connect the microtubules and microfilaments with each other. Structurally, to understand the arrangement of microtubules, microfilaments, and intermediate filaments, imagine an open umbrella. The central long handle with which you hold the umbrella is the microtubule, the lines on the umbrella, the frame of which are connected to this pole and are attached to the cloth of umbrella, are the microfilaments and the ones joining the pole and the lines on the umbrella, which open and close the umbrella, are the intermediate filaments. However, an open umbrella is semicircular, imagine the same as two semicircles forming a spherical cell.

The function of the cytoskeleton is analogous to the bony skeleton of our body. When the cell is suppressed or when it is on a surface it cannot stick to, it helps the cell to pull up and form a sphere. So, it provides anchorage and reinforcement to the cell as well as to many organelles within the cell. For example, the nucleus is anchored and held in its place inside the cell. During cell division, the microtubules and microfilaments help in moving chromosomes for separation along with other adjustments for cells to divide. The cytoskeleton also transports the organelles from place to place within the cytoplasm. For example, lysosomes might move to a food vacuole for digestion by gliding over the cytoskeletal network.

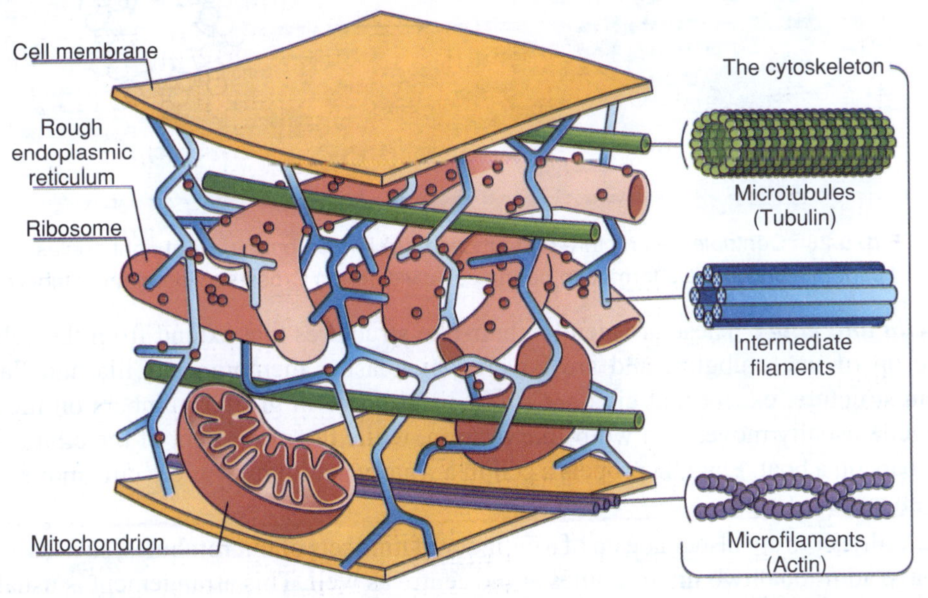

Fig. 5.27 Fibres and overall cytoskeletal network: (a) Microtubule; (b) Microfilament; (c) Structural arrangement of microtubule, microfilament, and intermediate filaments

Centrioles and spindle In the central part of an animal cell, there is another structure called centriole, which is also made up of microtubules. The centriole is located just outside the nuclear membrane in a microtubule-organizing centre called centrosome. Each centriole is composed of two sets of nine microtubules each, placed at right angles to each other. Each set of microtubules has a barrel-shaped cylindrical structure made up of microtubule triplets (Figure 5.28).

Functionally, the centrioles are responsible for organizing the microtubules into a complex of fibres called spindle. This has great significance during cell division as the spindle fibres attached

to each chromosome help the chromosomes to separate properly. Centrioles are also involved in the formation of the basal body in flagella and cilia.

A curious fact is that plant cells do not have distinct centrioles; however, microtubule assembly into spindle fibres still happens outside the nuclear envelope, which is clearly visible during cell division.

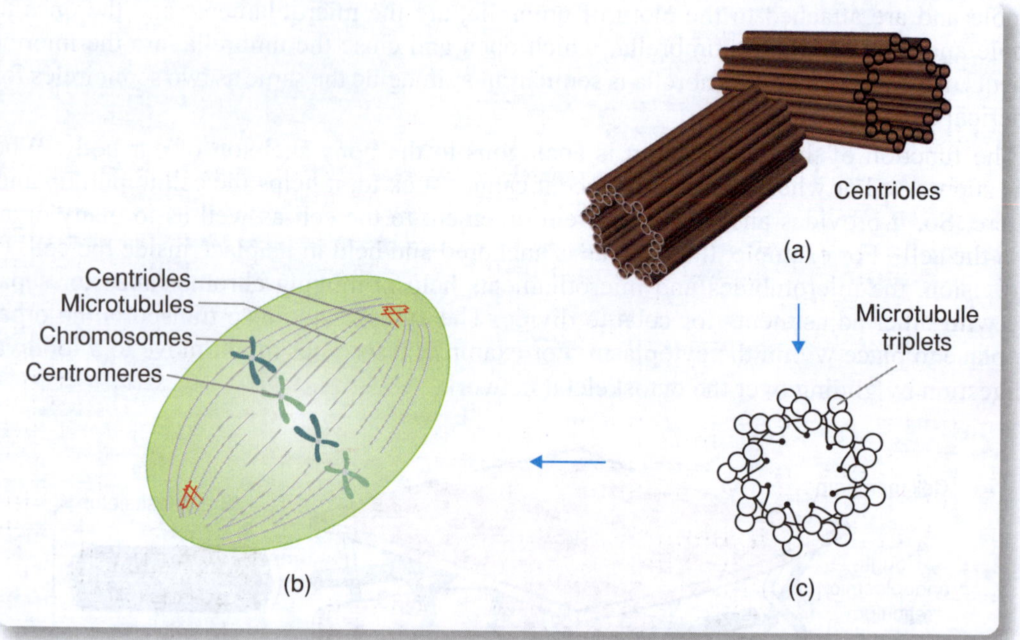

Fig. 5.28 Centrioles—structure and function; (a) Microtubule 9 + 9 at right angles; (b) Role of centriole and spindle formation during cell division; (c) Cross-section of microtubule triplets

Flagella and cilia These are slender, hair-like structures that extend from the cell surface, also made up of microtubules, and are covered by plasma membranes. Cilia and flagella are identical in structure, except that cilia are shorter and occur in greater numbers on the cell. The longer flagella usually moves in a whip-like fashion, while the movement of the cilia is back and forth, like oars on a boat. Flagella propels a cell in a watery medium, whereas cilia moves the fluids over the cell surface (Figure 5.29).

Structurally, both are also made up of a cylinder of nine sets of microtubules, such as a centriole, but they have additional two microtubules at the centre as well. This arrangement is usually called the 9+2 arrangement. A system of spokes and rings stabilizes this 9 + 2 array. This array that arises from a centriole and near the plasma membrane is called the basal body.

Examples of cells having cilia or flagella are: (a) Human sperm cell contains flagellum for motility; (b) protozoan paramoecium is covered with thousands of cilia; (c) human cells in lungs contain cilia that line air passage and sweep out mucus-containing bacteria and other contaminants; and (d) cilia in oviducts conduct eggs from ovaries to the uterus.

Prokaryotes such as bacteria and archaea also contain flagella. Functionally, their role is similar to eukaryotic flagella and helps in locomotion, but components of flagellar machinery, structure, anchorage, and nature of movement are different.

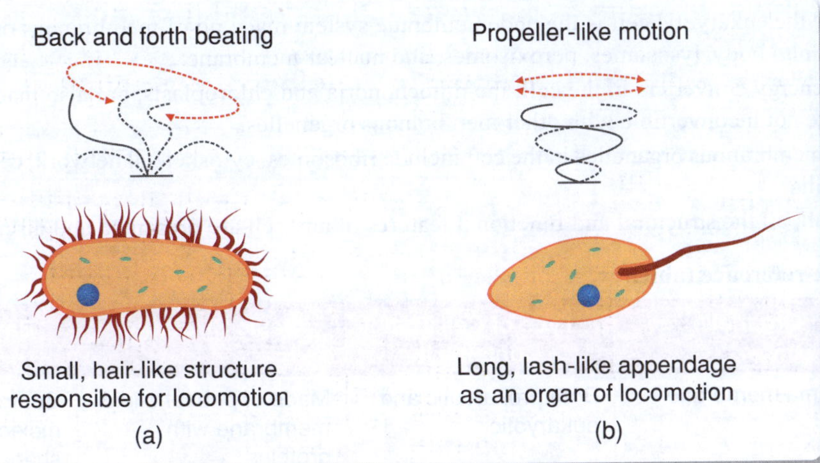

Fig. 5.29 Structure of: (a) Cilia; (b) Flagella

Interesting Fact 5.5

Differences Between Prokaryotic Flagella and Eukaryotic Flagella

Prokaryotic (Bacterial)	Eukaryotic
Smaller and simpler in structure	Larger and complex in structure
Made up of protein flagellin (53 kDa sub-unit)	Made up of Tubulin (9 + 2 microtubule arrangement)
Rotatory movement	Bending movement
Proton-driven	ATP-driven

Other specific structures in eukaryotic cells In addition to all the listed organelles mentioned here, plant cells also contain an external cell wall that supports and protects plant cells and a central vacuole, which has diverse roles in storage, structural support, and cell growth. Similarly in multicellular animals, there are cell-adhesion molecules that hold cells and tissues together along with cell junctions that reinforce cell adhesions. An extracellular matrix organizes the cell exterior and helps in protection and support.

SUMMARY

- Cell is the structural and functional unit of living organisms.
- Prokaryotic cells have a simple structure and are smaller than eukaryotic cells.
- Specialized organelles are not found in prokaryotic cells; only outer limiting membrane, ribosomes, cytoplasm, genetic material, and energy currency are common factors in any type of cell.
- Plasma membrane has a Fluid Mosaic model and is the semi-permeable barrier that gives the cell its shape and controls the transport of material.
- Different transport mechanisms operate in cells such as diffusion, osmosis, facilitated diffusion, active transport, and phagocytosis.

- Inside the eukaryotic cell is the endomembrane system made up of membranous organelles such as ER, Golgi body, lysosomes, peroxisomes, and nuclear membrane.
- Two energy converters in the cell, the mitochondria and chloroplasts, are also made of membranes but are not inconvertible with other membranous organelles.
- Non-membranous organelles in the cell include ribosomes, cytoskeletal network, centrioles, flagella, and cilia.
- Overall, all the structural and functional features of any cell are presented and discussed.

Quick-reference table:

Organelle	Type of Cell That Contains It	Basic Structure	Basic Function
Plasma membrane	Both prokaryotic and eukaryotic	Made of phospholipid membrane with proteins	Controls transport of molecules and gives shape to cell
Genetic material	Both prokaryotic and eukaryotic	Non-membranous, made of DNA; with some proteins	Hereditary information; has genes for protein synthesis and characters
Ribosome	Both prokaryotic and eukaryotic	Non-membranous; made of protein and RNA	Site of protein synthesis
Cilia and flagella	Both prokaryotic and eukaryotic	Prokaryotic composed of single type of protein Eukaryotic contains microtubules in 9 + 2 arrangement	Helps in movement and locomotion
Cytoskeleton	Eukaryotic	Made of microtubules, microfilaments, and intermediate filaments; each made of different protein	Structural support, aids in shape and movement of organelles
Fully formed nucleus with nuclear membrane enclosing nucleolus	Eukaryotic	Double membrane with nucleoplasm	Contains genetic material and is the site of ribosome manufacture and storage
Endoplasmic reticulum	Eukaryotic	Membranes forming sheets and canals	Surface for chemical reactions and intracellular transport
Golgi apparatus	Eukaryotic	Stacks of single membranes	Protein modifications and packaging into vesicles
Vacuoles and vesicles—lysosomes; peroxisomes	Eukaryotic	Membrane-enclosed sacs containing enzymes	Breakdown by enzymes

[Contd.]

[Contd.]

Mitochondria	Eukaryotic	Double-membrane organelle with folds called cristae and mitochondrial matrix	Site of aerobic cellular respiration for release of energy from nutrients
Chloroplast	Eukaryotic	Chlorophyll-containing structure with grana and stroma	Photosynthesis in plant cells, capture of light energy, and production of carbohydrates
Centriole	Eukaryotic	Two clusters of nine microtubule triplets at right angle	Formation of spindle during cell division

EXERCISES

A. Objective-Type Questions

1. Which of the following is not a membranous organelle?
   (a) Nuclear membrane (b) Mitochondria
   (c) Ribosome (d) Endoplasmic reticulum
2. Plasma membrane is made up of __________.
   (a) Phospholipids, proteins, cholesterol, and glycomoieties
   (b) Cellulose
   (c) Only phospholipids
   (d) Phospholipids and proteins but no glycoproteins
3. Which of the following statements is true about plasma membrane?
   (a) Is static in nature (b) Is hard and rigid
   (c) Is non-permeable (d) Fluid-like in consistency
4. Which of the four is an example of active transport?
   (a) Diffusion (b) Osmosis
   (c) Facilitated diffusion (d) Phagocytosis
5. Which of the following organelles does not have any association with ribosomes?
   (a) Cytoplasm (b) Endoplasmic reticulum
   (c) Nucleolus (d) Centriole
6. Hydrolytic enzymes in the cell are housed in which organelle?
   (a) Endoplasmic reticulum (b) Chloroplast
   (c) Lysosome (d) Golgi apparatus
7. Which of the following organelles is present in both prokaryotes and eukaryotes?
   (a) Mitochondria (b) Cytoskeleton
   (c) Chloroplast (d) Flagella
8. Structurally, a centriole contains __________.
   (a) Microtubules made of tubulin protein
   (b) Membrane made up of phospholipid
   (c) Ribosomes made up of RNA and protein
   (d) Chromatin made up of DNA

9. Chloroplasts are found only in __________ cells.
   (a) Human cells
   (b) Rose cells
   (c) *Escherichia coli* cells
   (d) Paramoecium cells
10. Difference between rough and smooth ER is the presence of __________.
   (a) Membrane
   (b) Ribosomes
   (c) Lysosomal vesicles
   (d) None of the above

B. Short-Answer Questions

1. In food preservation, how does adding excess salt help in preventing spoilage by bacteria?
2. What is one feature that is common in bacterial, plant, and fungal cells but is not found in animal cells?
3. If an electron micrograph of a eukaryotic cell shows a large number of mitochondria and ER, what kind of cellular activities would you anticipate that would require a huge number of these two organelles?
4. Give day-to-day examples of osmosis that you see in real life.

C. Self-Study Questions

1. Prepare different plant, bacterial, and other materials and view them one by one under a light microscope. Observe the cell structure and try to enlist visible cell organelles.

USEFUL WEBLINKS

1. https://www.etsu.edu/uschool/faculty/tadlockd/documents/apbio_06_detaillectout.pdf
2. https://biobeat.nigms.nih.gov/2021/03/take-a-tour-of-your-cells-organelles
3. https://www.nature.com/scitable/topicpage/what-is-a-cell-14023083/

VIDEO LINKS

1. https://sciencewiz.com/portals/cells/tour-inside-the-cell/a-tour-of-the-cell-more-advanced/

CHAPTER 6
Genetics of Life

LEARNING OBJECTIVES

After reading this chapter, you will be able to:

- Understand genetic material—structure, organization, and genomes
- Decode the central dogma of life— genes to mRNA to protein; DNA to DNA
- Analyse the process and outcomes of replication
- Delineate the steps involved in protein synthesis and detailed understanding of transcription and translation
- Get acquainted with the types and implications of mutations on DNA

CHAPTER TAKEAWAY **Everything about genes and genetic material**

This chapter presents the structure, function, outcomes, and impacts of the genetic material on the life forms on Earth.

PRELUDE

Since early times, a simple observation that children tend to resemble their parents, both, physically and in character was intriguing to many! However, an understanding of what and how the 'seed material' from both the parents transferred and mixed to produce the offspring was unclear. In the 18th Century, the idea of exploring this phenomenon began with a lot of experimentation in the form of breeding experiments. Over a period of time, a breakthrough in knowledge pertaining to genetic material, genes, heredity, and associated links was achieved. The molecular basis of heredity and the 'DNA as the chemical substance in cells, which carries the instructions for reproducing parental traits to children came much later in the 19th Century, when the 3-D structure of DNA was established by Watson and Crick in 1953. Since then, the focus of research has gained momentum and huge strides have been made in the past 70 years, leading to revolutionary discoveries in the field of molecular genetics and genetic engineering applications, benefiting mankind. In this context, in the next two chapters, we will gain basic knowledge of the genetics of life and understand inheritance in detail.

About DNA, Genes, and Genomes

The biochemistry of nucleic acids, such as DNA and RNA, and their basic structure and function have already been introduced in Chapter 4. Nucleic acid plays two important roles in the cell: it passes genetic information to the next generation of organisms and determines characteristics of organisms by synthesis of proteins. In almost all living organisms DNA molecules are the storehouses of genetic information contributing to heritable traits.

Discovery of DNA as Genetic Material

Early exploration into inheritance and understanding the composition and contents of the nucleus began in the 1860s by Gregor Mendel and Friedrich Miescher, respectively. Their observations set the stage for our molecular understanding of life.

The actual understanding of DNA as genetic material came much later in the 20th century. In 1928, Frederick Griffith proved that conversion of a non-pathogenic 'R' strain of *Streptococcus pneumoniae* to virulent 'S' strain was due to the phenomenon of transformation, and he called the agent the 'transforming principle'. The chemical nature of the 'transformation agent' was proved as DNA among the three candidate molecules: proteins, RNA, and DNA by Avery, Macleod, and McCarty in 1944 through their mice experiments with *S. pneumoniae*. The final series of experiments by Hershey and Chase confirmed conclusively that DNA was the hereditary material. These sets of experiments when taken together cemented the transformation by the genetic uptake and established DNA as the carrier of genetic information (Figure 6.1(a)).

Griffith Avery Macleod McCarty Hershey Chase

(a)

Wilkins Franklin Chargaff Watson Crick

(b)

Fig. 6.1 Pioneering contributions of scientists in understanding the DNA as genetic material

The next race was to establish the structure of DNA, which was completed in 1953 when Watson and Crick elucidated the structure and opened up a new domain of molecular biology. To achieve this, they developed the model based on information available from the research outcomes of various other scientists, i.e., X-ray diffraction data of Maurice Wilkins and Rosalind Franklin for DNA fibres; Chargaff's measurements of nitrogenous bases and ascertaining that the number of purines is equal to the number of pyrimidines to propose base-pairing rules known as Chargaff's rules. Information on the sugar–phosphate backbone and the polynucleotide chain with

polarity was also available. All this information led to the double-helix model for DNA wherein the two polynucleotide chains wound around each other in a right-handed helix with A=T and G≡C complementary base pairing. The model also proposed the anti-parallel strands running in opposite directions with turns; a full turn accommodates 10 nucleotides and is of 3.4 nm. For elucidating the DNA structure, Watson and Crick shared the Nobel Prize with Maurice Wilkins in the year 1962 (Figure 6.1(b)).

Experimental Insight 6.1 Historical Developments on DNA

1928 Griffith experiment

Griffith injected both S (virulent, mucous coat) and R (non-virulent, no capsule) strains into mice. The S strain kills mice whereas the R strain does not. Then, heat-killed S strain was injected and the mice lived. When the R strain was mixed with heat-killed S strain, the mice should have lived, but they died. He postulated that something has transformed the non-virulent R strain (maybe from S strain), and he called this as transforming particle.

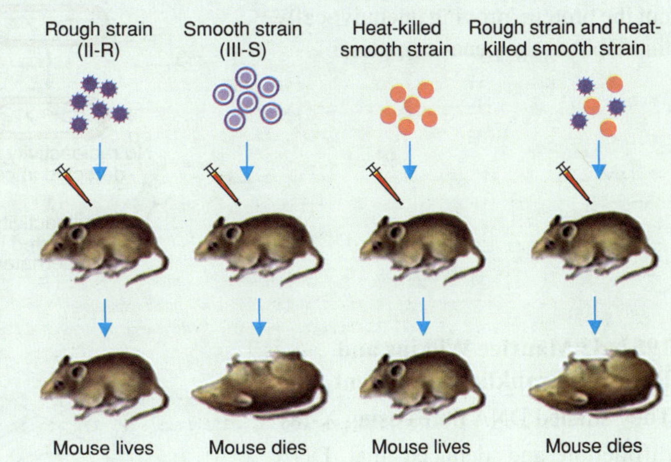

1944 Avery, Macleod, and McCarty experiment

They isolated the cell extract from S strain and separated the lipids and sugars. They treated it with proteases (destroys proteins), RNAses, and DNAses (destroy RNA and DNA, respectively). They added the treated extracts to R strain and virulent S strain appeared, proving that the transforming particle was DNA and not proteins or RNA that transforms bacteria by an enduring inheritable change.

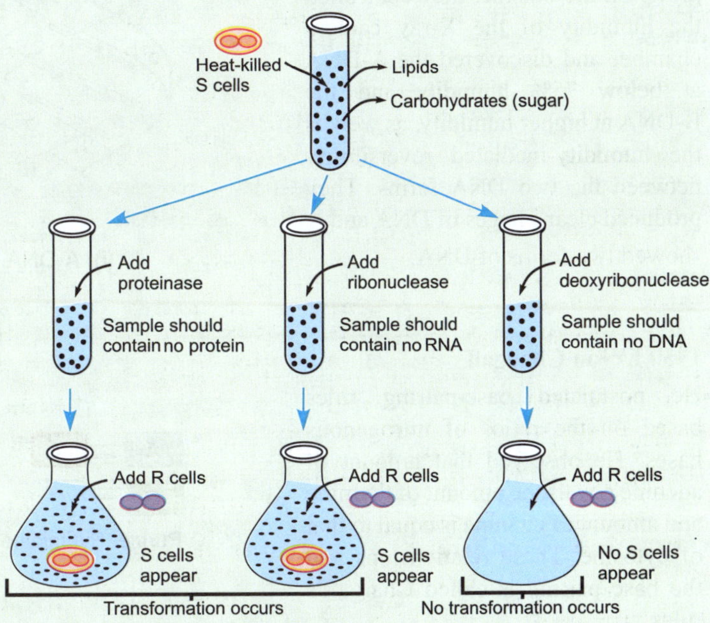

1952 Hershey and Chase experiment

Bacteriophages are viruses that infect bacteria and produce new virus particles. To identify genetic material, they labelled bacteriophage DNA and proteins radioactively and studied the phage lifecycle. DNA entered the cell and appeared in progeny phages but not the protein, proving unequivocally that DNA is the genetic material.

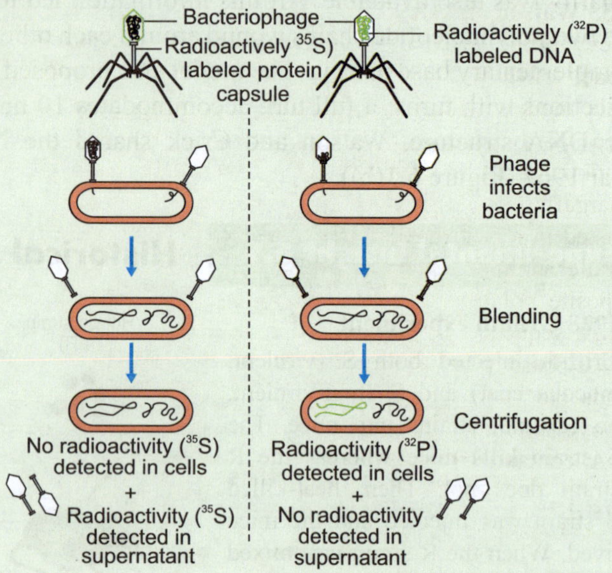

1952–53 Maurice Wilkins and Rosalind Franklin experiment

They studied DNA fibres using X-ray diffraction and deduced that DNA had two chains with sugar–phosphate group on the outside; they controlled the humidity of the X-ray camera chamber and discovered the A-DNA at below 75% humidity and the B-DNA at higher humidity, as well as the humidity-mediated reversibility between the two DNA forms. They produced clear images of DNA and showed two forms of DNA.

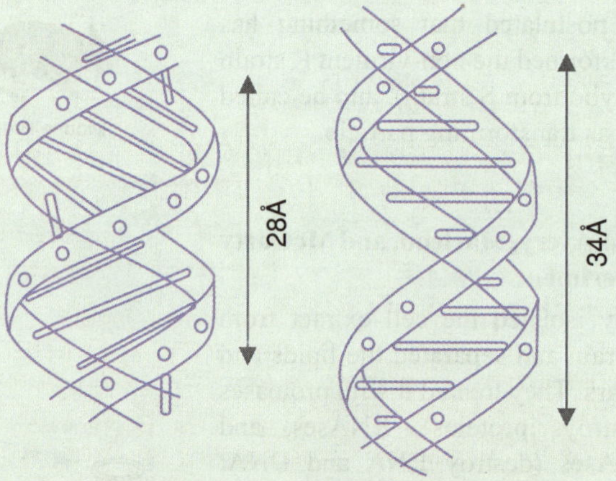

1950 Erwin Chargaff

He postulated base-pairing rules based on the ratios of nitrogenous bases. He observed that amount of adenine equals the amount of thymine and amount of guanine is equal to that of cytosine. These relationships and the base pairing is called Chargaff's rules.

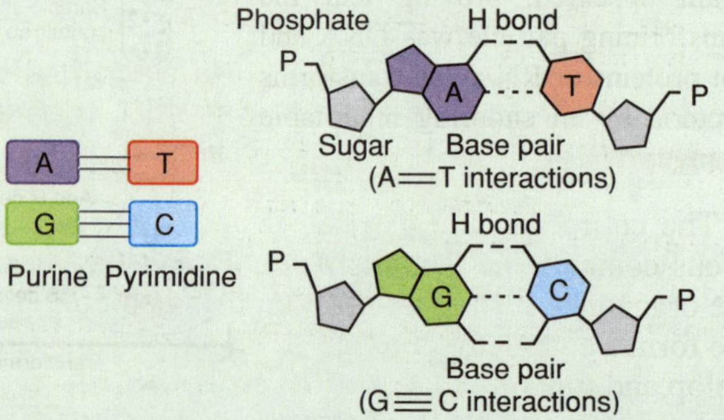

1953 Watson and Crick
They proposed the double-helix model of DNA. They constructed scale models with four nucleotides; used X-ray data of Wilkins and Franklin and biochemical data of Chargaff: Double-stranded molecule twisted in a right-handed helix, anti-parallel complementary strands with opposite polarity; 3′ and 5′ ends of DNA; sequence of DNA; A=T and G≡C base pairing as one purine binds to one pyrimidine; major groove and minor groove, one turn of 34 A′ contains 10 nucleotides.

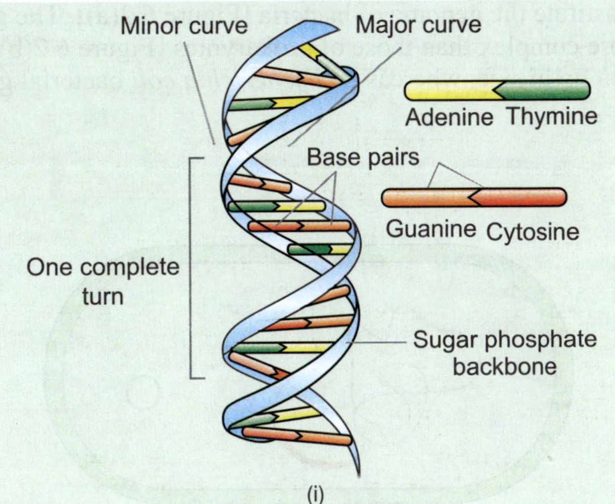

(i)

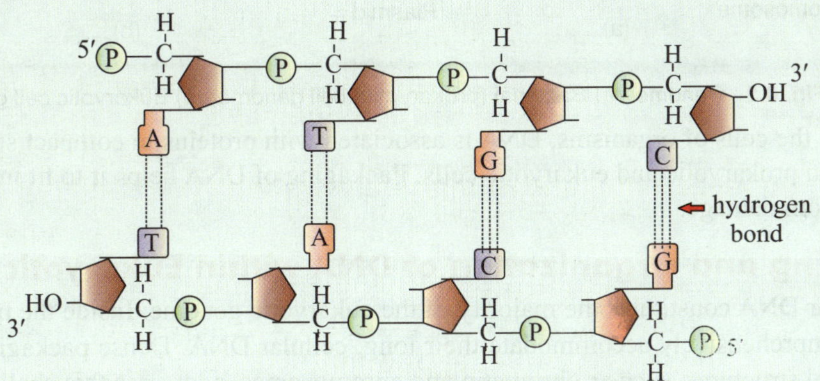

(ii) Double-stranded polynucleotide chain

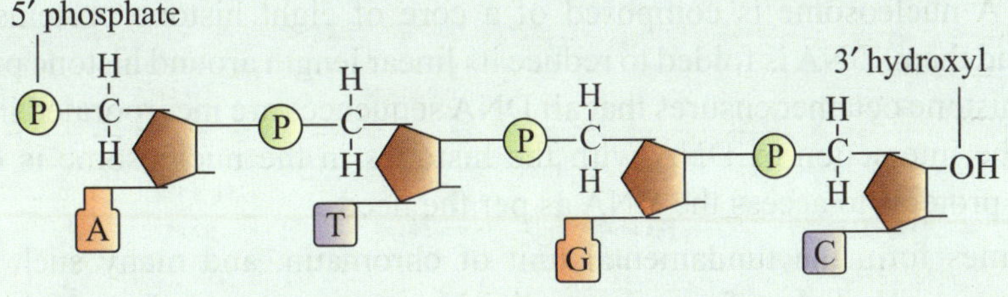

(iii) A polynucleotide chain

Genome of Organisms

The entire genetic material of an organism is called its 'genome'. In eukaryotic cells, the nucleus contains the genetic material of the organism and is called *nuclear DNA*. Additionally, some organelles such as mitochondria and chloroplast may also contain their own DNA. Together, these form a complete set of DNA, which contains all the information required for the organism to develop and function.

In prokaryotes such as bacteria, at the centre of the cytoplasm, a single circular DNA called *bacterial chromosome* is housed, and this nucleoid structure is not enclosed by a nuclear membrane. A few bacteria contain certain extra chromosomal material called *plasmids*. These together

constitute the genome of bacteria [Figure 6.2(a)]. The genomes of eukaryotes are much larger and more complex than those of prokaryotes [Figure 6.2(b)]. The human genome is estimated to be of 3 bn bp in size, whereas an *Escherichia coli* bacterial genome is about 4 mn bp.

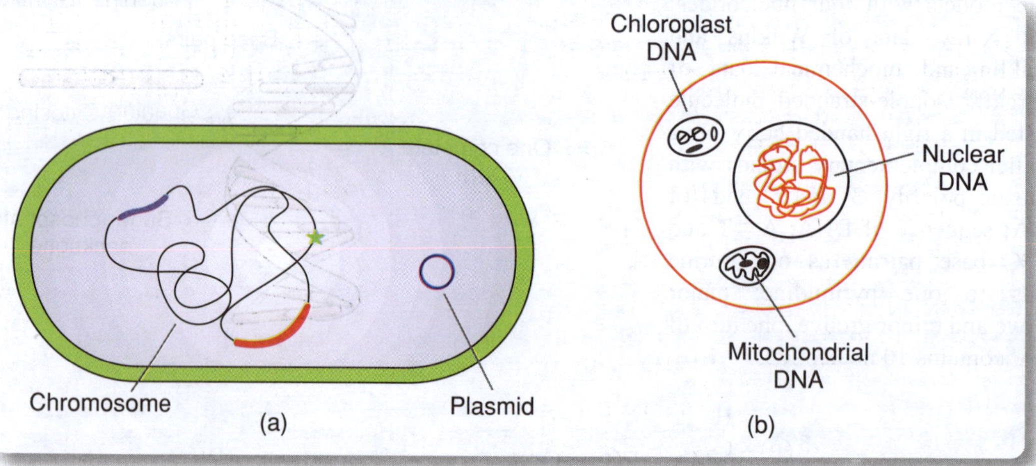

Fig. 6.2 Genome: (a) Bacterial (prokaryotic) cell genome; (b) Eukaryotic cell genome

Within the cells of organisms, DNA is associated with proteins in compact structures. This is true for both prokaryotic and eukaryotic cells. Packaging of DNA helps it to fit into cells and also protect it from damage.

Packaging and Organization of DNA within Eukaryotic Cells

Nuclear DNA constitutes the majority of the eukaryotic genome. Inside the nucleus, the cells need to comprehensively accommodate their long, cellular DNA. Dense packaging of DNA into nucleosomal structures, such as chromatin and chromosomes, addresses this challenge.

Compaction of DNA is a result of its association with proteins called histones to form nucleosomes. A nucleosome is composed of a core of eight histone proteins, and the DNA is wrapped around them. DNA is folded to reduce its linear length around histone proteins. Bending of DNA around histone octamer ensures that all DNA sequences are incorporated into the nucleosome assemblies. The interaction of DNA with the histones in the nucleosome is dynamic, allowing DNA-binding proteins to access the DNA as per the need.

Nucleosomes form the fundamental unit of chromatin, and many such nucleosomes with linker DNA are compacted to form chromatin. Nucleosomes are often depicted as 'beads on a string' to represent chromatin. Loosely arranged nucleosomes and DNA is called *chromatin*, which is found in resting cells. However, during certain stages of a cell's life, such as cell division, further compaction of the chromatin may take place, and it results in the formation of condensed forms of chromatin called *chromosomes* (Figure 6.3).

The entire genetic material of eukaryotes is arranged in multiples of chromatin fibres or chromosomes. *The number of chromosome is unique to each organism*. If the genetic material is present as single (one) copy of (arranged in) chromosomes, then the organism is said to be haploid; if there are two sets of the genetic material arranged in chromosomes, then the organism is said to be *diploid*. Every cell maintains a characteristic number of chromosomes. Prokaryotes, such as bacteria, have a single haploid chromosome, whereas the majority of eukaryotes are diploid.

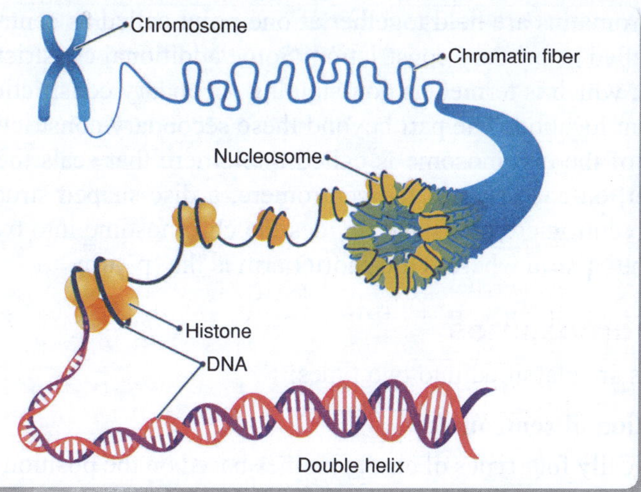

Fig. 6.3 Organization of DNA inside the Eukaryotic nucleus

So, DNA is organized as chromosomes inside the nucleus of cells. This DNA fulfils all the requisite functions of the genetic material—stores and transmits genetic information from one generation to the next, replicates itself, and controls the synthesis of proteins.

STRUCTURE OF CHROMOSOME

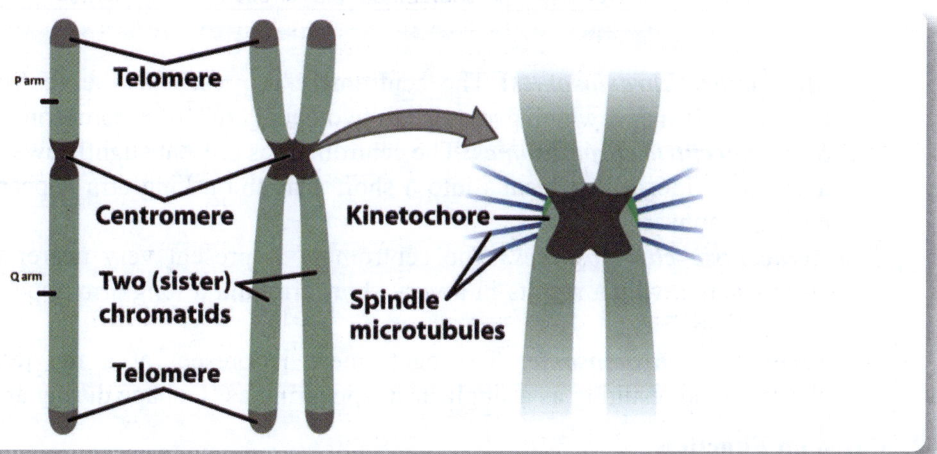

Fig. 6.4 Structure of chromosome

The nucleus contains thread-like structures called chromosomes, which have the ability to transmit genetic information from parents to offspring. In humans, DNA is approximately 2 metres in length, evenly present in its 23 pair of chromosomes. A haploid set of chromosomes is present in the gamete, which is half of the somatic cell that results from meiosis. Further, it is composed of two halves called as chromatin, which contains DNA, basic protein (histones), some non-histone protein, and RNA. The darkly stained and condensed region of chromatin containing tightly packed DNA, which is genetically inactive in nature, is called as heterochromatin. The lightly stained region with loosely packed DNA region is termed a euchromatin, which is genetically

inactive. These chromatins are held together at one point called as centromere, which appears as a narrow region called as primary constriction. Some additional constrictions may be present in a few chromosomes, which is termed as non-staining secondary constriction or nucleolar organiser (NOR) at a constant location. The part beyond these secondary constrictions is termed a satellite. The terminal part of the chromosome is called a telomere that seals the end, which prevents its shortening. Further, on each side of the centromere, a disc-shaped structure is present called as kinetochores. The centromere holds and divides the chromosome into two equal parts; the longer arm is known as the 'q' arm whereas the shorter arm is the 'p' arm.

Types Of Chromosomes

Chromosomes are classified into two types:

1. Based on position of centromere

There are basically four types of chromosomes based on the position of the centromere, which are as follows:

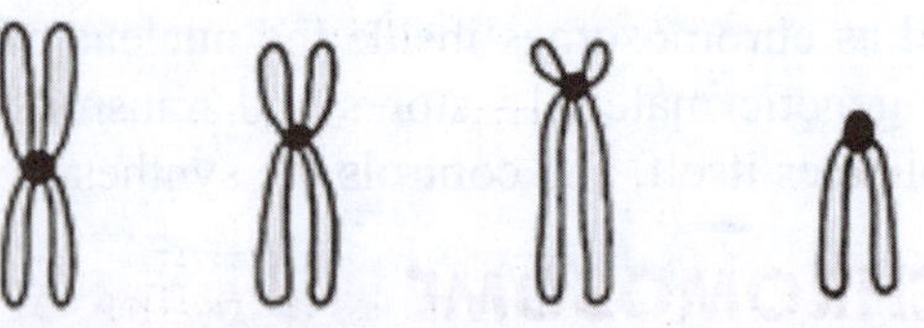

(i) ***Metacentric chromosomes:*** The centromere is present at the centre of chromosomes having a 'V' shape, which gives rise to two equal arms of chromosomes during anaphase.

(ii) ***Submetacentric chromosomes:*** The centromere is present slightly away from the centre of a chromosome, which results into a short arm and a long arm appearing as a 'L' shape during anaphase.

(iii) ***Acrocentric chromosomes:*** The centromere is present very nearer to one end of the chromosome, which results in a very short arm and a long arm appearing as a 'J' shape during anaphase.

(iv) ***Telocentric chromosome:*** The centromere is present at a the terminal end of the chromosome resulting as a single arm, appearing as 'I' shape during anaphase.

2. Based on Function

There are two types of chromosomes based on functions.

(i) ***Autosomes:*** They determine all the characters except gender. There are 44 (22 pairs) autosomes in humans.

(ii) ***Sex chromosomes (Allosomes):*** They determine the gender of the individual. There are two (1 pair) allosomes in humans. X and Y are the two sex chromosomes in humans.

Centromere is a constriction that joins the sister chromatids in a tetrad chromosome structure. Telomeres are repeats present at the ends of chromosomes that help to end replication problem (Figure 6.4).

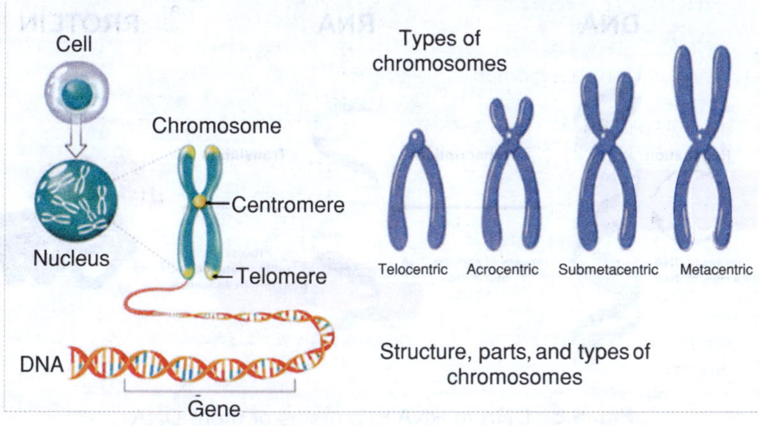

Fig. 6.5 Chromosome structure and types

The activities of DNA are regulated by specific functional units within the chromosomes called 'genes'. Each gene has a specific location on the chromosome called its 'locus'. The genome is a collection of all the genes of an organism. Genes are responsible for biological and physical traits and, hence, act as the basic unit of heredity passed on from parents to offsprings. Genes achieve their effect by directing the synthesis of proteins; hence, they are also defined as the regions of DNA that carry instructions for the synthesis of proteins.

There is a difference between prokaryotic and eukaryotic genes. Eukaryotic genes contain intronic regions, which are the non-coding part of genes. Exons are coding regions of the gene. No such introns are present in prokaryotes; hence, the entire gene is coded into proteins, justifying 'one gene–one protein or enzyme' concept. In eukaryotes, the introns are spliced out after synthesis of mRNA during transcription. So different ways of splicing introns (alternative splicing) can sometimes produce more than one protein from the same gene.

Interesting Fact 6.1

DNA terms differ on level of structural organization

DNA	Genome	Nucleosome	Chromatin	Chromosome	Gene
Any polymer of nucleotides in a double helical structure carrying genetic information	Collection of entire genetic material or DNA of an individual organism	DNA wrapped around protein in a 'bead in a string' arrangement	Loosely arranged nucleosomes inside the nucleus of a resting cell	Tightly condensed more compact chromatin	Fragment of DNA that codes for a protein Functional unit of heredity

Central Dogma of Biology

The DNA is the 'brain' behind all the activities of the cell. It instructs various processes leading to multiple outcomes. The central dogma explains the flow of instructions, wherein the information present in the DNA is converted into functional products. The sequential information, nucleotide-by-nucleotide, of DNA codes helps to create new copies of DNA or proteins inside the cell. For more copies of DNA, the genetic material is duly replicated. For protein synthesis, the DNA code is copied into mRNA, which then 'translates' it into a sequence of amino acids called protein. The two steps involved are transcription and translation (Figure 6.6).

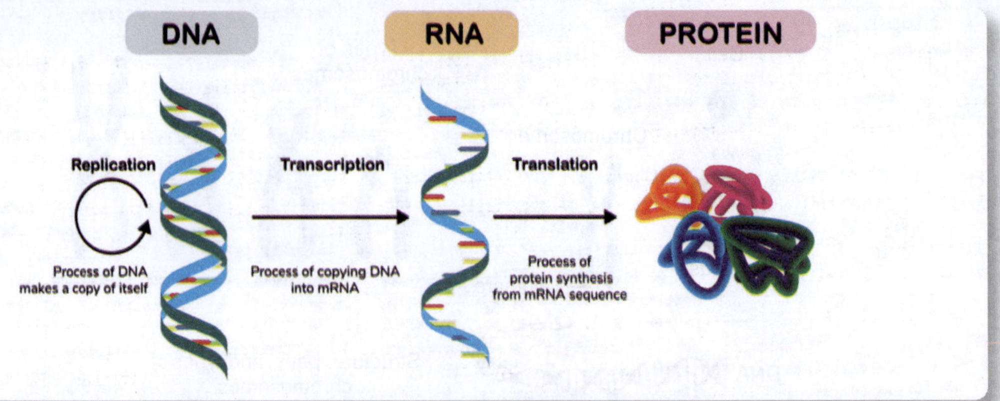

Fig. 6.6 DNA to RNA to proteins or more DNA

DNA Replication

Whenever new cells are formed by cell division, it is necessary to create additional copies of the genetic material so that every cell has its own genetic material. How the DNA gets duplicated before the cell divides was a big question once the DNA structure was elucidated. However, the structure of DNA itself and the complementary base pairing observation itself gave some clues about one strand acting as a template for the synthesis of the other strand.

DNA replication is the process by which a cell makes another copy of its genetic material DNA [Figure 6.7]. The replication is conserved in all cells, and the basic steps and general process remain the same. This DNA copying process is carried out by specific enzymes involved in various steps (Figure 6.9).

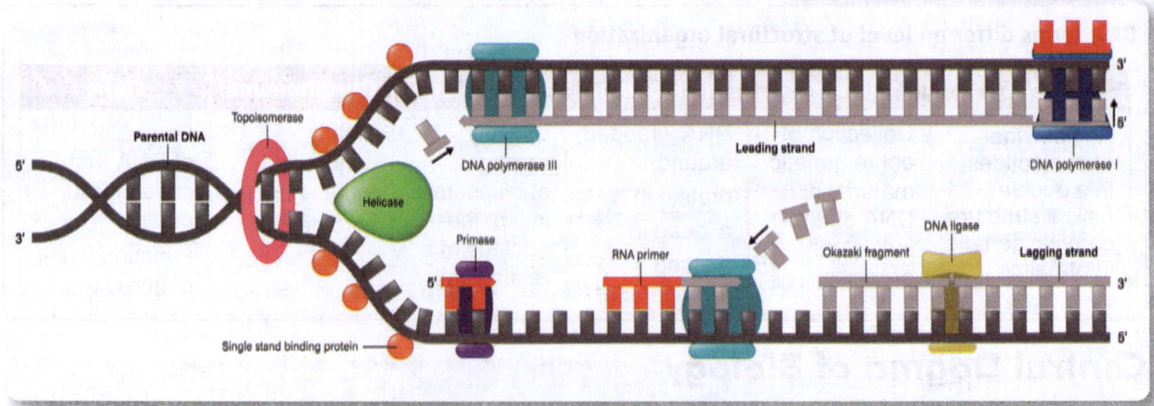

Fig. 6.7 Steps in DNA replication

Step-by-step mechanism of DNA replication with key players is as follows:

(a) The double-stranded DNA is first unwound at specific locations called 'Origin of Replication' and the two strands of DNA are separated. This exposes the template strands for new DNA synthesis. This unwinding is done by enzymes called *helicases*.

(b) The unwinding produces a Y-shaped molecule called *replication fork*. Once the two strands are separated, *single-strand binding (SSB) proteins* coat them and hold them in position and prevent them from pairing again. This unwinding leads to more coiling downstream (especially circular bacterial DNA), which is taken care of by *topoisomerases* that release the stress.

(c) A short stretch of a few nucleotides called *primer* binds to the complementary region of DNA. Primers are made up of RNA and are synthesized by the enzyme *primase*. Adding nucleotides one by one to synthesize new strand will lead to abortive initiation, as there is no force to hold the nucleotides initially, so the complementary base pairing of primers provides the stable platform and facilitates new nucleotide addition without aborting it repeatedly. The RNA primer also marks the starting point of new DNA synthesis, which is later easily identified and replaced, so that any mismatches at the beginning can be taken care of. Therefore, primers are essential.

(d) DNA polymerase now starts, adding new nucleotides based on complementarity to the template strand. Nucleotides enter as per base-pairing rules, e.g., if it is adenine in the template, thymine will be added, guanine will pair with cytosine, and so on. Thus exposed nitrogenous bases of original DNA guide the addition of appropriate nucleotides to form the new strand. Addition of new nucleotides always happens at the 3′ OH end and the DNA strand is synthesized in the 5′-3′ direction.

(e) As the two strands of DNA are antiparallel, only one template (3′-5′) will be able to add nucleotides in 5′-3′ direction. As unwinding happens, this strand will keep proceeding forward and the new strand will be synthesised continuously in the direction of unwinding of the double helix. The other template (5′-3′) has to go in the opposite direction of unwinding, so the exposed regions are in short stretches, and DNA polymerase synthesizes this strand discontinuously as short fragments called *Okazaki fragments* (after the scientist who detected them). The continuously synthesized template strand is called *leading strand* and the discontinuously replicating strand is called *lagging strand*.

(f) All these processes begin at a point in the template DNA called *origin of replication*. In prokaryotes, this process starts only at one place forming a replication fork, whereas in eukaryotes, the huge DNA is replicated with multiple origins of replications [Figure 6.7]. So, the process starts in several different places along the DNA, and when these multiple replication forks meet each other, a combined new strand of DNA is synthesized.

(g) Primers are then removed and the gaps are sealed by DNA ligases. The completion of DNA replication yields two copies of DNA double helices.

The DNA-copying process is quite accurate and almost error-free. *Proof-reading* and *editing* of new DNA for correct base pairing ensures that there is only an error of one nucleotide for every 2×10^9 nucleotides, which is quite small.

Experimental Insight 6.2 Models of DNA Replication

Scientists have proposed three different models for DNA replication, as follows:

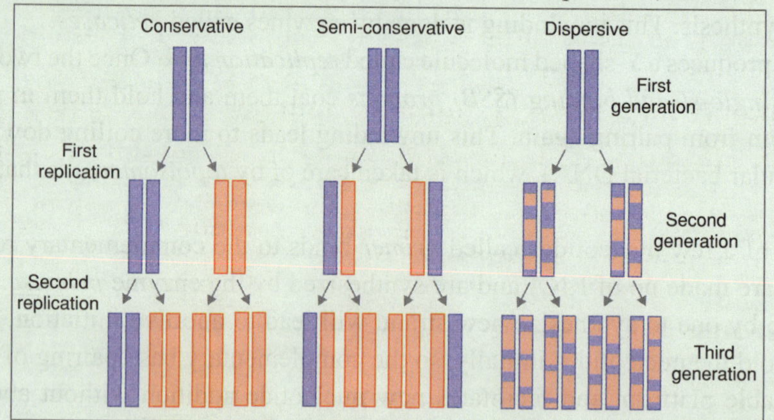

Source: Adapted from https://www.labxchange.org/library/items/b:LabXchange:b70119f9-581c-3443-b890-85d26b8db50d:html:1

In the conservative model, parental DNA strands (blue) remained associated in one DNA molecule while new daughter strands (red) remained associated in newly formed DNA molecules. In the semi-conservative model, parental strands separated and directed the synthesis of a daughter strand, with each resulting DNA molecule being a hybrid of a parental strand and a daughter strand. In the dispersive model, all resulting DNA strands have regions of double-stranded parental DNA and regions of double-stranded daughter DNA.

DNA Replication is Semiconservative

Meselson and Stahl, in 1958, proved the semi-conservative mode of replication by distinguishing parental DNA strands from the newly formed daughter strands.

E. coli bacteria was grown for several generations in the medium containing the heavy isotope of nitrogen (N^{15}). This labelled the parental DNA with heavy isotope. Then, the bacteria was transferred to N^{14} light medium. For the next few generations, all the newly synthesized DNA would incorporate lighter isotope N^{14}. From all these daughter generations of organisms, they isolated the DNA, mixed it with caesium chloride, and centrifuged it at high speed. DNA molecules move to positions where the DNA density equals that of CsCl solution and forms bands. Based on this experiment, the following results were found (Figure 6.8).

When the parent DNA with N^{15} replicated, after one generation, they did not see the heavy band. Instead, they observed a single band that was lighter than N^{15}. Parental bacteria grown in 'heavy' nitrogen had N^{15} on both strands. As the new lighter band appeared, they reasoned that all DNA molecules had a hybrid of one N^{15} strand and one N^{14} strand; hence, it was lighter than N^{15}/N^{15}, pointing towards semi-conservative mode of DNA replication. After two generations, they observed another even lighter band along with the hybrid one. This is because, subsequently, when they separate, the new strands formed are from N^{14}; so the percentage of N^{14} keeps going up and there are bands of N^{14}/N^{14} contributing to the even lighter band. In three generations and beyond, the lightest band becomes the predominant one, and in each successive generation, the proportion of heavy strands is reduced by half, and the 'heavy/light' band gradually fades.

These results also disproved conservative and dispersive models.

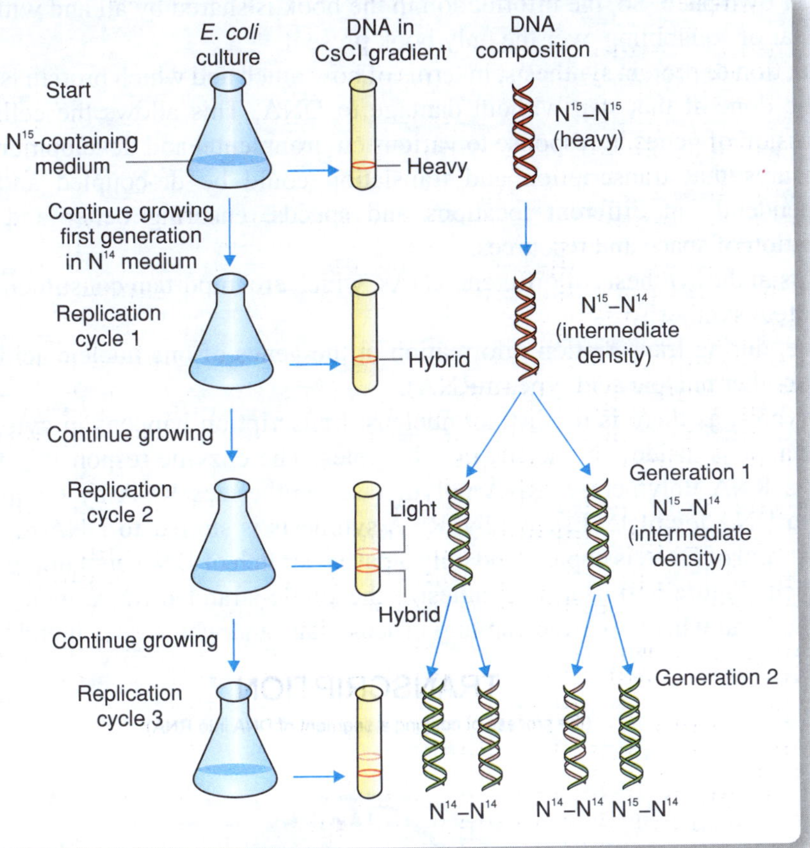

Fig. 6.8 Observations for semiconservative DNA replication in Meselson and Stahl experiments

Protein Synthesis

Inside the cell, information for synthesis of proteins lies in the genes present in the DNA. This information is coded as a specific sequence of nucleotides. The process of protein synthesis converts this nucleotide information into amino acid language for protein synthesis.

There are two steps in protein synthesis, as follows:

1. *Transcription* Nucleotide information from the DNA is converted to mRNA sequence.
2. *Translation* The mRNA information is translated to amino acid information to form protein.

Transcription Every protein inside the cell has a corresponding gene coding for it. The genome of an organism contains many genes coding for different proteins simultaneously. It is the process of making an RNA copy of the gene for a particular protein. DNA is only a carrier of information; action to convert it into protein starts by making the RNA copy in transcription. Why is this transcription process important? The following are the reasons for it:

- Instead of making proteins from a single-source DNA, having an RNA copy of each gene helps in parallel processing of all genes, without compromising the source DNA and also ensuring speed and accuracy. It is like making multiple copies of individual chapters of a reference book present in the library and making it available for a greater number of students. From the same book each student can photocopy and study individual chapters

at their own pace. So, the information in the book is shared by all and without much wear and tear or squabbling over the only book present.

- Regulation of protein synthesis, in terms of how much and which protein is to be produced, can be done at this step without damage to DNA. This allows the cells to control the expression of genes in response to various environmental and developmental signals.
- It ensures that transcription and translation could be de-coupled and could happen independently at different locations and speeds, ensuring better and more efficient utilization of space and resources.
- It helps in the synthesis of different RNAs, which are important constituents in the process of protein synthesis.

In essence, during transcription information in the genes of one nucleic acid type (DNA) is converted to another nucleic acid type (mRNA).

In prokaryotes, as there is no distinct nucleus, transcription happens in cytoplasm, whereas transcription happens inside the nucleus in eukaryotes. The enzyme responsible for carrying out transcription is RNA Polymerase. RNA polymerase synthesizes RNA by complementary base pairing using a DNA template. Essentially, RNA synthesis is similar to DNA replication, except that only one strand of DNA is copied, and only a partial stretch of DNA containing the gene for the protein is copied (Figure 6.9). But, to synthesize the single-stranded RNA, which strand of DNA should be copied and where does the copying process start and where does it end?

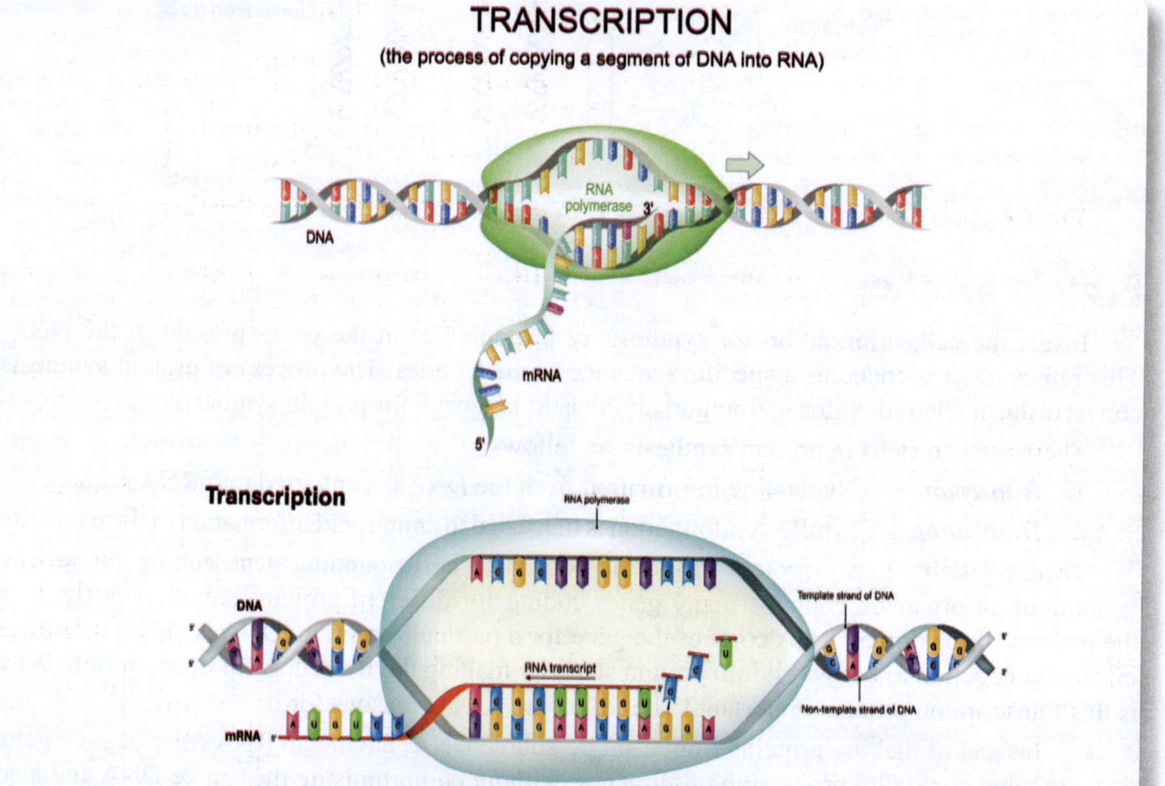

TRANSCRIPTION

(the process of copying a segment of DNA into RNA)

DNA

RNA polymerase

3'

mRNA

5'

Transcription

RNA polymerase

DNA

Template strand of DNA

RNA transcript

Non-template strand of DNA

mRNA

Transcription

Fig. 6.9 Transcription by RNA polymerase

Out of the two DNA strands, one strand contains the information about the protein. This strand is called the coding strand. RNA has to be a copy of this information. Hence, to synthesize this RNA, the other strand, i.e., the non-coding strand, is copied and is called the *template strand*. To begin transcription, RNA polymerase binds to a nucleotide sequence at the beginning of every gene called the *promoter region*. The promoter serves as an indicator for RNA polymerase to start binding and begin transcription. Then, from here onwards, the RNA polymerase unzips the double-stranded DNA, travels along the gene, and adds RNA nucleotides in the 5'-3' direction, complementary to the DNA strand used as a template. Thus, transcription results in the production of a single-stranded RNA, which has the information similar to the coding region of the gene, except that it has U instead of T. When it reaches the end of the gene, the RNA transcript is released. This RNA copy is called messenger RNA (mRNA) as the information on this RNA is used as a message to code the proteins in the next translation step of protein synthesis.

mRNA Processing in Eukaryotes

The Eukaryotic mRNA synthesized by transcription called pre-mRNA undergoes different processing events before it is converted into a final mRNA for translation. These processing events protect the mRNA and ensure a longer half-life than prokaryotic mRNA.

5′ capping In the pre-mRNA, 7′ methyl guanosine is added to the 5′ end that protects the mRNA, and this is called as 5′ capping. The cap protects the 5′ end of pre-mRNA from attack by ribonucleases and is recognized for assembling the ribosomes on mature RNA for translation.

3′ Poly A tail At the 3′ end a string of poly A residues is added by the enzyme Poly A polymerase. The poly (A) tail also protects the mRNA from degradation, aids in the export of the mature mRNA to the cytoplasm, and is involved in binding proteins involved in initiating translation.

It is important to note here that eukaryotic genes contain certain non-coding regions within the gene called introns. Exons are the coding regions that contribute to the protein information. So, eukaryotic genes are a mixture of coding regions called exons and intermittent non-coding regions called introns. Splicing out these introns is necessary to get a continuous coding parts or exons for translation. So, in eukaryotes, after transcription, pre-mRNA is formed, which is then converted into a final mRNA by splicing of introns. This final mRNA now has only the information for coding of proteins from exons and is then used for the next step of translation (Figure 6.10). Splicing also opens up the possibility of joining exons in different combinations to produce different mRNA transcripts. This is called alternative splicing and is a mechanism in eukaryotes for one gene to contribute to more than one protein—alternate forms of proteins.

Transcription also produces other RNAs, such as t-RNA and r-RNA, which have different roles in protein synthesis and are not translated further.

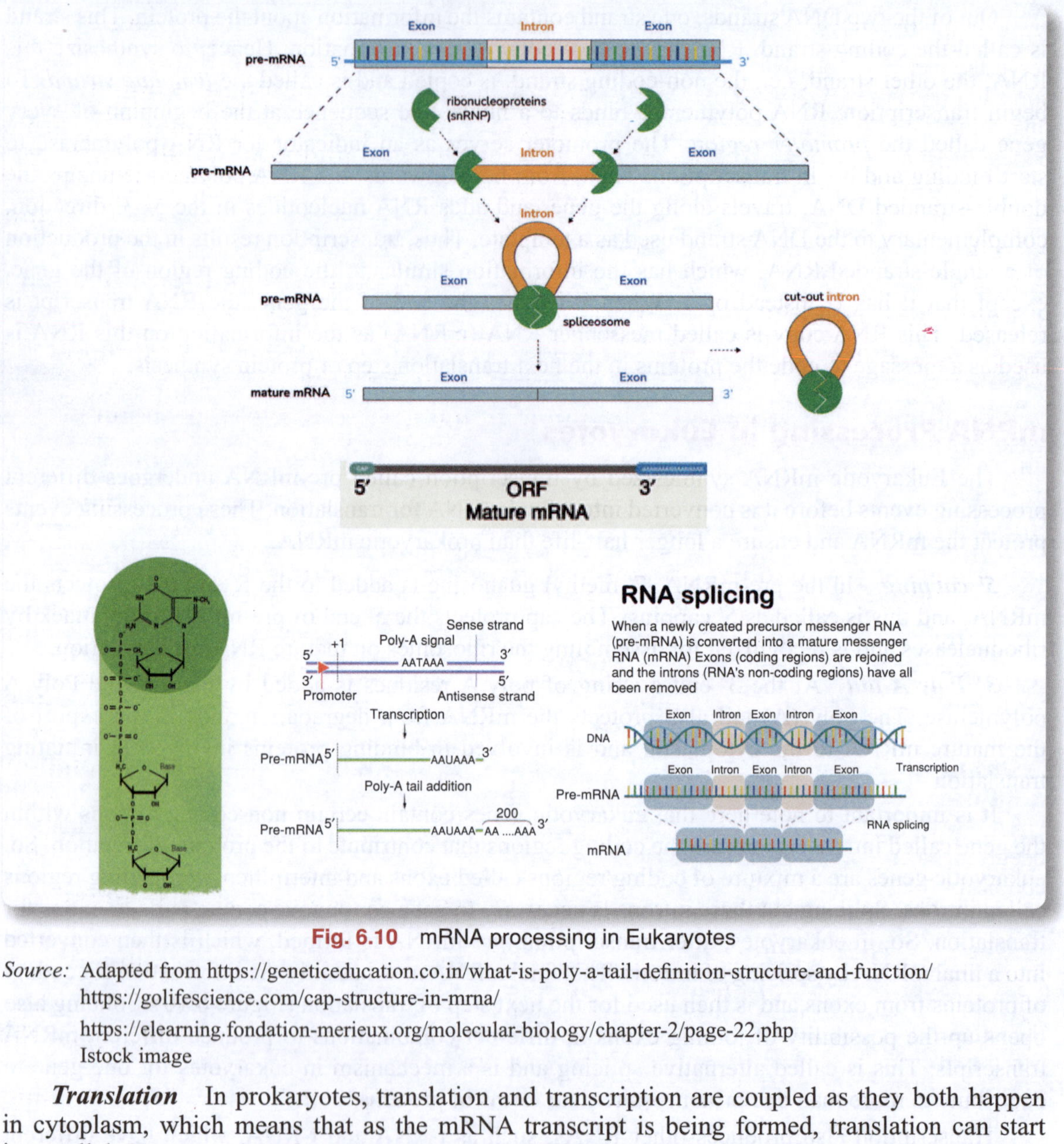

Fig. 6.10 mRNA processing in Eukaryotes

Source: Adapted from https://geneticeducation.co.in/what-is-poly-a-tail-definition-structure-and-function/
https://golifescience.com/cap-structure-in-mrna/
https://elearning.fondation-merieux.org/molecular-biology/chapter-2/page-22.php
Istock image

Translation In prokaryotes, translation and transcription are coupled as they both happen in cytoplasm, which means that as the mRNA transcript is being formed, translation can start in parallel. However, in eukaryotes, transcription happens inside the nucleus, and the mRNA is transported outside to the cytoplasm where translation occurs. Translation of mRNA is carried out with the help of ribosomes present in the cytoplasm and ER.

Translation is the use of nucleotide information encoded in the mRNA to assemble the protein monomers, i.e., amino acids, and thus translate the nucleotide language to amino acid language. There are four bases in nucleic acids: ATGC in DNA and AUGC in RNA. The combination of these bases needs to code for 20 amino acids that make up the protein.

The dictionary that helps to translate the nucleotide language to protein language is called the genetic code. The code uses combinations of three nucleotides in the mRNA to code for proteins, i.e., every amino acid is coded by three-letter codes of mRNA. This is because one-letter code could code only four unique amino acids; two-letter code could code for only $4^2 = 16$ amino acids; three-letter code for $4^3 = 64$ amino acids and provides 64 unique codes for 20 amino acids, which will serve the purpose well. This also indicates that as the number of codes is more than amino acids, some amino acids will be coded by more than one three-letter codes as the number of codes is more than amino acids. These three-letter codes are called *codons*. Further, a four-letter codon, $4^4 = 256$ codes for 20 amino acids and will lead to too high degeneracy and, hence, not useful (Figure 6.10).

First letter	Second letter: U	Second letter: C	Second letter: A	Second letter: G	Third letter
U	UUU, UUC } Phe UUA, UUG } Leu	UCU, UCC, UCA, UCG } e	UAU, UAC } UAA P UAG P	UGU, UGC } C UGA P UGG	U C A G
C	CUU, CUC, CUA, CUG } Leu	CCU, CCC, CCA, CCG } P	CAU, CAC } CAA, CAG } G	CGU, CGC, CGA, CGG } A	U C A G
A	AUU, AUC, AUA } e AUG e	ACU, ACC, ACA, ACG } h	AAU, AAC } A AAA, AAG } L	AGU, AGC } e AGA, AGG } A	U C A G
G	GUU, GUC, GUA, GUG }	GCU, GCC, GCA, GCG } A	GAU, GAC } A GAA, GAG } G u	GG, GGC, GGA, GGG } G	U C A G

Fig. 6.11 Genetic code for translation of mRNA into proteins

Codons are written in 5'-3' direction as per convention. Out of the 64 codons, three codons UAA, UAG, and UGA do not code for any amino acid; instead, they act as 'stop' codons, and their occurrence prevents further translation of the mRNA. These three codons are also called *nonsense* codons or terminator codons. So, in all, there are 61 codons for 20 amino acids. Some amino acids, such as methionine and tryptophan, have only one codon each, whereas most have more than one codon. A few such as serine, arginine, and leucine have as high as six codons representing them. Same amino acid coded by different codons is called *degeneracy*. Genetic code is 'commaless' and universal, which means the letters are sequentially read without break; and with a few exceptions, the same codons specify the same amino acid in all living organisms signifying its universality. So, the genetic code acts as a robust translational framework, wherein four alphabets are used to produce three-letter codes called codons, which specify amino acid additions during protein synthesis.

The process of translation is carried out by *protein-assembling machines* called *ribosomes*. Once the final mRNA reaches the cytoplasm, ribosomes are assembled onto it. Ribosomes consist of two parts: the smaller sub-unit and the larger sub-unit made up of a combination of ribosomal rRNA and ribosomal proteins. First, the smaller sub-unit attaches, followed by the larger sub-unit to assemble the ribosome onto the mRNA. Apart from mRNA and the ribosomes, the third critical component in protein synthesis is the tRNA, which has a clover-leaf structure and is responsible for bringing the amino acid to the ribosome for its addition to the polypeptide chain. The tRNAs contain anticodons that base-pair with the mRNA codons. The tRNA carries the correct amino acid cargo, corresponding to the anticodon, which, in turn, is specific to the codon. So, tRNA is already charged with an amino acid by the process of aminoacylation (Figure 6.12).

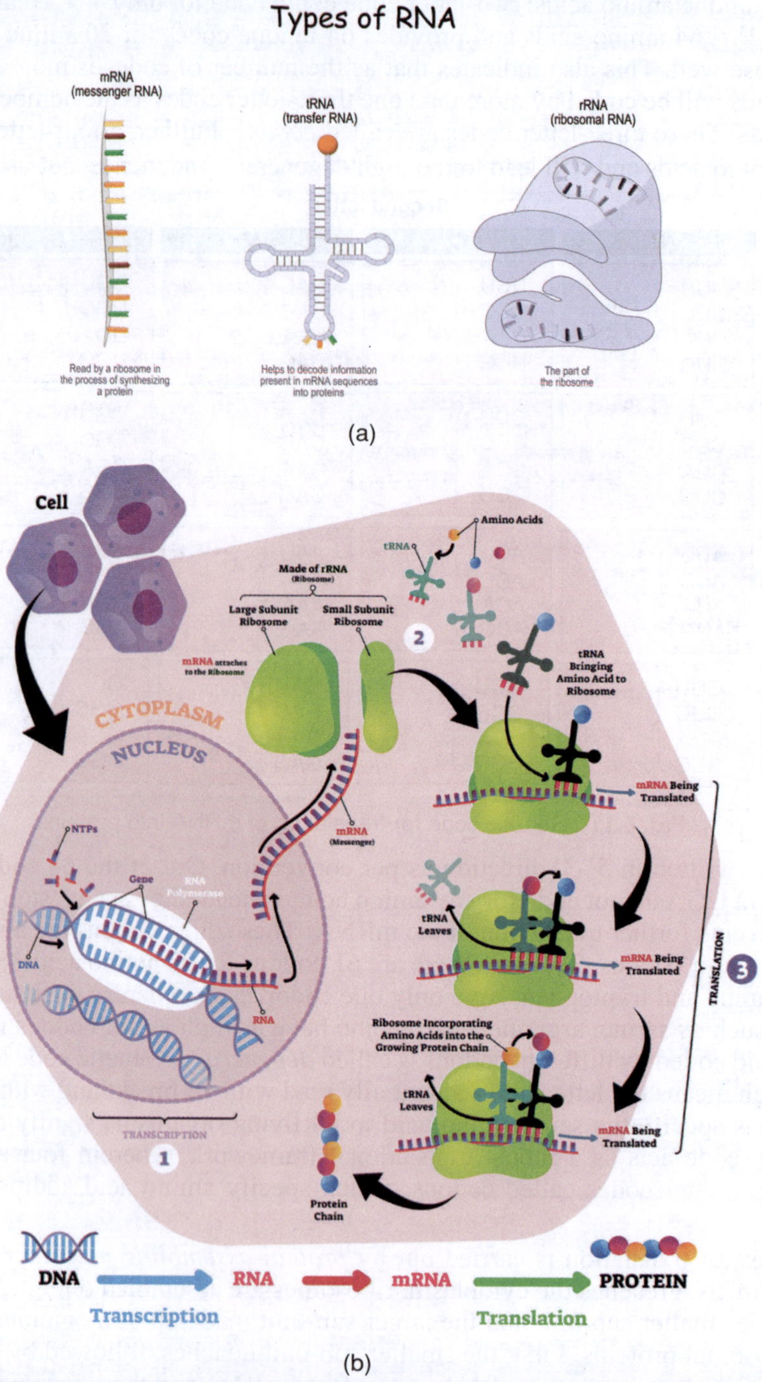

Fig. 6.12 Components of translation and basic features: (a) Types of RNA; (b) Protein synthesis

There are three major stages of translation: initiation, elongation, and termination (Figure 6.12). During *initiation*, the components of translation assemble on the start codon of mRNA with the help of a ribosome-binding site, just ahead of the start codon. The small sub-unit of the ribosome is directed to bind here, followed by the larger sub-unit to complete the ribosome. AUG is the start

codon, which codes for methionine. Initiation is handled by proteins called *initiation* factors in all organisms. The ribosome covers two codons or six nucleotides at a time. The incoming initiator tRNA with anticodons for start codon, enters the A-site—aminoacyl site—of ribosome with the amino acid and then moves to the P-site. The next tRNA with the amino acid then enters the A-site and forms a peptide bond with the initiator tRNA at the P-site. The initiator tRNA then exits through the E-site. This way, continuously during elongation, the assembled complex reads the string of codons one at a time and joins specific amino acids. At the P-site, the tRNAs hold the growing polypeptide and are called peptidyl tRNA once they move from the A-site and then exit through the E-site. Elongation factors facilitate this entire process of extension of the polypeptide chain with appropriate amino acids based on codon–anticodon recognition. *Termination* takes place when the A-site of the ribosome arrives at one of the stop codons on mRNA; UAA, UGA, or UAG. This stop

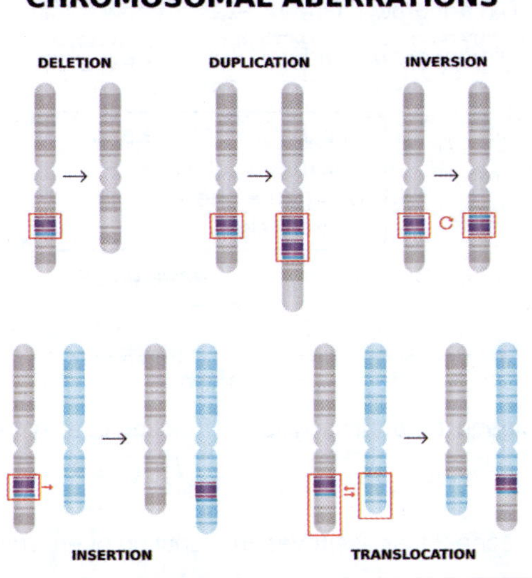

Fig. 6.13 Chromosomal abberations

Experimental Insight 6.3 Genetic Code

1. **How did we know genes code for proteins?** Beadle and Tatum, in 1941, while working with haploid organisms, *Neurospora crasa*, a bread mold, established a direct relationship between genes and enzymes. This organism is able to synthesize all the amino acids and chemicals needed for growth and, hence, easily grows in minimal media. They exposed the mold spores to X-ray and found that they failed to germinate and were unable to grow in the minimal media. This was because X-ray had mutated some of the genes, and the enzymes became non-functional. Complete media containing all nutrients allowed the spores to grow. Beadle and Tatum isolated arginine nutrient/auxotropic mutants, which are organisms that could not synthesize arginine and need to be supplied in media for their growth. Arginine synthesis is a multi-step process in the fungus requiring various enzymes. So, mutations by X-ray had impacted genes of this pathway, affecting the enzymes, leading to blocking of arginine synthesis pathway. Similar mutant studies with other nutrient mutants helped them to demonstrate the link that genes control the formation of enzymes, which are proteins. Based on their studies, they proposed the 'one gene–one enzyme' hypothesis, which was then updated to 'one gene–one polypeptide' hypothesis (Table 6.1).

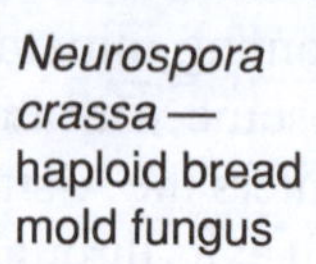

Prototrophic — can synthesize all nutrients and grow in minimal media → X-ray treatment → Fails to grow in minimal media — X-ray has mutated genes — affecting enzymes — nutrient synthesis impacted → Many nutrient pathways affected — but focus on arginine → Isolated and identified arginine-pathway mutants — steps in synthesis of arginine affected; hence, no growth in minimal media — supplementation with arginine and other precursors growth is seen → Link between gene and enzyme established — one gene–one enzyme hypothesis

Table 6.1 Beadle and Tatum Experiments

Bread mold	Minimal medium (MM)	MM + Ornithine	MM + Citruline	MM + Arginine
Wild type	Grew	Grew	Grew	Grew
Mutant 1	Did not grow	Grew	Grew	Grew
Mutant 2	Did not grow	Did not grow	Grew	Grew
Mutant 3	Did not grow	Did not grow	Did not grow	Grew

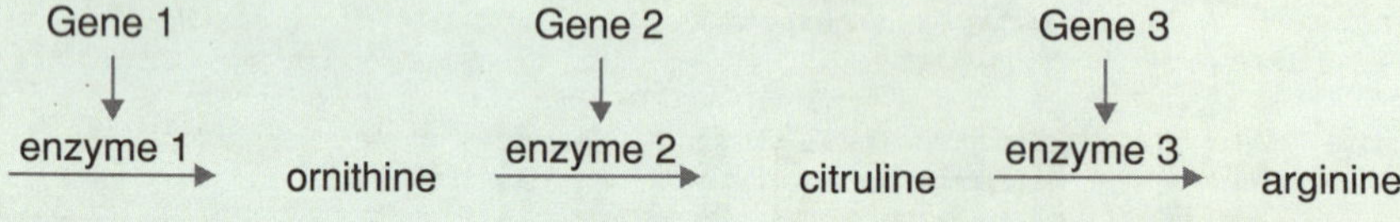

Mutant 1 had a mutation in gene 1 that destroyed the function of enzyme 1.

enzyme 1 (crossed out) → ornithine — enzyme 2 → citruline — enzyme 3 → arginine

Mutant 2 had a mutation in gene 2 that destroyed the function of enzyme 2.

enzyme 1 → ornithine — enzyme 2 (crossed out) → citruline — enzyme 3 → arginine

Mutant 3 had a mutation in gene 3 that destroyed the function of enzyme 3.

enzyme 1 → ornithine — enzyme 2 → citruline — enzyme 3 (crossed out) → arginine

2. **How was the genetic code identified?** Nirenberg pioneered the study on cracking the genetic code. In a 1961 experiment by Nirenberg and Matthaei, they used ribosomes + *E. coli*-free system with Poly-U synthetic RNA (synthesized with polynucleotide phosphorylase) with radio-labeled amino acids to observe what kind of polypeptide was made. They deciphered the first codon to be identified as UUU for phenylalanine (Phe). Other di, tri, and tetra polynucleotides were synthesized by H.G. Khorana, and homopolymer nucleotides AAA, CCC, and GGG were identified as codons for lysine, proline, and glycine, respectively.

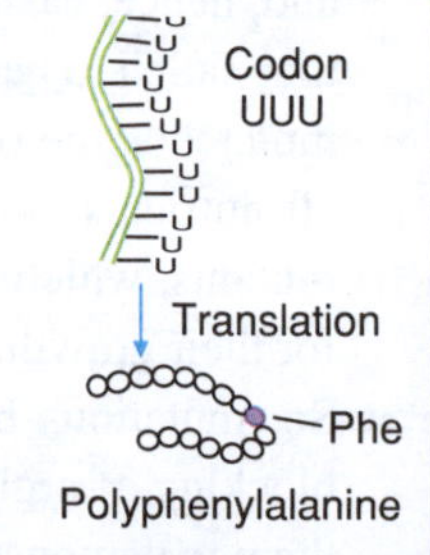

Nirenberg and Leder in 1964 subsequently designed the experiment, which pretty much broke the rest of the genetic code. They added individual amino acids to separate test tubes containing tRNAs, in effect causing the synthesis of specific aminoacyl-tRNAs. They then mixed their amino acid–bound tRNAs with isolated ribosomes and synthetic triplet codons of all 64 combinations.

They established that which codon corresponds to which amino acid and determined codon–amino acid identities, thus cracking the genetic code. Ultimately, 61 of the codons were matched to specific amino acids: Since 61 codons specify 20 amino acids, the genetic code was degenerate. The three stop codons were discovered by Charles Steinberg and Richard Epstein, and UAG was named as 'amber' after their friend Harris Bernstein. The remaining two stop codons were then named 'ochre' and 'opal' for UAA and UGA, respectively.

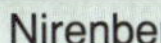

Nirenberg

Matthaei

Leder

codon does not have a corresponding tRNA, so the ribosome stalls. This is read by a release factor, which cleaves the protein from P-site, terminates the assembly, and ends the translation process. The protein will then fold inside the cell, obtain its optimal structure, and become functionally active.

Differences between Prokaryotes and Eukaryotes

In all the three aspects, DNA replication, transcription, and translation, differences exist between prokaryotic and eukaryotic cells. A few of those differences are presented as follows (Tables 6.2 and 6.3):

Table 6.2 Prokaryotic and Eukaryotic Differences

S. No	Parameter	Prokaryotic Cell	Eukaryotic Cell
1.	DNA Replication	Circular DNA	Linear chromosomes
		One origin of replication per chromosome	Multiple origins of replication per chromosome
		Occurs in cytoplasm	Occurs inside nucleus
		One replication fork	Multiple replication forks
		DNA Pol III main replicating enzyme	Many polymerases- α, β, γ, δ, ε DNA Pol δ major polymerase
2.	Protein synthesis-transcription and translation	Coupled transcription and translation in cytoplasm	Transcription in nucleus while translation in cytoplasm
		Single RNA Polymerase	Three different RNA Polymerases
		Poly cistronic mRNA transcripts having coding sequence for specific metabolic pathway	Monocistronic- coding sequence for only one protein or polypeptide
		No introns in genes, so no processing of mRNA	Presence of introns in genes pre-mRNA converted to mRNA with post-transcriptional modifications like splicing, 3′ poly A tail and 5′ cap.
		First amino acid for start codon is N-formly methionine.	First amino acid for start codon is methionine.
		Few initiation factors, Elongation and release factors	More in number and type
		70S ribosome with 30S and 50S subunits	80S ribosome with 60S and 40S subunits

Mutations in DNA

Changes in DNA sequences are called mutations. These alterations to DNA can impact the cell and lead to heritable changes for the cell. There are various reasons why DNA mutations happen. Errors in DNA replication can cause mutations. Exposure to certain agents, such as radiation, carcinogens, viruses, and some drugs, can also cause mutation the DNA. Alterations in DNA sequences can happen anywhere along the length of DNA. When in the protein-coding regions or regulatory sequences of gene expression, they affect the proteins of the cell. However, not all mutations are harmful; some changes may even prove beneficial to the organism.

Any change in the genetic material is a mutation. These changes could be at a single nucleotide level or major aberrations at chromosomal level. Based on the nature of change, mutations could be classified into the following types:

1. ***Point mutations*** This is a change seen in a single nucleotide of the DNA sequence. Although only one nucleotide is changed, it can manifest itself in various ways as follows (Table 6.3):

Table 6.3 Types and Impact of Point Mutations

Name of the Point Mutation	Impact at Codon Level in mRNA	Impact at Amino Acid Level in Polypeptide or Protein
Original DNA without any mutation	Wild-type mRNA sequence and codons UCC UAC AGA GCA UUC UGA	Polypeptide sequence and amino acid Serine — Tyrosine — Arginine — Alanine — Phenylalanine - STOP
Missense point mutation: UUC changed to CUC	Altered mRNA sequence UCC UAC AGA GCA **CUC** UGA	Altered: One codon; one amino acid change Serine — Tyrosine — Arginine — Alanine — **Leucine** — STOP
Nonsense point mutation: UAC changed to UAA	Altered mRNA sequence UCC **UAA** AGA GCA CUC UGA	Altered: Termination after UAA; truncated protein Serine — **STOP**
Silent mutation: GCA changed to GCC	Altered mRNA sequence UCC UAA AGA **GCC** CUC UGA	No impact; same amino acid alanine Serine — Tyrosine — Arginine — Phenylalanine — STOP
Insertion with frameshift: one nucleotide G added after first codon	Altered mRNA sequence UCC **GUA CAG AGC AUU CUG A...**	From second codon, complete frameshift; new amino acids; and stop codon lost, so longer protein Serine — **Valine** — **Glutamine** — **Serine** — **Isoleucine** — **Leucine** — continues as STOP codon lost
Deletion with frameshift: one nucleotide U deleted after first codon	Altered mRNA sequence UCC **ACA GAC CAU UCU GA...** continues as STOP codon lost	From second codon, complete frameshift; new amino acid; and stop codon lost, so longer protein Serine — **Threonine** — **Glutamic acid** — **Histidine** — **Serine** — continues as STOP codon lost

Interesting Fact 6.2

Simplified understanding of different mutation impacts

Let's mutate words...!

A sentence with 3-letter words	Nature of point mutation
THE RED BAG HAS RED BAT.	Original meaningful sentence with six words
THE RED BAG HAS **REM** BAT	Missense mutation — D substituted with M in the 5th word — changed sentence — one word change
THE RED BAG. HAS RED BAT.	Nonsense mutation — full stop inserted in-between after the 3rd word — half sentence — incomplete meaning
THE RED BAG HAS **REd** BAT.	Silent mutation — no change in meaning — alphabet different
THE **BRE DBA GHA SRE DBA** T...	Insertion leading to frameshift mutation — complete change in meaning — B added after first word and original words retained
THE **EDB AGH ASR EDA AT**...	Deletion leading to frameshift mutation — complete change in meaning — R removed after first word and original words retained

(i) *Missense mutation* Substitution of one nucleotide with another, which can lead to codon change, resulting in a wrong amino acid getting incorporated in the protein. So missense substitution mutations of one nucleotide will impact one amino acid. For example, if the codon changes from UUC to CUC, it will incorporate leucine amino acid instead of phenylalanine.

(ii) *Nonsense mutation* Substitution of one nucleotide can result in introduction of a stop codon, which would cause the ribosome to stop protein synthesis. The protein will be terminated too soon and a functional protein will not be formed. For example, nucleotide substitution can convert UAC to UAA; hence, instead of incorporating tyrosine, the polypeptide chain is terminated.

(iii) *Silent mutation* Sometimes there will be a nucleotide change due to mutation, but it will not impact the protein as the codon change will still result in the same amino acid to be incorporated. This will happen as there are some amino acids with more than one codon. Substitution of one codon with another for the same amino acid will result in silent mutation, and no impact on the protein will be seen. For example, if GCA changes to GCC or GCG, it would still incorporate alanine, as all these codons specify alanine as per genetic code.

(iv) *Single nucleotide insertion and deletion* If one nucleotide is inserted in between or one nucleotide is deleted, there will be a complete shift of the reading frame of mRNA as the genetic code is non-overlapping. The three-letter codons from the point of insertion or deletion will be changed completely, and the protein will be a different one with a new lot of amino acids. The same situation will hold if two nucleotides are added or deleted. However, if three nucleotides are added or deleted, then the impact will be confined to only one additional amino acid being added or one missing amino acid. But the reading frame will be restored from the next codon onwards.

2. ***Chromosomal aberrations*** These are major changes in the DNA, which impact at the chromosomal level. Segments or parts of the chromosome are altered; hence, many genes may be impacted and could have a significant impact on the organism's life and development. The different types of chromosomal aberrations that impact at the structural level include the following (Figure 6.14):
   - *Inversions* In this kind of mutation, a part of the chromosome is broken and gets reattached in a flipped or inverted orientation. The genes between the breakpoints are impacted and the sequences are read in opposite orientation.
   - *Translocations* In this type of mutation, a broken part of the chromosome attaches itself to a different location in the same chromosome or to a different chromosome. This kind of erroneous integration can also happen when pieces from two different chromosomes trade places with each other. Translocation is reported in certain disease conditions such as Down's Syndrome, schizophrenia, and muscular dystrophy.
   - *Duplications* This mutation occurs when a portion of the chromosome is replicated and reattaches adjacent to the original sequence. This is called tandem duplication. Sometimes the duplicated chromosome can be separated far away by non-duplicated regions, and it is said to be displaced. Duplications can lead to copy number variations of genes and lead to gene dosage issues.
   - *Deletions* These are aberrations in which a part of the chromosome is lost or destroyed, leading to loss of genetic information.

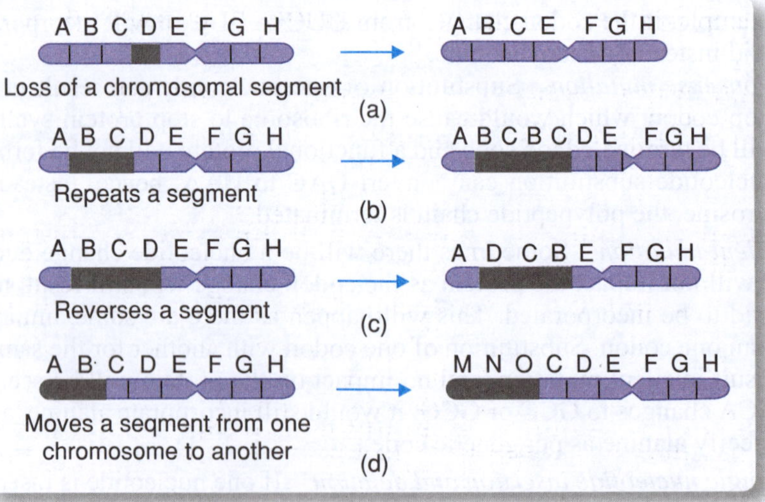

Fig. 6.14 Types of chromosomal aberrations: (a) Deletion; (b) Duplication; (c) Inversion; (d) Translocation

Source: Adapted from http://www.sarthaks.com/1419756/what-are-different-types-of-chromosomal-aberrations

3. ***Mutations caused by viruses*** Some viruses insert their RNA into the host DNA, and the presence of these new viral sequences may interfere and impact the genes and DNA sequences of those areas in the host. These can be classified as insertion mutations. Such

insertions in a few cases, such as the Human Papilloma virus (HPV), can lead to cancers. Thus, viruses act as mutagenic agents.

Mutations and inheritance

If the mutations are seen in the sex cells of the individuals, they tend to be inherited. Somatic mutations affect the individual whereas germline mutations could be carried forward to the next generation. How DNA changes are inherited will be a part of the next few chapters.

Interesting Fact 6.3

Point mutation and sickle-cell anaemia

- Sickle-cell anaemia is a genetic disorder due to a point mutation in beta haemoglobin gene.
- Mutation leading to change of codon from GAG codon to a GUG codon in the 6th position encodes valine instead of glutamic acid.
- This mutation produces abnormal haemoglobin called sickle haemoglobin (HbS) that causes RBCs to sickle when exposed to low oxygen level.
- Sickle-shaped RBCs get trapped in small blood vessels, which can cause pain and damage organ.

	Normal	Sickle Cell
DNA	GAG CTC	→ mutation → GTG CAC
RNA	GAG	GUG
Protein	GLU	VAL
Haemoglobin		

Source: Adapted from https://evolution.berkeley.edu/dna-and-mutations/a-case-study-of-the-effects-of-mutation-sickle-cell-anemia/

Experimental Insight 6.4 DNA: the ultimate data storage solution

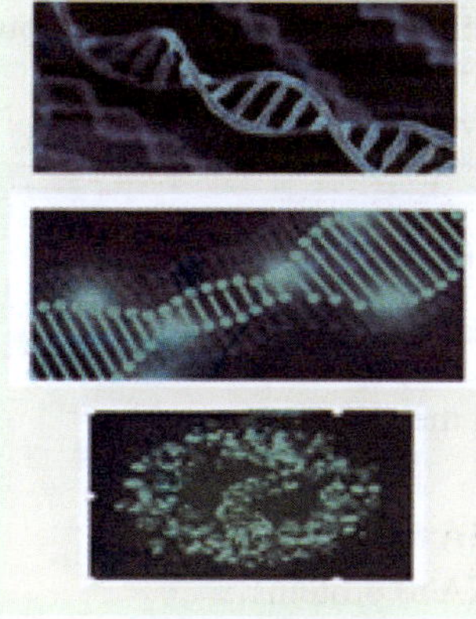

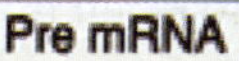

Pre mRNA

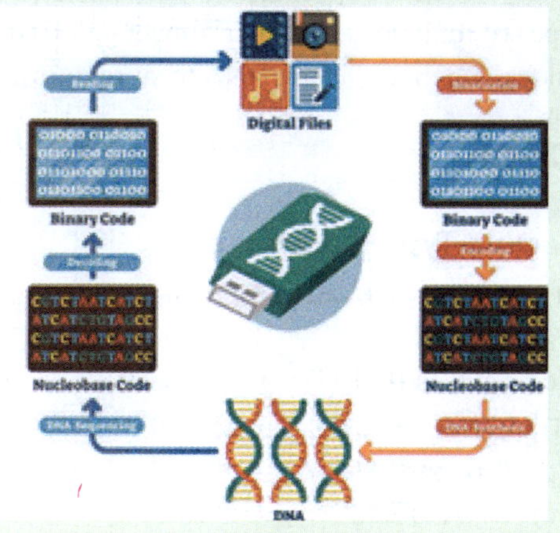

To store large data efficiently and inexpensively, one of the modern-day solutions that has emerged is to archive information in DNA molecules. Instead of large reels of magnetic tape, data storage in DNA is more energy efficient, long lasting, and less expensive. DNA properly encapsulated with a salt remains stable for decades at room temperature and should last much longer in the controlled environment of a

data centre. DNA does not require maintenance, and files stored in DNA are easily copied for negligible cost. DNA can store large volume of data in smaller space—zettabytes worth of data can be stored in a DNA that will be the size of a small ball, and this is a great boon for space saving. The Adaptive DNA Storage Codex (ADS Codex) of Los Alamos National Laboratory translates data files from the binary language of computers (zeroes and ones) to the four-letter code of DNA. As we are aware, the four nucleotide bases of DNA store information and act as a manual for building every organism on Earth. Specific sequences of A, C, G, and T create a code for various proteins and activities and form the blueprint in the form of a genome. ADS Codex tells exactly how to translate the zeros and ones into sequences of four-letter combinations of A, C, G, and T, and vice versa. Apart from this, a user friendly method has been developed recently—the DNA bricks method with methylated DNA and Barcode method were also developed for data storage. Molecular storage in DNA is a wonderful attempt, but error rates are high as there could be insertions, deletions, frame shifts, etc., whereas the error can only be reversal of zero and ones in digital storage.

Detecting errors are also tougher in DNA storage. With data storage needs continuing to grow exponentially, DNA as a storage tool shows great promise due to its amazing data density of storing 215,000 terabytes in one gram of DNA.

The additional reading section below gives the references relating to data storage in DNA for further referencing and student interest.

SUMMARY

- Details of genetic material – its structure, organization, and genomes were studied.
- The central dogma of biology was explained.
- DNA replication methodology, enzymes, and the outcomes were elaborated upon.
- Protein synthesis – transcription and translation were discussed.
- Eukaryotic mRNA processing is touched upon.
- Types of mutations and their impact on proteins and living systems were highlighted.
- A novel attempt of using DNA to store information as a modern day development was touched upon.

EXERCISES

A. Objective-Type Questions

1. 1. Which of the following is not true for all double-stranded DNA samples?
   (a) A/T = 1 (b) (A + C)/(G + T) = 1
   (c) (A + T)/(C + G) = 1 (d) All the above
2. Transcription copies genetic information from
   (a) Proteins to DNA (b) RNA to DNA
   (c) DNA to RNA (d) RNA to proteins
3. A deletion of a single base in a protein-coding gene will most likely result in
   (a) Duplication (b) Faulty RNA polymerase
   (c) Frameshift (d) Absence of tRNA
4. A nucleotide change in DNA without any change to the sequence of amino acids in protein is called as ________.
   (a) Chromosomal aberration (b) Nonsense mutation
   (c) Missense mutation (d) Silent mutation

5. A-site, P-site, and E-site are parts of ________.
   (a) RNA polymerase (b) Nucleosome
   (c) Gene (d) Ribosome
6. Eukaryotic genes contain certain non-coding regions called ________.
   (a) Exons (b) Codons
   (c) Introns (d) Anticodons
7. ___________ has a clover-leaf structure with an anti-codon arm.
   (a) Chromosome (b) Ribosome
   (c) tRNA (d) mRNA
8. Which of the following does not denote DNA?
   (a) Chromatin (b) Gene
   (c) Chromosome (d) Histone
9. In a genetic code, ________ is the number of codons that code for 20 amino acids.
   (a) 16 (b) 61
   (c) 64 (d) 3
10. UGA, UAA, and UAG are known as __________ .
    (a) Initiator codons (b) Stop codons
    (c) Anti-codons (d) Exons

B. Short-Answer Questions

1. In both replication and transcription, the DNA template is copied. Give two differences between the two processes.
2. For a moment, assume that DNA replication occurs continuously on both the strands of DNA. Which of the following will not be a part of DNA replication in such a scenario: primase, helicase, DNA polymerase, Okazaki fragments?
3. How did Meselson and Stahl prove that DNA replication is semi-conservative?
4. Transcription and translation are coupled in prokaryotes but not in eukaryotes. Explain.
5. All the three RNAs have a link with ribosomes. What is the link? Which two RNAs connect due to base pairing and in which process?

C. Self-Study Questions

1. Read, in general, about the theories proposed by Charles Darwin and Jean B. Lamarck, and try to see the connection with selection of genes (characters) under the influence of the environment (conditions) in evolution of organisms.

FURTHER READINGS

1. Taylor, Ian. Dark DNA: How scientists could soon unlock the mystery of your genome. BBC ScienceFocus magazine, June 9, 2024.

USEFUL WEBLINKS

1. https://my.clevelandclinic.org/health/body/23064-dna-genes--chromosomes
2. https://www.mybiosource.com/learn/what-is-dna/
3. https://my.clevelandclinic.org/health/body/23064-dna-genes--chromosomes
4. https://www.mybiosource.com/learn/what-is-dna/

5. https://www.sciencefocus.com/the-human-body/dna
6. Do it yourself DNA data storage https://www.nature.com/articles/d41586-024-03312-6
7. DNA: The Ultimate Data-Storage Solution
8. https://www.scientificamerican.com/article/dna-the-ultimate-data-storage-solution/

VIDEO LINKS

1. https://study.com/academy/lesson/video/overview-of-the-genetic-code.html
2. https://study.com/learn/lesson/video/what-is-central-dogma-of-molecular-biology-central-dogma-theory-of-genetics.html
3. How to store data on DNA? https://www.youtube.com/watch?v=ZuB3yB7QWCk

CHAPTER 7

Inheritance and Mendelian Genetics

LEARNING OBJECTIVES

After reading this chapter, you will be able to:

- Get acquainted with the entire modern terminology associated with genetics and inheritance
- Appreciate the beginning of genetics by Mendel, his experiments, and his laws of inheritance
- Factor inheritance ratios in monohybrid (single factor) and di-hybrid (double factor) crosses
- Enlist modified Mendelian patterns such as co-dominance and incomplete dominance
- Compare multiple allelism and polygenic inheritance
- Understand linkages, X-linked inheritance, and other phenotypic influences

CHAPTER TAKEAWAY **Laws of inheritance and different types of inheritance patterns**

This chapter presents a comprehensive coverage of the laws and patterns of genetics and inheritance from the beginning (Mendelian) till today.

PRELUDE

The ability to reproduce is one of the most fundamental features of living organisms. Reproduction ensures the continuation of species, and traits are transmitted from generation to generation. While roses produced roses, dogs produced dogs, and humans produced humans in every generation, what is interesting to note is the variations we see among individuals and populations. Children from the same parents are not identical except for monozygotic twins. Siblings vary from one another and are also different from the parents in some ways. Inheritance of traits is hence a fascinating topic to explore. Genetics is the study of inheritance and heritable variations. As Mendel pioneered the understanding of inheritance and his work set the stage for modern genetics, this field of study is rightfully called 'Mendelian Genetics'. The field expands much beyond Mendel and includes many more concepts, which are also covered here.

Fundamentals of Genetics

When a cell divides, chromosomes divide, and the DNA is passed on to the daughter cells through the chromosomes during nuclear division. We have already seen that chromosomes contain specific genes—each gene contributes for a characteristic. Thus, genetic material when transmitted helps in transmitting characteristics as well.

At the organism level, reproduction is an important aspect of the human life cycle. It helps to transmit genetic information and, in turn, characteristics from one generation to another in living organisms. Reproduction is of two types: (a) asexual reproduction and (b) sexual reproduction (Figure 7.1).

Asexual reproduction requires only one parent, which divides to produce two identical organisms that are genetically identical to the parent. Binary fission in bacteria, fragmentation in amoeba, budding in Hydra, and so on, are examples of asexual reproduction. In these organisms, before the parent cell splits into two, it duplicates its genetic material/chromosomes and, thus, each daughter cell receives an identical copy as the parent cell.

Sexual reproduction, on the contrary, requires two parents to donate genetic information towards creating a new individual. In sexually reproducing organisms, the genome is diploid ($2n$)— two copies of genetic information—one from each parent. These two copies have the *complete set* of genes necessary for coding the organism's characteristics, but they *need not be identical*. They may contain different versions of the genes called *alleles*.

To produce a diploid ($2n$) individual by sexual reproduction, one set of chromosomes comes from the female gamete (n) and another set from the male gamete (n), and fertilization of these two gametes leads to a diploid zygote ($2n$), which then grows into a new individual. Sex cells such as sperm and egg are haploid and contain only one copy of every gene. The alleles in egg and sperm combine to form a new genome ($2n$) that is different from either of the parents. So, the gametes are haploid (carry one copy of the genetic information), and these haploid gametes are produced by meiosis. Every diploid individual produces haploid gametes for fertilization and, hence, only one copy of the individual genome is transmitted. Whichever of the two copies is present in the gamete that eventually fertilizes decides the genetic makeup of the new offspring.

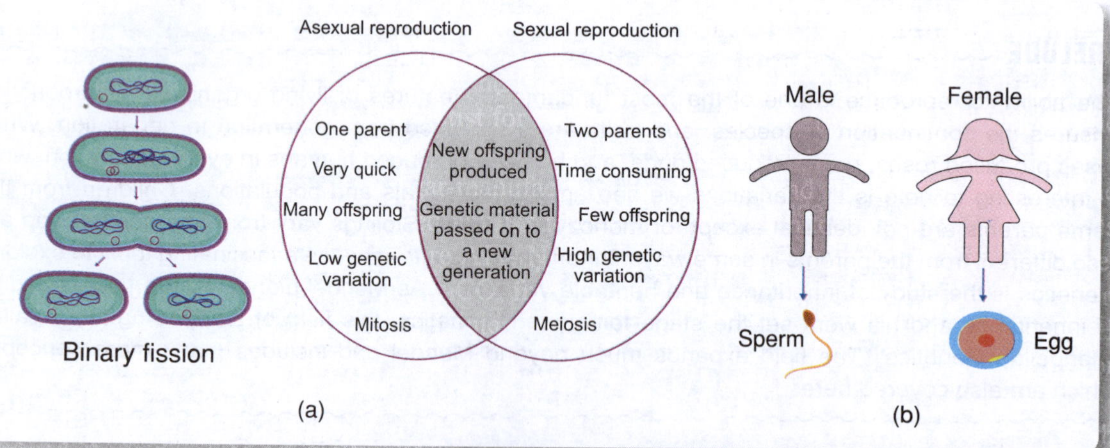

Fig. 7.1 Types of reproduction: (a) Asexual; (b) Sexual

Source: Adapted from https://www.nagwa.com/en/explainers/647147683954/

Terminologies Associated with Modern Genetics and Inheritance

Genome It is a set of all the genes necessary to code for all the characteristics of and produce a functionally fit organism (Figure 7.2).

Diploid An organism with two sets of genetic material (2*n*). Each chromosome is present in two copies, so each gene has another counterpart in the same cell. For example, human diploid number of chromosomes = 46; 23 pairs; 22 pairs of autosomes and one pair of sex chromosomes.

Haploid The genome of a haploid organism has only one copy of genetic material—only one copy of each chromosome and gene. For example, human gametes—male sperm and female egg—are haploid; Male ants, bees, and wasps develop from unfertilized eggs and are haploid, whereas their female counterparts develop from fertilized eggs and are diploid.

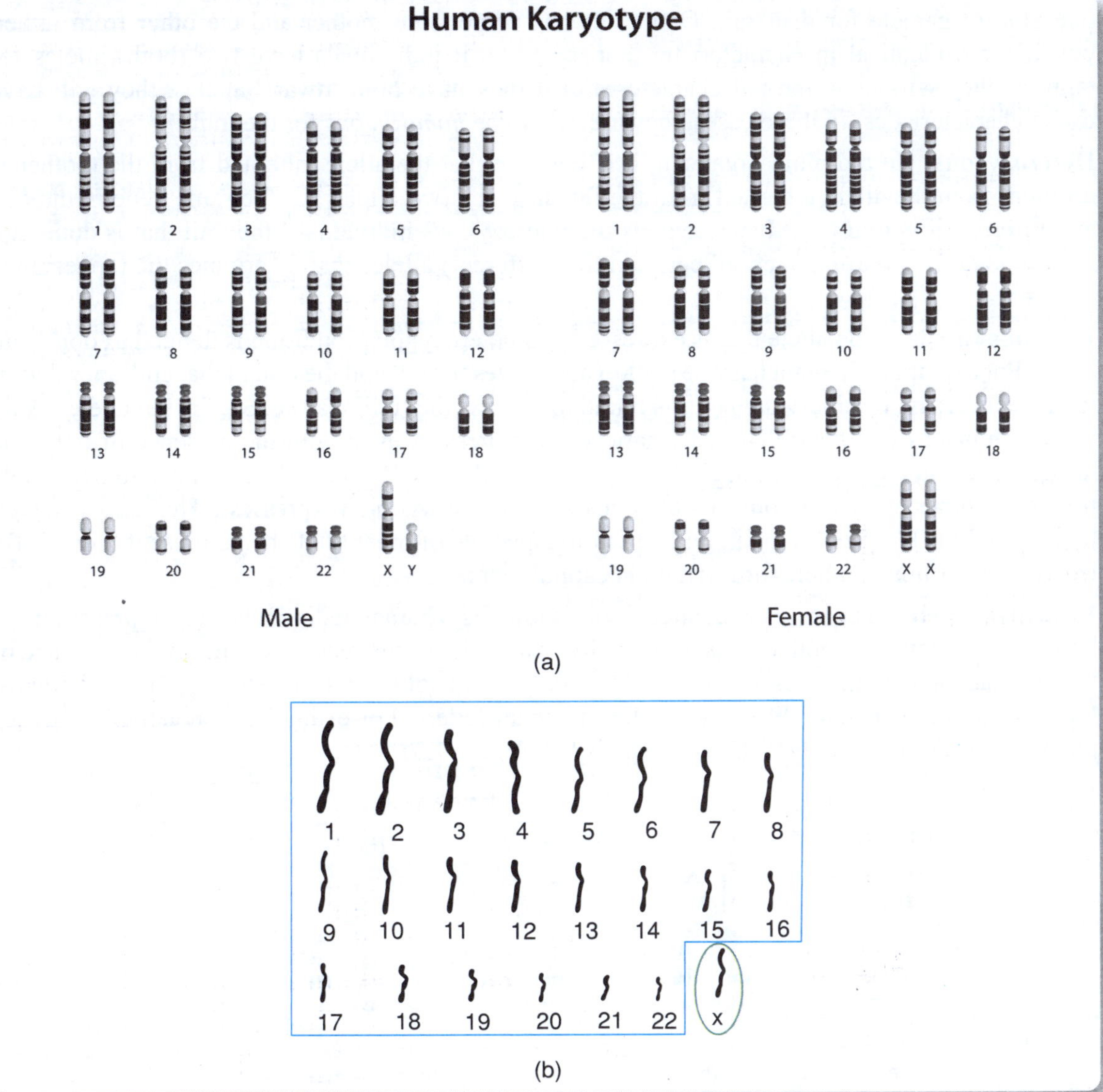

Human Karyotype

Male

Female

(a)

(b)

Fig. 7.2 Human genome: (a) Replicated but haploid; (b) Haploid gametes

Alleles Diploid organisms have two copies of genes/genetic material. The two copies can be different. These different versions of the gene are called alleles.

Genotype The genetic makeup of the two versions of the gene, as present in the cells of an individual, is called their genotype. Thus, there are the two copies identical or different. The alleles possessed by an individual denote the genotype. It is one type or two different types. Either both alleles are the same or are different.

Phenotype It is a measurable characteristic or distinctive trait displayed by the individual as a result of his/her genetic makeup, i.e., genotype. For example, 'tall plant' or 'dwarf plant' is a phenotypic trait. *The phenotype is a result of genetic expression.*

Homozygous In a diploid organism, if both copies carry the same identical allele of a gene, they are said to be homozygous. Homozygotes contain the same alleles at a single locus and produce one kind of gamete for that trait. One allele has come from mother and the other from father, but they are identical in characteristic. For example, if individuals have two (both) alleles for tallness, they will show the tall characteristic; if they have both 'dwarf' alleles, they will have dwarf characteristics. In these cases, they are said to be *homozygous* for the trait.

Heterozygous In a diploid organism, it is possible that the allele inherited from the mother is different from the allele inherited from the father. If the two alleles for the same gene in the cell are different, it is said to be heterozygous. In heterozygous individuals, the trait that is dominant is expressed. It is the 'interaction' between the (different) alleles that determines the (appearance of) trait.

Dominant allele The allele that is expressed in a heterozygous condition is defined as dominant allele. For example, if a plant has one allele T that codes for tall and the other t that codes for dwarf, then the heterozygous individual (Tt) is found to express the 'tall' feature. So, the tall allele is said to be dominant over dwarf. The tall character is expressed, even with the presence of only one (single) allele in the heterozygous individual, as this allele is dominant. Infact, the dominant allele masks the other allele, and only its own characteristic is visible (expresses). That is, whenever, by convention, T is present, plant will be tall irrespective of other allele being present 't or T'. By convention, dominant alleles are written in capital letters.

Recessive allele The recessive allele will show its characteristic only when present in a homozygous state, i.e., both copies of the individual code for the recessive trait. In the presence of the dominant allele, the recessive trait will be masked and not express in heterozygous individuals. By convention, recessive alleles are written in small letters. For example, a phenotype of dwarf plant will be produced only if the genotype is 'tt' (Figure 7.3).

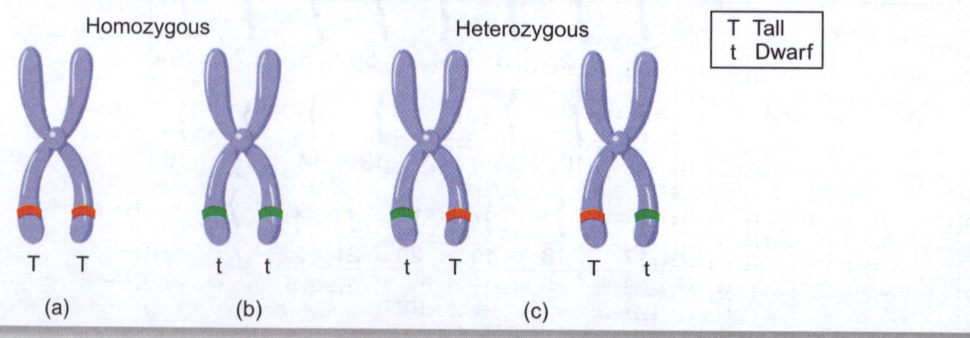

Fig. 7.3 Genotype, phenotype, homozygous, heterozygous, dominant, and recessive alleles

Mendel's Laws of Inheritance and His Pioneering Work

Gregor Mendel (1822–1884), an Augustinian monk, experimented with garden pea (*Pisum sativum*) and carried out controlled matings between plants with different traits. He hand-pollinated the pea flowers by taking anthers from one plant and placing them on the stigma of another, thus, achieving cross-fertilization in a controlled manner. His observations and results provided the first insights on the inheritance of traits and their patterns.

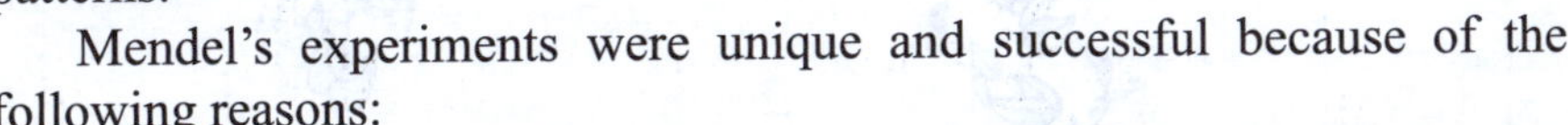

Mendel's experiments were unique and successful because of the following reasons:

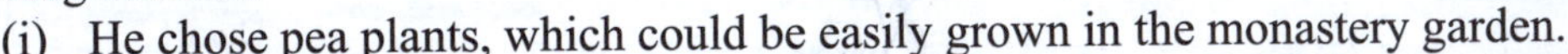

(i) He chose pea plants, which could be easily grown in the monastery garden.

(ii) Pea plants usually self-pollinate and produce both pollen and eggs in the same plant. When self-pollination occurs over several generations, it develops a pure line of plants with homozygous genetic makeup, without change of traits from one generation to the other.

(iii) He studied discrete, non-overlapping characters, one character at a time. He obtained plants that had alternative forms of these characters and showed differences in their traits.

(iv) As part of his experimental design, he cut the anthers, prevented self-pollination, and carried out cross-pollination with a purposeful intent.

(v) He then studied the distribution of the traits in the next generation to arrive at his laws or conclusions on inheritence.

Mendel selected seven such traits to study, as shown in Table 7.1.

Table 7.1 Mendelian traits

Character	Dominant Trait	Recessive Trait
Seed shape	Round	Wrinkled
Seed colour	Yellow	Green
Flower colour	Violet	White
Pod shape	Full (inflated)	Constricted
Pod colour	Green	Yellow

Contd.

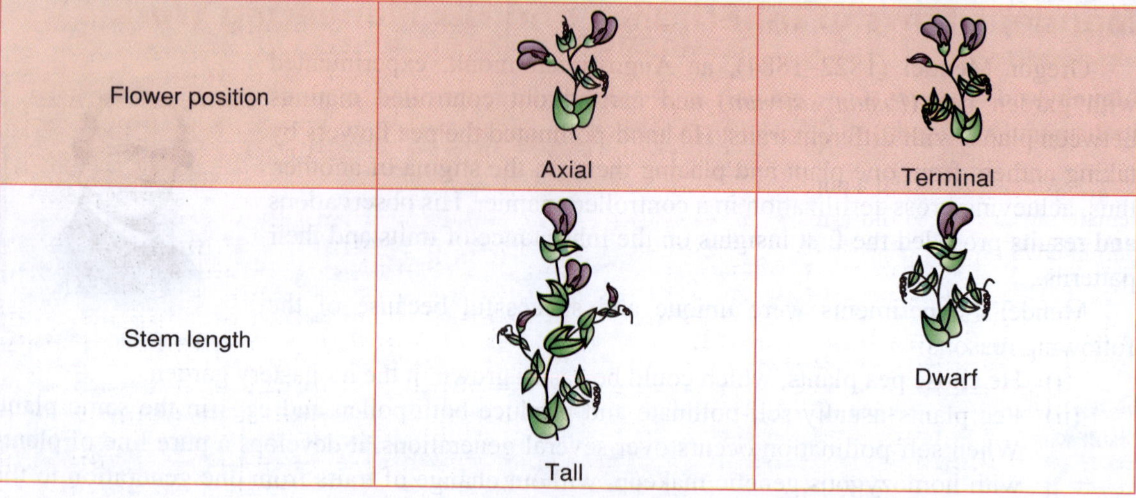

Flower position	Axial	Terminal
Stem length	Tall	Dwarf

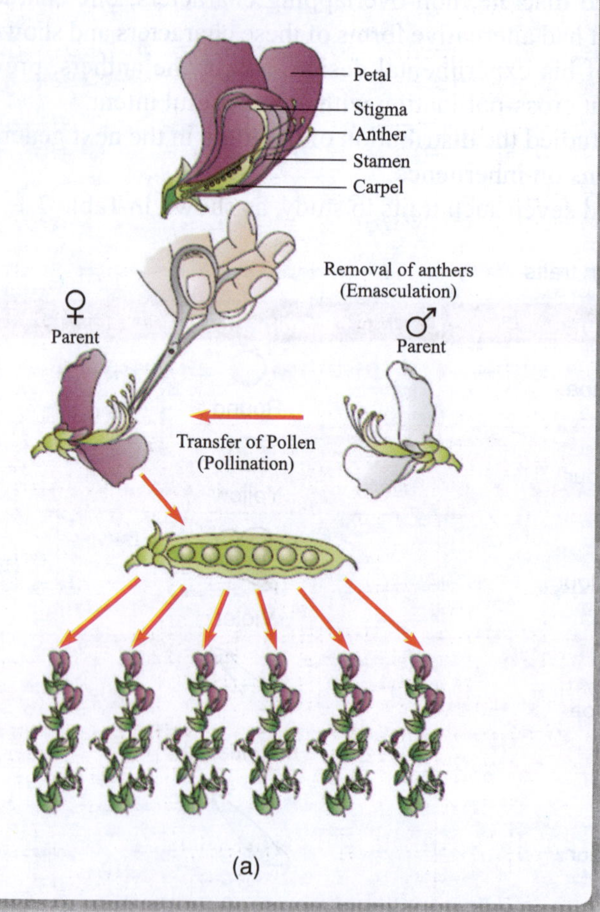

Fig. 7.4 Mendel´s experiments with pea plant—monohybrid cross—flower colour: (a) Steps in making a cross in pea

Experimental Insight 7.1 Mendel's Experiments

Monohybrid or single-factor cross Let us understand the outcome of one trait— flower colour—and its distribution in the offspring, which led Mendel to formulate his laws (Figure 7.4).

Step 1: Mendel fertilized a pure-line white-flower male plant with the pollen of a pure-line purple-flower female plant. He also did a reciprocal cross, i.e., pure-line white-colour female plant with a pure-line purple-colour male plant.

White (Male) × Purple (Female) and Purple (Male) × White (Female)

These plants that were used for the cross were called parental (P) generation.

Step 2: The first filial (F1) generation of offspring was produced. In both the cases (crosses), the same result was obtained. All the F1 seeds produced plants with purple flowers.

Step 3: It was as if the white colour had vanished.

Step 4: Then, he used these F1 purple plants to cross-fertilize each other (self-fertilize).

F1 Purple × F1 Purple

Step 5: Now, it produced the second (F2) generation of offspring. In this generation, the white colour reappeared. Mendel counted close to 1000 plants. Plants with purple flowers were around 700+ and those with white colour were around 220 only. Thus, the ratio of purple:white plants rested at around 3:1.

He repeated his experiments with all thc seven traits, studying one character at a time, and this monohybrid ratio led him to propose laws that were called Mendel's laws of inheritance. Mendel published his work in 1865; however, the full significance of his work was realized much later. It is also important to note that he did not know about DNA or even chromosomes during his lifetime. As a mathematical deduction based on counting plants and understanding their phenotype, he deduced the genotypes and proposed the inheritance patterns.

Mendel's experiments and the results of the experiments are as follows (Table 7.2):

Table 7.2 Mendel´s Results on All Seven Traits and Outcomes

Trait	Dominant	Recessive	Numbers in F_2	Ratio
Seed				
Seed shape	Round	Wrinkled	5474:1850	2.96:1
Seed colour	Yellow	Green	6002:2001	2.99:1
Whole plants				
Flower colour	Purple	White	705:224	3.15:1
Flower position	Axial	Terminal	651:207	3.14:1
Plant height	Tall	Short	787:277	2.84:1
Pod				
Pod shape	Inflated	Constricted	882:299	2.95:1
Pod colour	Green	Yellow	428:152	2.82:1

Mendel's Laws of Inheritance

Mendel formulated three laws based on his studies: Law of Dominance, Law of Segregation, and Law of Independent Assortment.

The first two laws are obvious from monohybrid crosses, and third law is interpreted from dihybrid cross.

1. ***Mendel's Law of Dominance*** Adult plants such as 'pea' carry a pair of 'hereditary units' or 'factors' that govern the inheritance of each trait, and these factors can be different. When the purelines of contrasting traits (P1) are crossed, only one form of the trait appears in F1 (first filial) generation. The characteristic appearing in F1 generation (heterozygote) masks the other counterpart and, hence, is dominant over the other. Law of dominance states that *in a heterozygous individual, one trait exhibited by the hybrid offspring is dominant over the other.*

Tips

Now, we know the genetic basis of inheritance, about genes, and that the allele for the trait is the dominant allele and the other one is recessive, which will be expressed only when both the copies are recessive.

2. ***Mendel's Diploid Law of Segregation*** The pair of hereditary units or factors present for each trait in an individual gets segregated when the gametes are formed. Gametes contain only one form of the trait; half of the gametes carry one allele of the pairs and the other half carries the other allele of the pairs. When this gamete with one allele unites with the gamete of the other parent during fertilization, the offspring inherits these two alleles—one allele from each parent.

Law of segregation states that *during gamete formation, the hereditary factors for a trait segregate randomly, and each gamete receives only one of the pair*. However, each factor has an equal probability of entering any gamete.

Tips

Now, we know that these two alternate factors of a pair for a gene are alleles, and we know about allele segregation during gamete formation.

Di-hybrid or two-factor cross As a natural extension, Mendel then studied two characters simultaneously and designed experiments to test their inheritance patterns. In a dihybrid cross, the F1 is produced from a cross that involves two characters. When individuals with contrasting traits for two characters are crossed, the F1 will be a heterozygote for both the traits.

Mendel used two characters—seed shape and seed colour. Seed shape could be either round (R) or wrinkled (r). Similarly, seed colour can be yellow (Y) or green (y). In the parental generation, he chose parents—one who had dominant traits in both the characters and the other had recessive traits in both characters.

Step 1: Parental generation plants crossed. True breeding plant with round and yellow seeds (RRYY) with a pure-line plant with wrinkled and green seeds (rryy).

Round-yellow plant RRYY × Wrinkled-green plant rryy

Step 2: The F1 offspring was a heterozygote (di-hybrid) for both the characters. Gametes produced in the two parents after segregation were RY and ry.

RY × ry = RrYy (F1 heterozygote of two characters)

Step 3: A di-hybrid cross of two F1 double heterozygotes.

RrYy × RrYy

This produced four different types of gametes—RY, Ry, rY, and ry. Then, offsprings were produced in the F2 generation.

Step 4: In F2 generation, there were 16 possible genotype combinations of offspring from each cross. Four different phenotypes were produced in F2; the F2 phenotypic ratio was 9:3:3:1 (9 round-yellow: 3 round-green: 3 wrinkled-green: 1 wrinkled-green). This can be easily explained using a Punnett square as follows:

As per Mendel's count of 1000 plants, he obtained 315 round-yellow plants, 108 round-green, 101 wrinkled-yellow, and 32 wrinkled-green, justifying the ratio of 9:3:3:1. Based on these results, he deduced that the hereditary factors for the two characters assort independently during the formation of gametes. This also confirmed the law of segregation (3:1 ratio for round:wrinkled and yellow:green). Based on mathematical deduction, he showed that these two traits assorted independently (Figure 7.5).

3. ***Mendel's law of independent assortment*** This law states that *during gamete formation, the segregating pair of alleles for the two characters assort independently*. So, the allele for seed shape that the gamete receives (R or r) is not influenced by the allele for seed color (Y or y) and vice versa. The two events are completely independent of each other.

He further produced confirmatory evidence for this law by doing an additional experiment of test cross. He crossed the heterozygous F1 plant (RrYy) to a double homozygous recessive plant (rryy). He reasoned that if independent assortment is true, then the F1 should produce four different gametes: RY, rY, Ry, and ry with equal frequencies. When combined with the 'yr' gamete of homozygous recessive plant, all the four phenotypes should be observed in equal proportions (1:1:1:1). His experimental results of 55 round-yellow seeds: 51 round-green seeds: 49 wrinkled-yellow seeds: 53 wrinkled-green seeds were close to the expected ratios, thus strengthening his law of independent assortment.

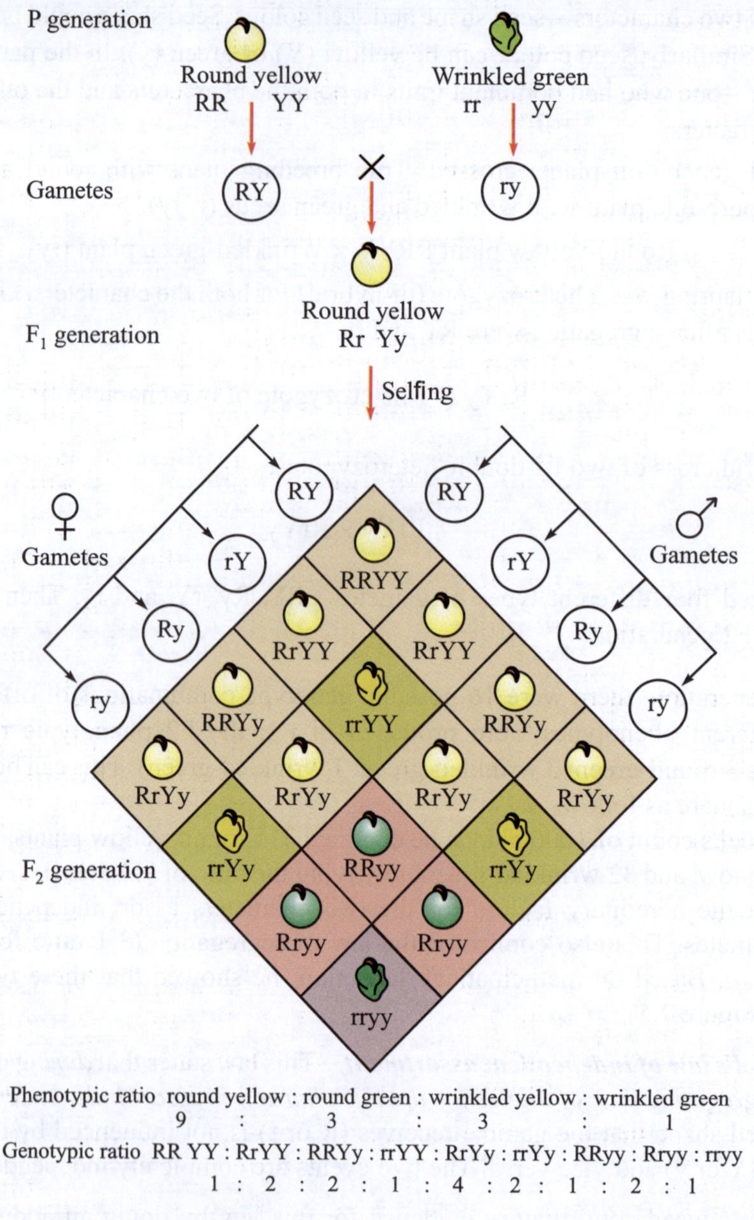

P generation

Round yellow
RR YY

Wrinkled green
rr yy

Gametes

RY × ry

Round yellow
Rr Yy

F_1 generation

Selfing

♀ Gametes: RY, rY, Ry, ry

♂ Gametes: RY, rY, Ry, ry

RRYY

RrYY, RrYY

RRYy, rrYY, RRYy

RrYy, RrYy, RrYy, RrYy

rrYy, RRyy, rrYy

Rryy, Rryy

rryy

F_2 generation

Phenotypic ratio round yellow : round green : wrinkled yellow : wrinkled green
9 : 3 : 3 : 1

Genotypic ratio RR YY : RrYY : RRYy : rrYY : RrYy : rrYy : RRyy : Rryy : rryy
1 : 2 : 2 : 1 : 4 : 2 : 1 : 2 : 1

Fig. 7.5 Mendel´s di-hybrid (two-factor) cross leading to the Law of Independent Assortment

Although Mendel's experiments were primarily with garden pea plants, his analysis and laws are applicable to all species. These laws provide coherent insights on the inheritance traits for single-character and double-character crosses.

Interesting Fact 7.1

How to solve Mendelian genetics problems??

1. First read the problem given. Note down the information that is available and the question that is asked.
2. From the given information, determine the parental genotypes from the problem through the phenotypes or genetic data given.
3. Determine the gametes that will be formed in both parents. Haploid genotype of the sex cells—male sperm and the female egg.
4. Set up a Punnett square to show all possible fertilization outcomes.
5. Determine the offspring genotypes and phenotypes.

A tall pea plant and a dwarf pea plant are crossed. Given below are two situations: (a) Both parents are homozygous; (b) Both parents are heterozygous

Find offspring genotype and phenotype.

***Solution*:**

(a) Let allele 'T' denote tallness, 't' denote shortness. Parental genotype is known from phenotype TT and tt, as they are homozygous for the trait. All gametes produced in one parent will have 'T' allele and the gametes of the other parent will be 't'. Punnett square is as shown below. All plants (100%) in next generation are heterozygous: Genotype Tt Phenotype tall.

(b) Both parents are heterozygous, so parental genotype is Tt. Gametes produced by both parents, 50% will have 'T' allele and 50% will have 't' allele. Punnet square is as shown below. Genotypic ratio of offspring is 1TT: 2Tt: 1tt; Phenotypic ratio is 3 tall 1 short.

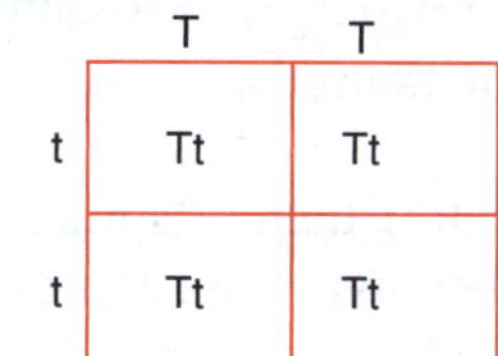

	T	T
t	Tt	Tt
t	Tt	Tt

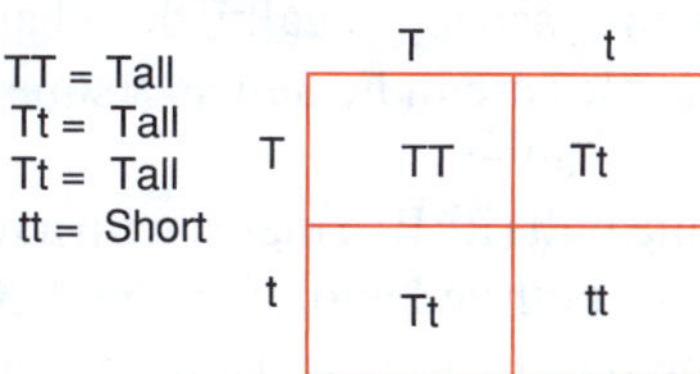

TT = Tall
Tt = Tall
Tt = Tall
tt = Short

	T	t
T	TT	Tt
t	Tt	tt

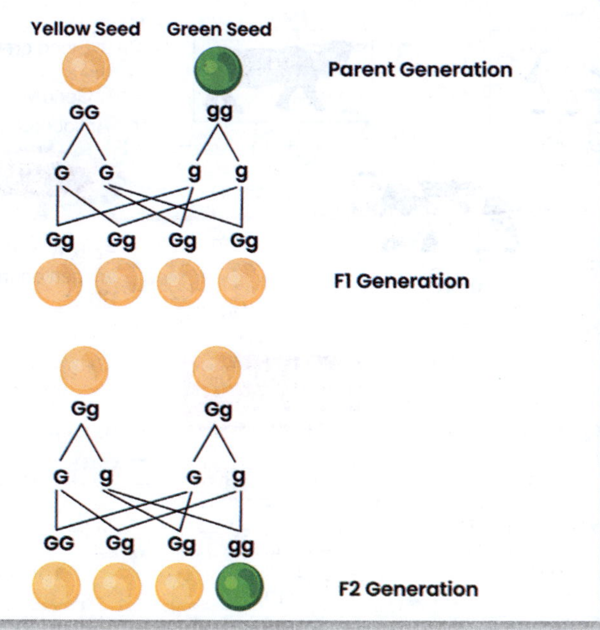

Monohybrid cross

Modified Mendelian Patterns

In the 1900s, research on genetics gained new momentum as it became clear that (i) each gene has a specific location on chromosomes called 'locus', (ii) chromosomes are the factors that are inherited—chromosomal theory of inheritance, and (iii) gamete formation happens by meiosis. Genetic research that followed in the 20th Century not only supported and validated Mendel's observations and laws but also revealed many variations on his basic principles in nature. Apart from the straightforward allelic interactions of dominant and recessive, there are inheritance patterns that do not conform to this dominant–recessive framework and are called *modified Mendelian patterns*.

Co-dominance

Normally, in Mendelian inheritance pattern, the heterozygote shows only one trait and completely masks the other trait. So, out of the two alleles, one is dominant and the other is recessive.

However, in co-dominance, none of the two alleles of a heterozygote shows absolute dominance, both of them express their observed phenotype to some degree. So, no allele is dominant or recessive, and both of them are equally dominant—hence called co-dominance. Conditions in which the heterozygote expresses both phenotypes simultaneously is called a co-dominant characteristic. Hence, the phenotype of the heterozygote will be very different from either of the homozygous parents (Figure 7.6).

To represent co-dominance, usually the character or trait is given a capital letter and the alternative forms of alleles are indicated as a superscript. For example,

(a) Hair colour in cattle:
When a white bull (R^WR^W) mates with a red cow (R^RR^R), their offspring are *roan*, meaning that they have both red and white hair. Both the phenotypes are expressed.

(b) Blood group AB in humans: Both the antigens A and B are expressed on RBC.

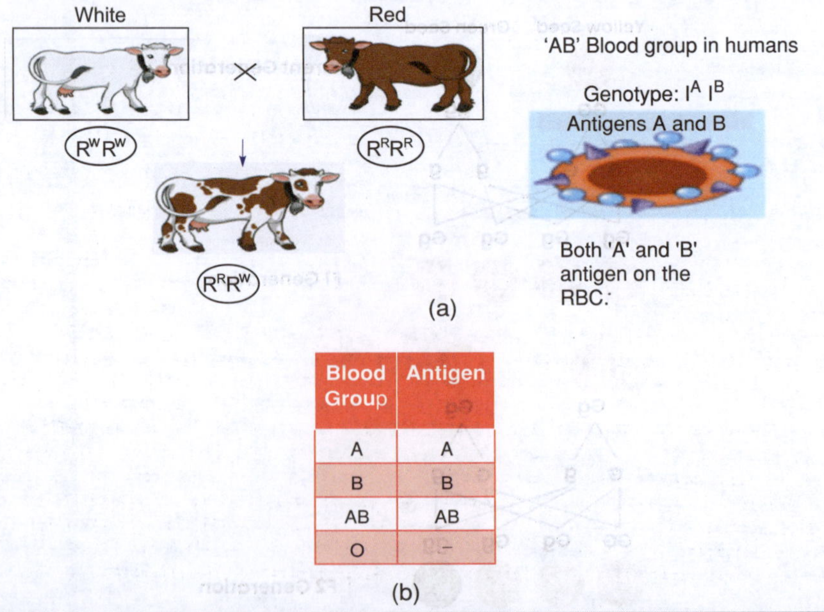

Blood Group	Antigen
A	A
B	B
AB	AB
O	–

Fig. 7.6 Examples of co-dominance: (a) Cow colour; (b) Blood group AB is codominant

Incomplete Dominance

In this inheritance pattern, the heterozygote shows a phenotype that is intermediate between the two homozygous phenotypes. While the homozygotes show absolute traits, the heterozygous phenotype appears blended. Here also, for representation, they are expressed as superscripts. This is different from codominance in the sense that both the alleles may be expressed and mixed to produce intermediate phenotype or the presence of one allele may have half the pigment and the other allele produces no pigment, causing a lighter phenotype, which looks like blending. So, the presence of the other allele makes it to express incompletely; hence called incomplete dominance.

For example, flower colour in 4 o'clock plant (*Mirabilis jalapa*) and snapdragon *Antirrhinum majus* flowers (Figure 7.7).

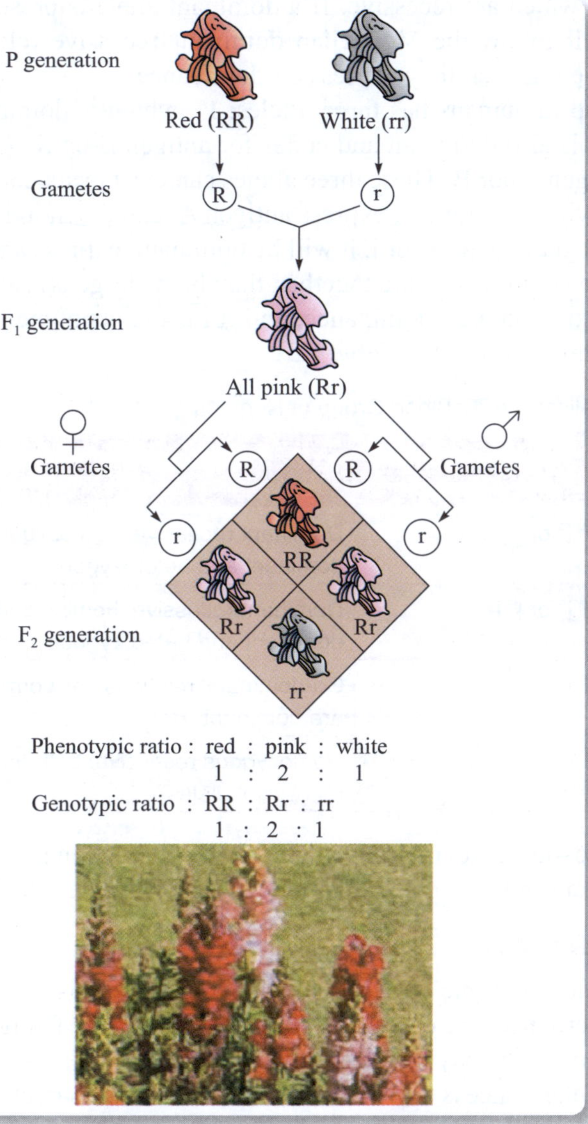

Fig. 7.7 Incomplete dominance in Snapdragon

Multiple Allelism

Till now, our discussion has focused on traits that have two different alleles expressing a trait. While it is true that in the cells of diploid organisms, within an individual, there can be only two versions of a particular gene—only two different alleles. However, for a particular trait in a population, collectively there can be more than two alleles. To simplify, the same characteristic can be expressed by multiple versions of the gene. This phenomenon of multiple alleles for a particular trait in a population is called *multiple allelism*. However, out of these multiple alleles, each individual can have any of the two combinations possible within the cells.

For example, the human ABO blood group system is the best example of multiple allelism. It is also interesting to note that depending on the combination of alleles, the pair can behave in a dominant–recessive relationship or co-dominance. This is purely based on which ones are dominant alleles and which are recessive. If a dominant allele is present with a recessive one in a heterozygote, it will follow the Mendelian dominant–recessive relationship; however, if two dominant alleles come together, they express co-dominance.

ABO blood group in humans has three alleles: I^A, which is dominant and codes for antigen A on RBC; I^B, which is also dominant and codes for antigen B on RBC; i, which is recessive and codes for neither antigen A nor B. These three alleles can exist in six combinations. If I^A is together with either I^A or i, it will be dominant, express antigen A, and the individual will have blood group A. Similarly, if I^B is together with I^B or i, it will be dominant, express antigen B, and the individual will have blood group B; if I^A and I^B are together, then both antigens A and B are expressed, and the blood group is AB, expressing co-dominance; and if i is the genotype (i exists with i), then blood group is O (neither antigen A nor B) (Table 7.3).

Table 7.3 Multiple allelism—ABO Blood Group of Humans

Phenotype Blood Group	Possible Alleles and Genotype	Relationship of Alleles	Antigen on RBC
A	$I^A I^A$ or $I^A i$	Dominant recessive; homozygous dominant and heterozygous	A
B	$I^B I^B$ or $I^B i$	Dominant recessive; homozygous dominant and heterozygous	B
AB	$I^B I^A$	Co-dominant relationship; both alleles are dominant	A + B
O	ii	Homozygous recessive; both are recessive alleles	None

One characteristic–one gene–multiple alleles in the population is multiple allelism. Fur colour in mouse and eye colour in Drosophila are other examples of multiple allelism.

Polygenic Inheritance

'Poly' also indicates 'multiple'. So, is polygenic inheritance also the same as multiple allelism? **No.** In polygenic inheritance, the context itself is very different. Till now all characteristics were expressed by one gene—they may have multiple versions of alleles as we saw in multiple allelism. However, polygenic inheritance is the reverse: many genes–one characteristic. Some characteristics are determined by the interaction of several genes. Many traits and phenotypic characters are inherited by many alleles present at different loci known as *polygenic inheritance*.

Polygenic characters manifest as a gradation of types—a continuous range—rather than 'on' or 'off' effect. These characters show a great variety within a population. For example, (i) skin colour of humans—a range of colours from light to dark is seen; (ii) height in humans—people range from short to tall. Several different genes contribute to the same character, and individual genes provide a quantitative difference.

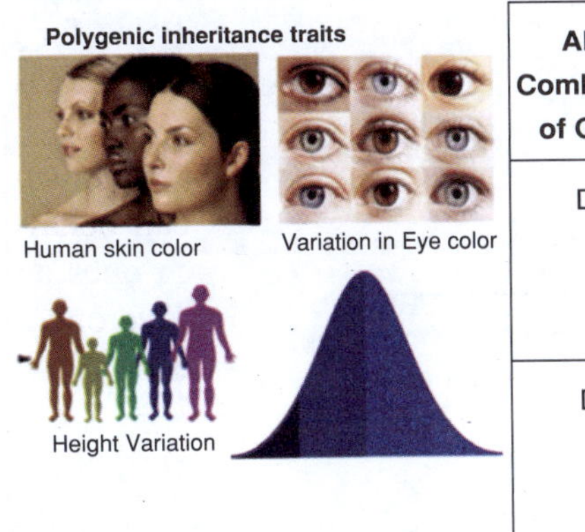

Allelic Combination of Gene 1	Allelic Combination of Gene 2	Allelic Combination of Gene 3	Genotype and Phenotype	Progressive Manifestation from Dark to Light
D^1D^1	D^2D^2	D^3D^3	6/6 alleles dominat-dark end of spectrum	
D^1d^1	D^2D^2	D^3D^3	5/6 alleles dominat dark; 1 light	
D^1d^1	D^1d^1	D^3D^3	4/6 alleles dominat dark; 2 light	
D^1D^1	D^1d^1	d^3d^3	3/6 alleles dominat dark; 3 light	
d^1D^1	D^1d^1	d^3d^3	2/6 alleles dominat dark; 4 light	
d^1D^1	d^2d^2	d^3d^3	1/6 alleles dominat dark; 5 light	
d^1d^1	d^2d^2	d^3d^3	6/6 alleles recessive; light end of the spectrum	

Fig. 7.8 Polygenic inheritance: (a) Eye colour; (b) Hair colour; (c) Skin colour; (d) Height

When multiple genes are involved in a characteristic and each gene has allelic differences, then a range of traits can be produced. Imagine a trait such as skin colour, which is controlled by three genes at three different loci, each gene has two alleles. So, six alleles can produce a range of variations from all six dominant to all six recessive. The combinations would be a spectrum, and that quantitative spectrum is a result of polygenic inheritance (Figure 7.8).

Ideally,

D = dark allele; d = light allele

However, in real life, we see many more shades or variations than just six, the reason being that polygenic inheritance is often *modified by the environment and is not a result of genetics alone.* A typical bell-shaped distribution is obtained with more concentration in the middle and less at the extremes (Figure 7.9).

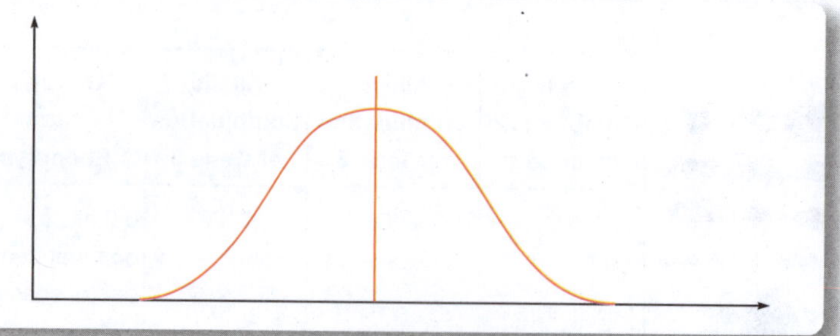

Fig. 7.9 Bell-shaped structure of polygenic inheritance

Pleiotropy

In polygenic inheritance, many genes affect one character, whereas in pleiotropy, one gene affects many characters. Although a single gene may produce one type of protein, it has a variety of phenotypic effects. '*Pleio*' means changeable. For example, metabolic disorder Phenylketonuria (PKU) and sickle-cell anaemia are pleiotropic in nature.

In normal individuals, the amino acid phenylalanine is converted to tyrosine by enzyme phenylalanine hydroxylase. If this gene is mutated, the enzyme is affected. So only one gene is impacted, but the absence of this enzyme can have multiple effects such as build-up of phenylalanine, which then gets converted to phenyl pyruvic acid that kills the nerve cells and causes mental retardation. Tyrosine is the substrate for other reactions such as the production of thyroxine and melanin; hence, absence of tyrosine can impact growth and also cause lack/reduced melanin production, leading to light skin (Figure 7.10). Hence, one gene–many impacts is pleiotropy!

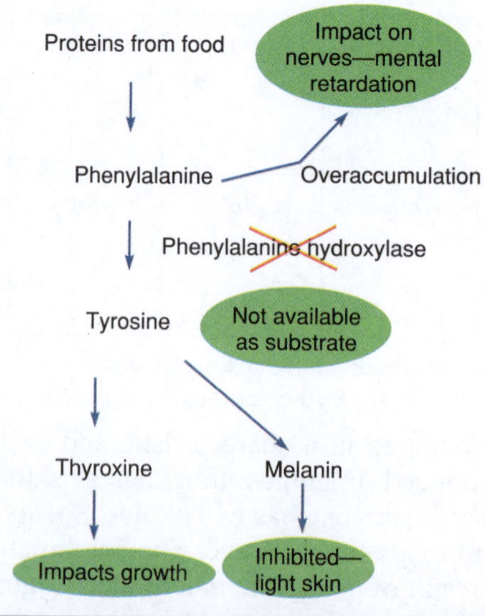

Fig. 7.10 Phenylketonuria

Similarly, in sickle-cell anaemia, mutation in the haemoglobin gene affects haemoglobin structure and function. This causes multiple effects such as sickling of RBCs; destruction of RBCs, causing anaemia and clumping of cells leading to interference in blood circulation and failure of blood supply to specific regions. These in turn can manifest as other serious conditions such as heart failure, weakness and fatigue, impaired mental function, paralysis, and so on (Figure 7.11).

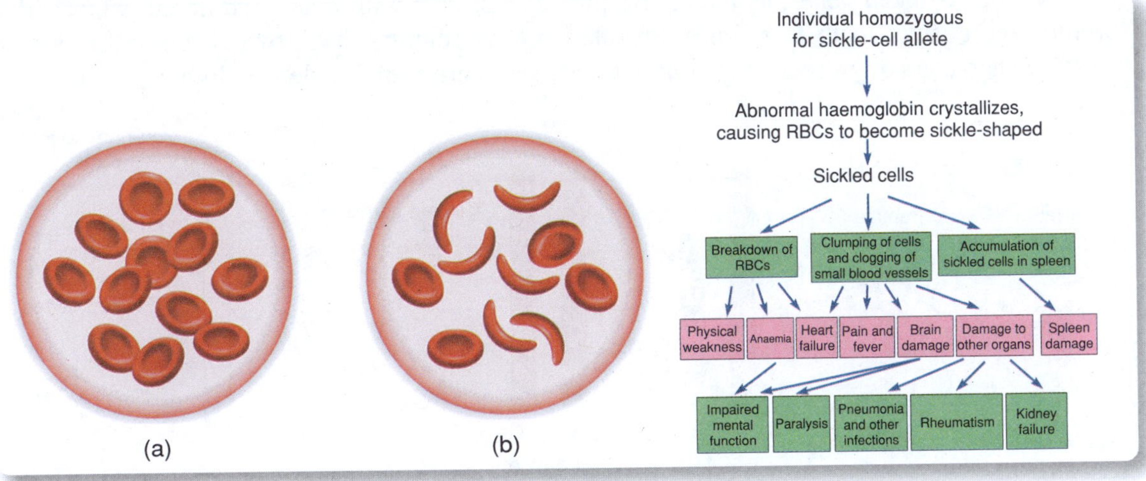

Fig. 7.11 Sickle-cell anaemia: (a) Normal RBCs; (b) Sickle-shaped RBCs

Linkage Groups

Mendel's laws of segregation and independent assortment work primarily if the genes are located on different chromosomes. This is because, in a diploid organism, the homologous chromosomes (two copies) of a diploid pair segregate independently, whereas individual chromosomes assort independently. For example, if humans have 23 pairs of chromosomes, segregation will be true for each pair of chromosomes, i.e., if there are two copies of Chromosome 1—these two copies will segregate independently; the same will be true for Chromosome 2, 3, and so on. One chromosome pair does not influence the other, and they assort independently (independent assortment), i.e., Chromosome 2 will not influence Chromosome 3, 4, or any other chromosome; they all are independent.

Every chromosome has many genes. If the genes are present on the same chromosome, they tend to be inherited together. Genetic linkage occurs when genes for different traits are located on the same chromosome, causing them to be inherited together more often. They are said to form *linkage groups*. The only way the genes within chromosomes can be separated is by 'crossing over' during gamete formation, which helps in mixing of traits from both parents. But if genes are present close to each other, crossing over is less likely, and they may stay together and be inherited as a group. This is because genes that are closer together on a chromosome are less likely to be separated during chromosomal crossover.

Sex-linked Genes

There are two types of chromosomes: autosomes and sex chromosomes. In humans, out of the 23 pairs, 22 pairs are autosomes and one pair is sex chromosome. Sex chromosomes carry sex-determining genes and contribute to sexual characteristics; hence, they are different in male and

female sexes. Autosomes, however, carry the same genes in both the sexes. In humans, males carry the X and Y chromosomes, wherein the Y chromosome has genes to determine maleness, while females carry two X chromosomes (XX).

Genes carried on X chromosomes are called X-linked genes, and genes found on Y chromosome are said to be Y-linked. During inheritance, as males have the XY genotype, there are no matching genes for the X-linked genes, as the X chromosome is large with many additional genes and Y chromosome is small with few genes. Females, on the contrary, have two X chromosomes, so recessive characters of X-linked genes affect male sex more than female sex (Figure 7.12).

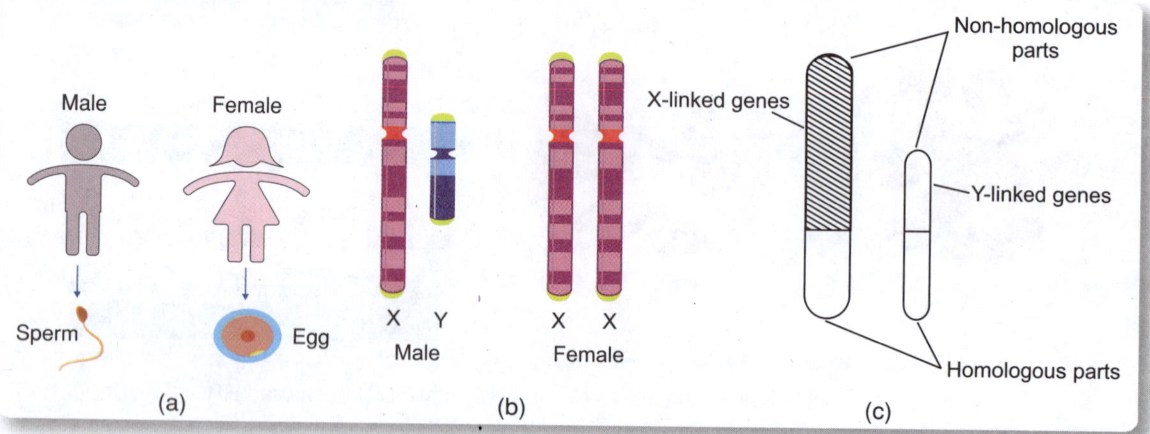

Fig. 7.12 Sex determination in humans and sex-linked genes: (a) Sex determination; (b) Sex chromosomes; (c) X and Y chromosomes

Sex-linked Inheritance

Sex-linked genes and alleles are represented as a superscript on the X or the Y chromosomes, e.g., X^N or X^n for dominant and recessive alleles, respectively. Genes on X chromosomes are said to be X-linked, and their inheritance pattern is interesting. The Y chromosomes are small and have very few genes as compared with X chromosomes.

Some X-linked genes can cause diseases in humans, such as colour blindness, haemophilia, and muscular dystrophy. Hence, the study of their inheritance patterns and gender-specific impact is critical to understand. Dominant–recessive relationship can be seen in X-linked alleles too, wherever two copies are present (as in females XX). However, if only one copy is present (as in males XY), then whichever allele is present, either dominant or recessive, will be expressed. There is a possibility of heterozygotes in females for some X-linked genes, but no counterpart in males and they express as hemizygous—one copy.

Different genes and their characteristics in X and Y chromosomes areas follow (Figure 7.13):

Let us understand the inheritance pattern. Mothers (XX) will transfer the X chromosome to, both, sons and daughters, whereas the father (XY) will always transfer the X chromosome to daughter and Y chromosome to son. So, daughters inherit one X chromosome from the mother and another X chromosome from the father, while sons inherit X chromosome from the mother and Y chromosome from the father.

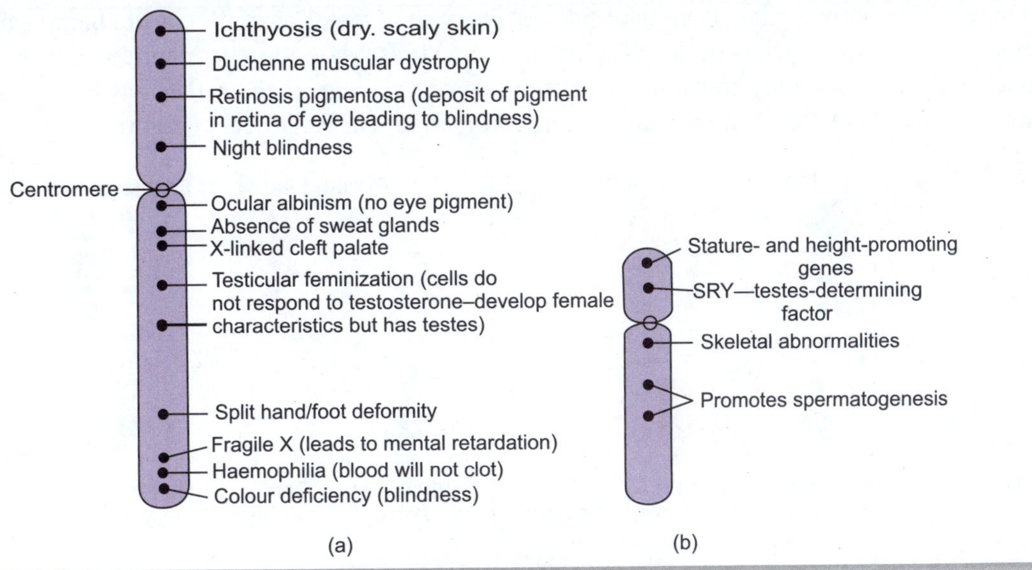

Fig. 7.13 Genes and characters: (a) X chromosome; (b) Y chromosome

Now, imagine a situation where in the mother is heterozygous for X-linked Haemophilia allele; her genotype will be $X^N X^n$—50% of her gametes will carry X^N normal allele and 50% of her gametes will carry X^n. Then, if her husband is normal, his genotype will be $X^N Y$. The cross between these parents will produce normal daughters, and 50% of sons will be normal and 50% will be haemophilic. The genotype of daughters will be homozygous dominant for normal allele ($X^N X^N$) or heterozygous carriers ($X^N X^n$). In either case, they will not inherit the disease. But, in case of sons, along with Y chromosome from father, those sons who have inherited X^N from mother will be normal and those who have inherited X^n from mother will be affected (Figure 7.14).

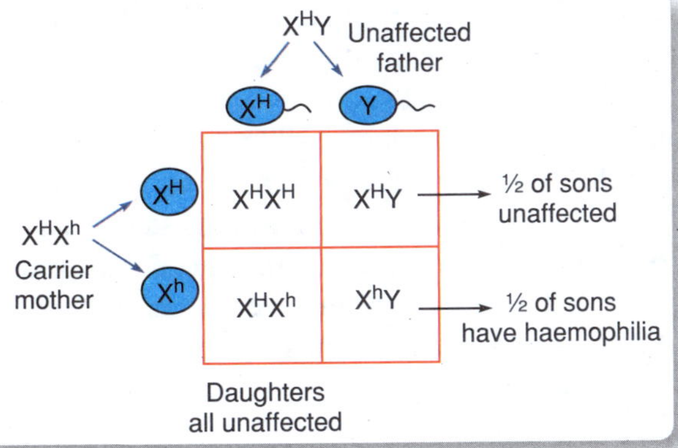

Fig. 7.14 Inheritance of X-linked genes

Source: Adapted from https://www.khanacademy.org/science/biology/classical-genetics/sex-linkage-non-nuclear-chromosomal-mutations/a/sex-linkage-sex-determination-and-x-inactivation

Recessive X-linked traits appear more often in males than in females.

With only one X chromosome, males are more frequently affected than females when they receive a 'bad' allele from the mother. On the contrary, both maternal and paternal X chromosomes must be affected for females to express the trait. So, even if they receive a bad allele, they will

often receive a normal allele from their fathers, preventing the disease allele from being expressed. Females are usually carriers of the X-linked recessive trait. Male-to-male inheritance is impossible because a father can only transmit a Y chromosome to a son. If the father has bad allele on X chromosome, there will be no impact on sons, *daughters will be carriers* (Figure 7.15).

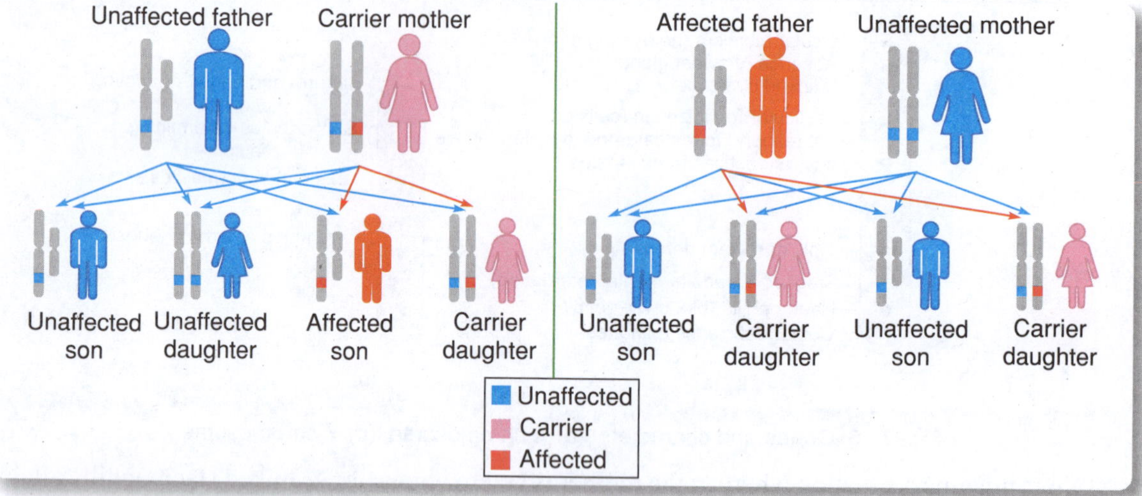

Fig. 7.15 X-linked inheritance depending on the affected gene present in mother or father

Source: Adapted from https://www.lecturio.com/concepts/sex-linked-inheritance/

Penetrance and Expressivity

Sometimes, presence of a specific allele for a trait does not mean it will certainly be expressed. This is true for some traits where there will be a difference in the proportion of individuals expressing the trait even though they may carry that allele. This phenomenon is called *penetrance*. If all individuals carry the gene and express the trait, it is called complete penetrance. For example, Neurofibromatosis type 1 (NF1) is an autosomal dominant condition that shows complete penetrance; in Huntington's disease, penetrance is 99%—almost everyone who carries the gene will develop the condition if they live long enough. On the contrary, if some people do not exhibit the features of the trait of a particular allele, the condition is said to have reduced or is incomplete penetrance. 'Incomplete' or 'reduced' penetrance means the genetic trait is expressed in only a part of the population. For example, Osteogenesis imperfecta has incomplete penetrance; there will be individuals who have *COLIA1* and *COLIA2* mutations responsible for the disease but have no symptoms. Many factors such as age, sex, environment, epigenetic modifiers, and modifier genes are linked to penetrance (Figure 7.16).

Expressivity refers to variation in phenotypic expression when an allele is penetrant. This means the allele expresses itself, but there will be a variation in severity of phenotype across individuals having and expressing the allele. The way the signs and symptoms show up in an individual will vary. For example, a dominant allele produces polydactyly in humans, but not all humans with the allele display the extra digits. So, it has reduced penetration, but there will be a variability even in individuals who express an extra digit—an extra digit may occur on one hand or feet, the digit could be full size, or it would be just a stub, and so on. This allele has reduced penetrance as well as variable expressivity.

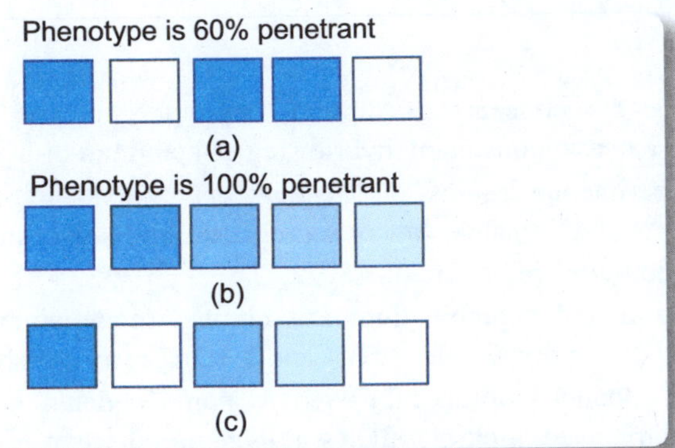

Fig. 7.16 (a) Incomplete penetrance; (b) Variable expressivity; (c) Both

Source: Adapted from https://www.frontiersin.org/journals/genetics/articles/10.3389/fgene.2022.920390/full

Interesting Fact 7.2

Pedigree of haemophilia in the royal families of Europe

Genetics is about family histories. Haemophilia A, a sex-linked trait associated with a defect in clotting factor VIII was the first human genetic trait to be found to follow a sex-linked inheritance pattern.

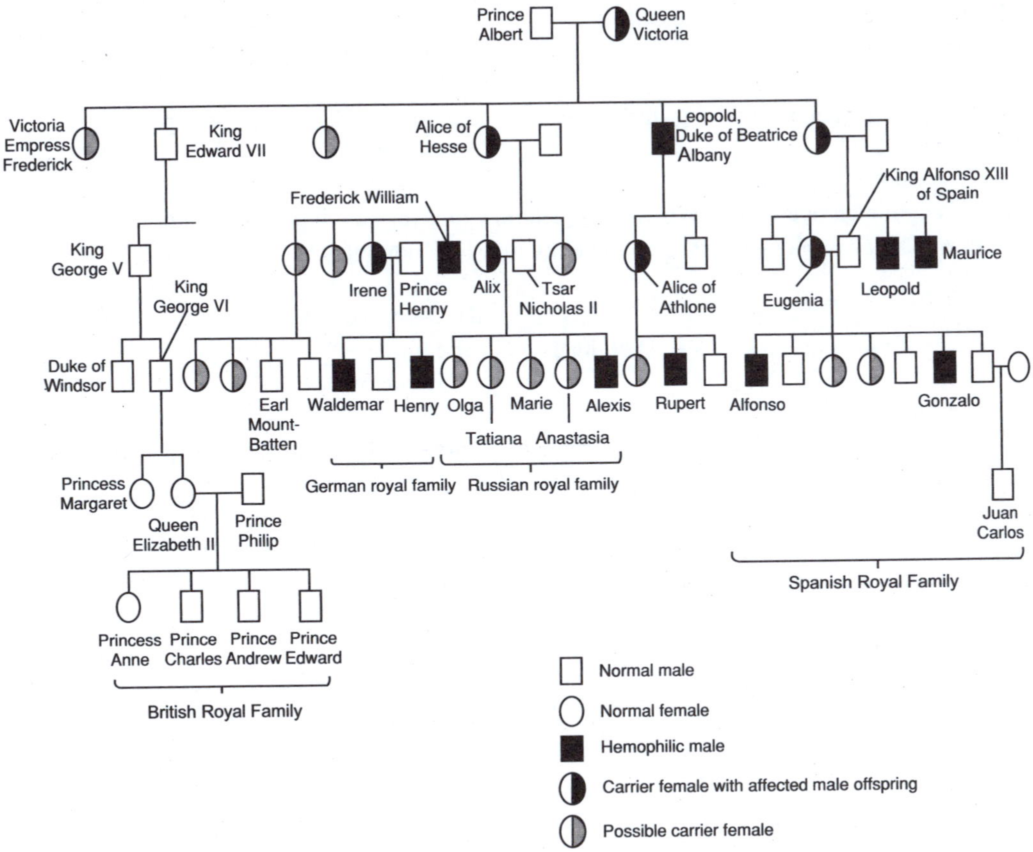

Prince Albert
Queen Victoria
Victoria Empress Frederick
King Edward VII
Alice of Hesse
Leopold, Duke of Albany
Beatrice
King Alfonso XIII of Spain
Frederick William
King George V
King George VI
Irene
Prince Henny
Alix
Tsar Nicholas II
Alice of Athlone
Eugenia
Leopold
Maurice
Duke of Windsor
Earl Mount-Batten
Waldemar
Henry
Olga
Tatiana
Marie
Anastasia
Alexis
Rupert
Alfonso
Gonzalo
German royal family
Russian royal family
Princess Margaret
Queen Elizabeth II
Prince Philip
Juan Carlos
Spanish Royal Family
Princess Anne
Prince Charles
Prince Andrew
Prince Edward
British Royal Family

Normal male
Normal female
Hemophilic male
Carrier female with affected male offspring
Possible carrier female

SUMMARY

- Understanding the concepts of diploid, haploid, genotype, phenotype, dominant, recessive, homozygous, heterozygous, monohybrid single-factor and di-hybrid double-factor crosses.
- Obtaining experimental insights on Mendel's experiments and how he proposed the three laws of inheritance: law of dominance, law of segregation, and law of independent assortment.
- How genes and traits are inherited!
- Study of one trait and a combination of two traits. Mendelian pattern in monohybrid and di-hybrid crosses, with F2 phenotype ratios of 3:1 and 9:3:3:1, respectively.
- Variations to Mendel's observations led to non-Mendeian inheritance patterns: co-dominance, incomplete dominance, multiple alleles, polygenic inheritance, and pleiotropy—the Principle and inheritance pattern in each case.
- Genes on the same chromosome are likely to be inherited together. Linkage groups and sex-linked inheritance.
- Chromosomes XX in females and XY in males. Gender-specific phenotypic outcomes based on X-linked inheritance.
- Penetrance and variable expressivity with examples.

EXERCISES

A. Objective-Type Questions

1. The different forms of a gene are called __________.
   (a) loci (b) homologous
   (c) gametes (d) haploid
2. __________ is a phenomenon by which a single gene has more than one phenotypic effect.
   (a) Polygenic inheritance (b) Dominant
   (c) Pleiotropy (d) Linked
3. Black is dominant over white. When black heterozygous pigs are intercrossed, approximately, what fraction of black progeny is expected to be homozygous dominant?
   (a) 1/3 (b) 3/3
   (c) 0/3 (d) 2/3
4. A mother of blood group O has an O child. The father could be ____________.
   (a) A, B, or O (b) Homozygous, A only
   (c) AB only (d) None of the above
5. The ABO blood group of humans is determined by three alleles in the population. How many total genotypes are possible for the blood group phenotypes?
   (a) 6 (b) 2
   (c) 4 (d) 3
6. The characteristic that is expressed in a heterozygous condition and masks the other character is said to be __________.
   (a) Recessive (b) Dominant
   (c) Co-dominant (d) Polygenic
7. The phenotypic ratio in F2 generation for a di-hybrid cross is ______________.
   (a) 3:1 (b) 9:3:3:1
   (c) 12:4 (d) 1:2:1

8. What is the probability (%) of the following parents, both recessive for plant height, short plant (tt) × short plant (tt), to produce a tall plant?
   (a) 100 (b) 0
   (c) 25 (d) 50
9. Identify the gamete that can be produced by a female with a genotype of AaBb?
   (a) AB (b) Aa
   (c) Bb (d) Ab
10. Which of these is not a law proposed by Mendel?
   (a) Law of concentration (b) Law of segregation
   (c) Law of independent assortment (d) Law of dominance
11. Co-dominance is seen in _______________.
   (a) Blood type A (b) Blood type B
   (c) Blood type AB (d) Blood type O
12. In an X-linked gene, if father is affected, then what will be the status of sons in the next generation?
   (a) All sons will be affected. (b) No son will be affected.
   (c) Half of the sons will be affected. (d) He will not have sons at all.

B. Short-Answer Questions

1. What is the difference between variable expressivity and pleiotropy?
2. Can reduced penetrance and variable expressivity be true for a single trait? Justify your answer with an example.
3. What is criss-cross inheritance? Which linkage group will show the same?
4. A dominant allele H governs short hair in guinea pigs and its recessive allele h governs long hair. Hair colour is specified by co-dominant alleles at independent loci—C^YC^Y is yellow, C^YC^W is cream, and C^WC^W is white. What would be the genotypic and phenotypic ratios of progeny if a di-hybrid short-cream pigs is crossed with genotype HhC^YC^W?

C. Self-Study Questions

1. Read, in general, about the theories proposed by Charles Darwin and Jean B. Lamarck, and, yet again after this chapter, try to see the connection with selection of genes (characters) under the influence of the environment (conditions) in evolution of organisms.

FURTHER READINGS

1. Schaefer, G.B., and James N. Thompson, Jr. Medical Genetics: An integrated Approach. McGraw Hill Education, 2013.

USEFUL WEBLINKS

1. https://atlasgeneticsoncology.org/teaching/30025/mendelian-and-atypical-patterns-of-inheritance
2. https://microbenotes.com/sex-linked-inheritance/
3. https://www.nature.com/scitable/topicpage/gregor-mendel-and-the-principles-of-inheritance-593/
4. https://accessmedicine.mhmedical.com/content.aspx?bookid=2247& sectionid=173744718

VIDEO LINKS

1. https://www.youtube.com/watch?v=ya7h-Y-9l8c
2. https://www.youtube.com/watch?v=aJIsoQhSAo8

CHAPTER

8

Cellular Metabolism

LEARNING OBJECTIVES

After reading this chapter, you will be able to:

- Get acquainted with metabolism and biochemical pathways
- Understand biocatalysts—enzymes—their basic characteristics, regulation, and function
- Become aware of the universal energy currency—ATP
- Appreciate the environmental balance of autotrophs and heterotrophs
- Know the catabolic cellular respiration in living organisms and anabolic photosynthesis in plants

CHAPTER TAKEAWAY **Recycling of 'energy' and 'matter' by cellular metabolism**

This chapter throws light on the relationshsip between body (organism), food, nutrition, and energy through the process of metabolism at the cellular level.

PRELUDE

A simple question, why do we eat food? Simple answer, to get nutrients. Why do we need nutrients? We all know that nutrients are the building blocks of our body and the source of energy. Energy from nutrients?? How? A good question! Let us try answering this pertinent question in this chapter. All organisms obtain energy from external sources as they cannot generate energy but only transform it. For the entire universe, one prominent source of energy is sunlight. But we cannot consume sunlight, can we? Right! So, how are our energy needs met? Another source of energy could be the chemical bond energy of molecules in nature or in other organisms. Some organisms in the universe can convert the energy from sunlight into organic molecules, and we, then, use the chemical bond energy of those organic molecules to fuel our energy needs. There is a continuous need for energy as biological activities such as growth, reproduction, communication, repair, and response to stimulus, so energy transformations are also happening continuously. Obtaining energy and building blocks is a life-long endeavour and constitutes one of the important characteristics of life—*metabolism*.

Metabolism

All the chemical reactions and associated energy changes that take place within living systems, in total, is called *metabolism*. To meet the energy needs of organisms for fuelling their various activities, such as growth, movement, reproduction, etc., harvesting energy from organic molecules is a necessary process.

To obtain energy from organic molecules, metabolic processes involve three steps: (i) nutrient uptake, (ii) nutrient breakdown and processing, and (iii) waste elimination.

Different organisms expend energy and obtain the required nutrients by different means; for example, plants absorb nutrients from soil, animals eat and swallow food, protozoans carry out phagocytosis, etc. These nutrients provide the organic raw materials to the body to be broken down. Once nutrients are inside the body of the organism, they are processed by a network of chemical reactions. These chemical reactions produce various intermediates that are used to manufacture new parts, repair, and reproduce. The breakdown of organic molecules also produces energy, which is used to fuel essential activities. Some parts of the nutrients are harmful or useless and are eventually eliminated as waste products. Heat is also a waste product that is produced as an outcome of metabolic activity. Waste elimination from the body also differs between different organisms; for example, it can be diffused out or shunted out through cell surfaces in some organisms, while others may have an elaborate waste disposal or excretory system to remove waste molecules.

Breaking down of large organic molecules (nutrients or food) into smaller compounds with the release of energy is called *catabolism.* The reverse of this is called *anabolism*, synthesizing large organic molecules (nutrients or food) from smaller inorganic ions. So, metabolism has two arms—anabolism and catabolism (Figure 8.1).

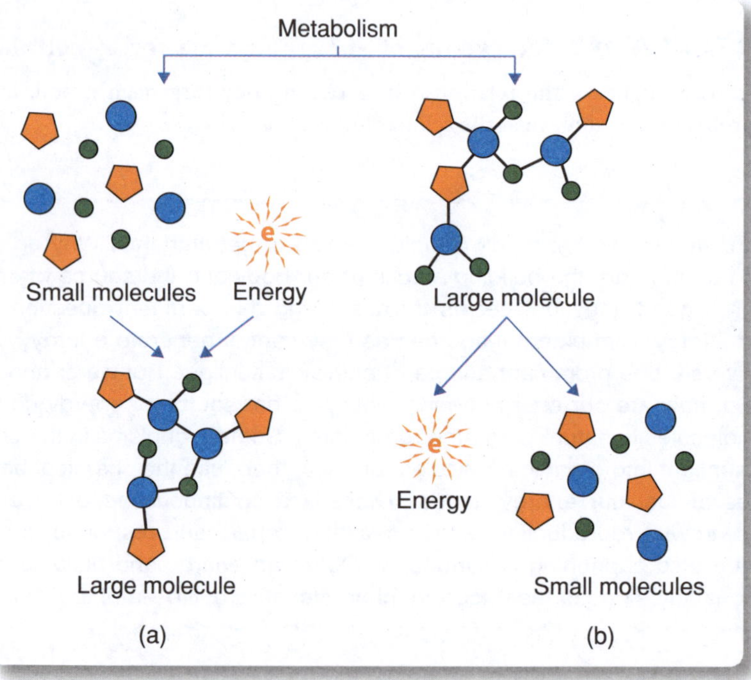

Fig. 8.1 Types of metabolic pathways: (a) Anabolism; (b) Catabolism

Interesting Fact 8.1

Metabolism

Table 8.1 Anabolism Versus Catabolism

S. No.	Anabolism	Catabolism
1	Process of building up—constructive process	Process of breaking down—destructive process
2	Smaller molecules are combined together to form larger complex molecules	Larger molecules are broken down into smaller molecules
3	Energy is formed and stored	Formed energy is released and used
4	Examples: Photosynthesis; glucose to glycogen Synthesis of proteins from amino acids	Examples: Glucose from carbohydrate Glucose to carbon dioxide and water

Depending on the capability of organisms to metabolize, they are classified into two groups as follows:

1. *Autotrophs* are organisms that synthesize their own food—*self-nourishing*. They use the energy from sunlight and convert the inorganic CO_2 from the atmosphere to organic molecules, such as glucose, by an anabolic process called *photosynthesis*. Hence, plants and some bacteria are called photoautotrophs. The food synthesized by photosynthesis is catabolised by them for their own energy requirements as well as it serves other organisms. Some other autotrophs use chemical energy and are called *chemoautotrophs*.
2. *Heterotrophs* are organisms that depend on autotrophs for their nutrient needs. They cannot synthesize their own food and are being nourished by other organisms. They catabolize the organic molecules received from other organisms (Figure 8.2).

Fig. 8.2 (a) Autotrophs; (b) Heterotrophs

So, some organisms carry out the synthesis of food by anabolic process, whereas all organisms require a catabolic process to break down their food and obtain energy. The energy 'need' in all organisms is common, although the source of energy is different! Autotrophs use the energy

from sunlight, synthesize organic compounds, and then break them down for obtaining energy. Heterotrophs, on the contrary, rely only on chemical energy from organic compounds obtained from other sources.

Table 8.2 Autotrophs Versus Heterotrophs

Autotroph	Heterotroph
Producers Produce their own food for energy; use photosynthesis or chemosynthesis	Consumers Consume other organisms to get proteins and energy
Sun → Grass →	Grasshopper → Toad
Plants, Some bacteria, Algae	Animals, Most bacteria, Fungi

The process of synthesis of food by autotrophs is 'photosynthesis', and the breakdown of food to yield energy by organisms (both autotrophs and heterotrophs) is 'cellular respiration' (Table 8.2). Photosynthesis is an anabolic process, whereas cellular respiration is a catabolic process.

Metabolic reactions occur in a step-by-step series of interconnected chemical reactions. These enzyme-mediated biochemical reactions lead to biosynthesis (anabolism) or breakdown (catabolism) of natural products within a cell or tissue. These multistep series of reactions are called *biochemical pathways* or *metabolic pathways*. There are hundreds of different pathways taking place within cells of organisms. Cellular respiration and photosynthesis are two such major pathways involved with energy *homeostasis*.

In eukaryotic cells, cellular respiration takes place in mitochondria, whereas chloroplast is the site of photosynthesis. In prokaryotes, as they lack mitochondria and chloroplast, these processes are carried out in the cytoplasm with the support of cell membranes and other specialized membranes.

Enzymes as Biocatalysts

Biochemical reactions involve interconversion of molecules with energy changes, i.e., substrates converted into intermediates and then to final end-products. Biochemical reactions in living organisms need to happen at rapid rates for organisms to survive, as they must obtain energy and building materials within a short time. Usually, conditions involving systems that can increase

facilitation in the rate of reaction, such as high temperature or pH, are not present inside living systems as high temperature can destroy structural components of cells, like proteins, and it will increase the rate of all reactions and not of specific reactions as desired. So, typically, living systems have a neutral pH and relatively low temperature. Hence, to speed up chemical reactions within a cell, catalysts are needed. Enzymes are biocatalysts, which accelerate biochemical reactions.

Most enzymes are proteins by nature (a few RNAs also have catalytic activity); they can be made of a single polypeptide or could be made of larger proteins with multiple identical (homomers) or different sub-units (heteromers). To understand biochemical reactions, it is important to understand the different characteristics and features of enzymes (Figure 8.3). Different characteristics of enzymes are described below:

- ***Nature of enzyme action*** Enzymes act by lowering the energy demands of the reaction. The initial energy investment required to start a reaction is called 'activation energy'. Enzymes help in lowering the activation energy so that the molecule occupies a transition state, where bonds are easily broken, and, subsequently, there is formation of new bonds. For example, hydrolysis of disaccharide sucrose into glucose and fructose by enzyme sucrase. Binding of sucrase to substrate sucrose forms an unstable intermediate, lowering the activation energy, and facilitates the split into glucose and fructose.
- ***Substrate specificity of enzyme*** The specific size and 3-D shape of each enzyme ensures binding of enzymes to specific substrates. The enzyme physically fits with the substrate and forms the enzyme–substrate complex (ES). This site of binding is called binding site or attachment site on the enzyme surface. The place on the enzyme where catalysis occurs is called the active site. Binding site and catalytic site are usually parts of the active site; binding site helps to bind and orient the substrate for catalytic activity. Active sites are usually shaped as pockets or grooves. The ES binding then converts the substrate to one or more products. Enzymes are released unchanged after the reaction and are cycled repeatedly through multiple reactions before destruction by wear and tear.

  Most enzymes bind to substrates very specifically—like a lock and key; whereas there are some enzymes that can induce a fit with the substrate by molding and adjusting itself to the substrate and are somewhat flexible. Both 'lock and key model' and 'induced fit hypothesis' are valid explanations about how enzymes interact with their substrates.
- ***Turnover number of an enzyme*** Turnover number defines the rate at which an enzyme converts its substrate. The number of substrate molecules that can be converted by a single enzyme molecule is turnover number and can range from a few molecules to several million molecules per second.
- ***Enzyme nomenclature*** Based on the specificity of substrates and the reaction they catalyze, each enzyme has a unique name made of three parts: the first part indicates the substrate, the second part indicates the reaction it catalyzes, and third part is a tag '-ase'. For example, DNA polymerase, where the substrate is DNA, catalyzes the polymerization of DNA and ends with suffix '-ase'. Some older names of enzymes like pepsin, trypsin, etc., have stayed, although they can also have formal names as indicated earlier.

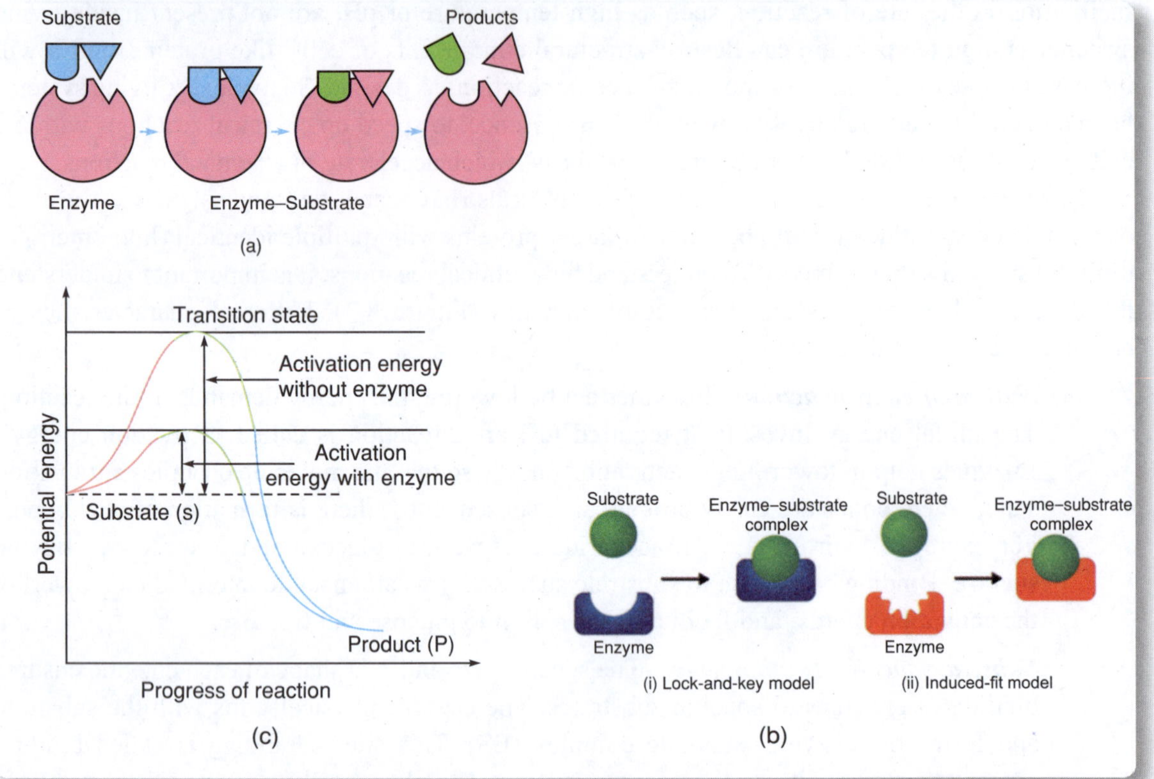

Fig. 8.3 (a) Nature of enzyme action; (b) Mechanisms/models of enzyme–substrate interactions; (c) Enzyme-catalyst—Activation energy

- ***Enzyme helpers*** Co-factors and co-enzymes are enzyme partners that act as enzyme helpers. Co-factor is a generic term for a broad category of molecules that can be inorganic or organic, and help enzymes. Cofactors are non-proteinaceous molecules without which an enzyme is unable to catalyse a reaction. Inorganic co-factors include metal ions such as Zn, Fe, Mn, Cu, etc. A type of co-factor, which is an organic molecule is usually referred to as co-enzyme NAD+, FAD, etc.

Vitamins help in the synthesis of co-enzymes. Riboflavin (B12) produces co-enzyme flavin adenosine dinucleotide (FAD), and niacin is metabolized to co-enzyme NAD+. Hence, vitamins are required in your diet as cells cannot synthesize these. These co-enzymes play a significant role in the transfer of electrons and protons in cellular respiration and photosynthesis (Table 8.3).

Table 8.3 Enzyme Helpers
(a) Cofactors vs Coenzymes

Metal Ion	Coenzyme
Inorganic in nature	Organic or metalo-organic compounds
Take part in catalysis	Do not take part in catalysis but help in carrying specific ions or functional groups

(b) Cofactors

Ions	Enzyme
Cu^{+2}	Cytochrome oxidase
Fe^{+3}	Catalase
K^{+1}	Pyruvate kinase
Mg^{+2}	Hexokinase

(c) Coenzymes

Coenzyme	Enzyme
Thiamine pyrophosphate	Pyruvate dehydrogenase
Coenzyme A	Acetyl CoA carboxylase
Biotin	Pyruvate carboxylase
Pyridoxal phosphate	Glycogen phosphorylase

- ***Environmental effects on enzyme activity*** Multiple factors impact enzyme activity and vary the turnover number of enzymes. Every enzyme needs an optimum temperature to function effectively. Very high and low temperatures affect enzymes by distorting their shape or preventing molecular motion and contact with substrate. Very low temperature slows down molecular motion and reduces formation of ES complexes. High temperature up to a certain point helps in increasing molecular motion and better ES contact, but beyond a certain temperature, the enzyme is denatured and its shape is lost; thus becoming inactive. For example, high fever affects digestive enzymes.

  Optimum pH ensures the right conditions for accumulation of hydrogen ions on certain parts of the enzyme. High or low pH may also affect the shape of enzyme as positively charged hydrogen ions may/may not accumulate as required on the exposed side-chain of enzymes, thus abrogating the function. The ions in solution may alter the enzyme's active site and overall shape of the enzyme. For example, in the digestive system, each enzyme has its own pH range of activity, e.g., amylase in saliva works at neutral pH, pepsin in stomach works at acidic pH (1.5–2), whereas trypsin in pancreas and small intestine works best at alkaline pH (8–9).

- ***Enzyme regulation*** Cells can adjust the activity of many enzymes upward or downward, depending on the state and need of the cell.

  The number and kind of enzymes produced are regulated within the cell by gene-regulator proteins. Gene-repressor proteins reduce enzyme synthesis, whereas gene-activator proteins increase enzyme synthesis depending on the need of the cell.

  Conditions within the cell such as pH can also favour one enzyme over the other. For example, one substrate acetyl is acted upon by three enzymes: citrate synthase for energy, malate synthase for malate, and fatty acid synthase for fatty acid (Figure 8.4(a)). Multiple enzymes competing for the same substrate is called *enzymatic competition*. Depending on the need of the cell, enzyme numbers and conditions will be regulated.

  Inhibitors can also be used to reduce enzyme activity: (i) competitive inhibition happens when inhibitors closely resemble substrates, and they fit into and occupy the active site, thereby blocking access to normal substrates and slowing down reaction rate. As inhibitor and substrate are competing for the same active site, this kind of inhibition is called *competitive inhibition* (Figure 8.4(b)). For example, methotrexate and organophosphates inhibit enzymes competitively. The binding of inhibitors can be reversible or irreversible.

Irreversible inhibition can only be overcome by synthesizing more enzymes. (ii) Non-competitive inhibitors do not resemble substrate, bind to other parts of enzymes, and change enzyme configuration, making active sites inaccessible to substrate leading to ineffective enzyme catalysis. If the non-competitive inhibitor binds to a site other than the active site, the inhibitor is called an *allosteric inhibitor*. For example, cyanide on cytochrome oxidase. Cyanide binds irreversibly to the copper of cytochrome oxidase, inhibiting its activity.
Negative feedback inhibition is a type of control by which the high level of the end product sends a feedback to negatively regulate enzyme activity of one or more of the previous enzymes in the pathway. This ensures that the production of end product drops and the enzyme activity is restored; this cycle continues (Figure 8.4).

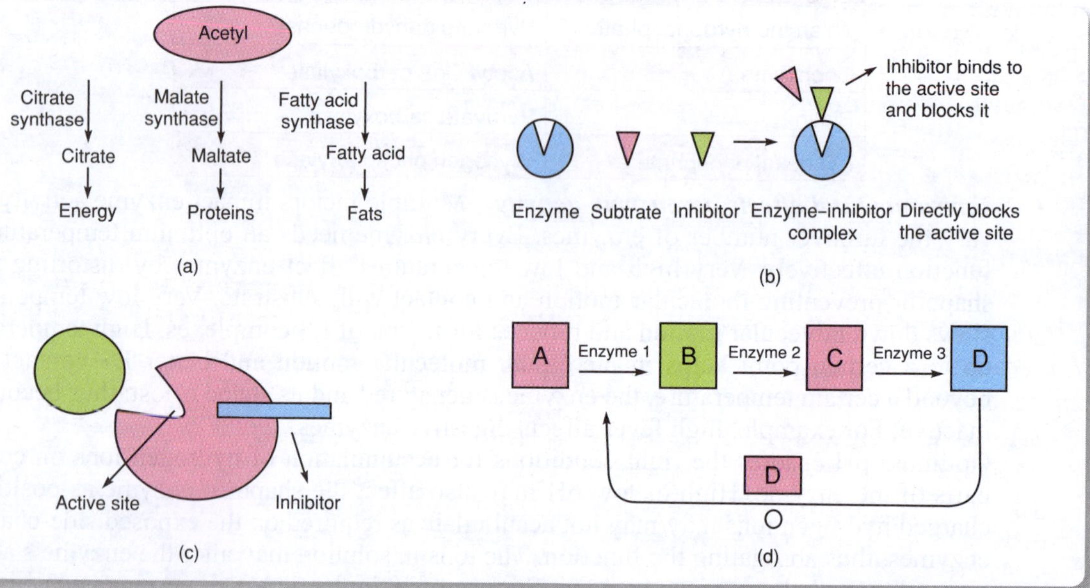

Fig. 8.4 Enzyme regulation: (a) Enzymatic competition; (b) Competitive inhibition; (c) Non-competitive inhibition (allosteric); (d) Negative feedback inhibition

Classification of enzymes based on functions: Enzymes can be categorized based on the kind of reactions that they catalyse. (i) Oxidoreductases – catalyse oxidation – reduction reactions, e.g., glucose oxidase.

(ii) Hydrolases – accelerate hydrolysis of substrates, e.g., Nucleases, proteases, proteinase K etc..
(iii) Lyases – eliminate groups and leave double bond in substrates, e.g., decarboxylases such as Pyruvate decarboxylase.
(iv) Isomerases – involved in conversion of isomers, e.g., glucose isomerase, triose phosphate isomerase.
(v) Ligases – synthesis by joining two molecular substrates, e.g., DNA ligase and some synthetases.
(vi) Translocases – help in moving molecules across membranes or from one location to another, e.g., mitochondrial translocases.
(ii) Transferases – transfer functional groups (like phosphate, methyl etc.) from one molecule to another, e.g., kinases, methylases, etc.

Many examples of these categories of enzymes are used in industries and laboratories for various in-vitro applications. Amylase for starch hydrolysis, proteases for protein digestions,

glucose isomerase for production of high fructose syrup, lysozymes for breaking bacterial cell walls, restriction endonucleases for cutting DNA in genetic engineering, Taq Polymerases in PCR, Luciferase for bioluminescence, Papain in meat tenderization, cosmetics, etc., are few notable examples.

RESTRICTION ENDONUCLEASE (RE)

Restriction endonucleases are enzymes known to cause internal cleavages in DNA molecules after identifying a particular nucleotide sequence known as a recognition or restriction site. It is also called a molecular scissor as it breaks down dsDNA into fragments by cutting them at specific sites. The restriction enzymes shield the bacteria from bacteriophages by cleaving their specific restriction sites by hydrolysing their DNA. Further, a restriction site is a sequence of 4–8 bases, which are mostly palindromic in nature (The sequence of bases happens to be the same when read both backwards and forward.)

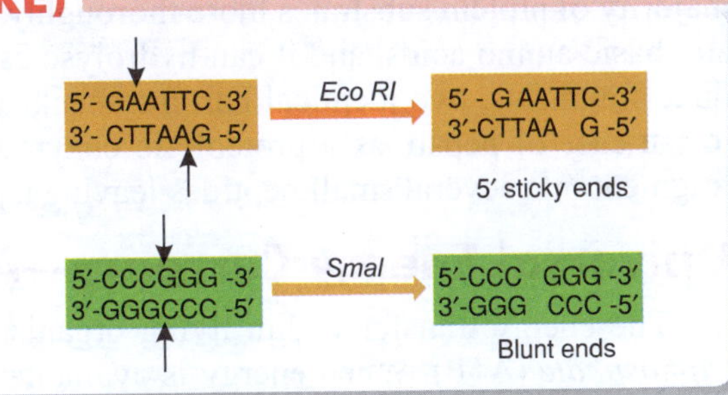

Restriction site of EcoRI and SmaI generating sticky and blunt ends

EcoRI is the best example of restriction endonuclease, which cleaves the DNA between G and A on each strand producing 'sticky ends', which can be later joined by DNA ligase. Two types of ends are formed after the cutting of RE, i.e., blunt and sticky ends.

(i) Blunt end- It is produced when the enzyme is cut in a straight manner between the bases resulting a paired bases on each side and does not contain any 3' and 5' overhangs.

(ii) Sticky end- It is generated by a staggered cut having unpaired bases with 3' and 5' overhangs as they ensure proper joining of fragments.

Cysteine Protease—Papain

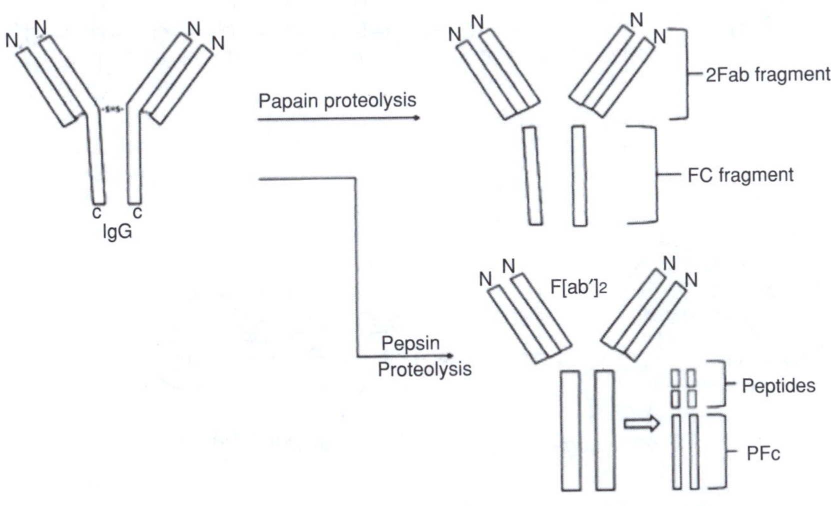

Fig. 8.5 The proteolytic action of papain on IgG with respect to pepsin

The first cysteine protease to be separated from papaya was papain (Papaya proteinase I). Its structure is determined by three-dimensional X-ray analysis. It is used commercially in different industries, including pharmaceutical, cosmetic, meat, textile, and food. Three disulfide bridges present to stabilize the protein along with it have the ability to resist high pH and temperature. Two separate structural domains with a gap between them make up its three-dimensional structure.

When compared with pancreatic proteases, papain is more efficient in breaking down the majority of protein substrates more thoroughly. It can cleave the peptide bonds of leucine, glycine, and basic amino acids, and it can hydrolyse esters. When it is treated with IgG, it is cleaved into three fragments, two identical Fab fragments and the third fragment called Fc fragment, whereas in the case of pepsin as a proteolytic enzyme, IgG is cleaved into F(ab)2 fragment and the Fc fragment into several small peptides leaving a pFc fragment.

Universal Energy Currency—ATP

The energy transfer within living organisms is handled by a molecule known as *adenosine triphosphate* (ATP). When energy is available, the living organisms store that energy by making ATP, and when energy is required then ATP is broken down to release energy.

ATP is a *nucleotide* that consists of adenine (nitrogenous base), ribose (sugar), and three phosphates. The hydrolysis of one or two phosphates from ATP releases energy. The release of terminal phosphate from ATP produces adenosine diphosphate (ADP), and this ADP can be further hydrolysed to stable adenosine monophosphate (AMP). These two bonds are hence called *high-energy phosphate bonds* as they are useful forms of energy. The energy released from ATP hydrolysis is used to catalyze biochemical reactions. This happens by coupling the energy of ATP hydrolysis with endergonic reactions by transferring a phosphate group from ATP to the target molecule by the process of phosphorylation. On the other side, ATP can be replenished and regenerated by addition to phosphate groups to ADP or AMP. Whenever energy is harvested from external sources, such as sunlight or through chemical reactions from the bond energy, it is stored by attaching a phosphate to ADP.

(a) ATP

Fig. 8.6 Energy currency ATP: (a) Structure of ATP; (b) Hydrolysis and energy production; (c) ADP

Balancing the energy-producing process to the energy-consuming process is an interplay of regeneration and hydrolysis of ATP through chemical reactions. So, functionally, ATP acts as a chemical currency—earned and used. It, thus, functions like a rechargeable battery, which has the right amount of energy to power various processes (Figure 8.6).

Breaking down of food particles (large organic compounds) to smaller molecules yields energy in the form of ATP. The first metabolic pathway for energy harvesting that we will focus on is cellular respiration (Figure 8.7).

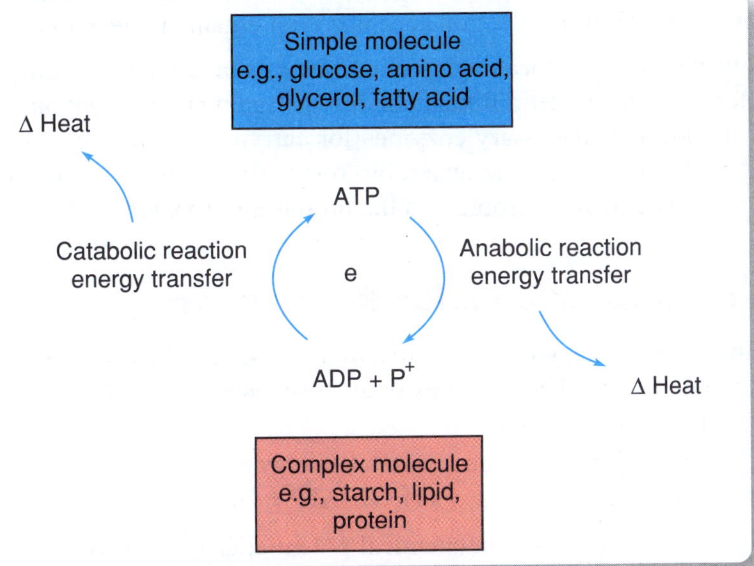

Fig. 8.7 ATP: Energy and functioning

Catabolic Cellular Respiration in Living Organisms

Cellular respiration is a process that controls the release of chemical energy from the bonds of large organic molecules for use in activities required to sustain life.

The catabolic release of energy from organic molecules is driven by a cycle of electron transfers through coupled reactions called *oxidation–reduction reactions* (redox reactions). In redox reactions, the electron donor is called a reducing agent, and the electron acceptor is an oxidizing agent. Loss of electron(s) from reducing agent is called 'oxidation', whereas the gain of electron(s) from oxidizing agent is called 'reduction'. This relocation of electrons in a redox reaction releases energy from the bonds of food/fuel and stores this energy in the new chemical bonds by synthesizing ATP.

Cellular respiration is analogous to combustion of hydrocarbons in gasoline to fuel the movement of automobiles. It provides energy as well as the greenhouse gas—carbon dioxide. In living cells, organic compounds act as fuel sources. Breakdown of food particles to release ATP is routed through this process. Hence, all organisms need to catabolize their nutrients and carry out cellular respiration, if they have to survive. So, both autotrophs and heterotrophs break down molecules by cellular respiration to obtain their energy needs.

We all know that respiration or breathing involves taking in oxygen and giving out carbon dioxide. Similarly, most cells use oxygen to breakdown complex molecules such as carbohydrates, proteins, and fats, which serve as fuels, and give out carbon dioxide. This is called cellular respiration. Depending on whether the cell uses oxygen or not, the catabolic degradation can be classified into the following:

- ***Aerobic cellular respiration*** Breakdown of complex organic molecules in the presence of oxygen. Oxygen acts as a terminal electron acceptor.
- ***Anaerobic respiration or fermentation*** The form of breakdown where oxygen is not involved. It is also called more accurately as fermentation. Other molecules act as electron acceptors, P, S, etc. if the electron acceptor is an organic molecule.

Some organisms, such as the single-celled yeast, can effectively carry out both aerobic respiration and fermentation under different conditions leading to different outcomes and products. Organisms that do not have necessary enzymes for carrying out the steps of aerobic respiration primarily indulge and go through the anaerobic route. But in most organisms, the fundamental process to generate ATP remains aerobic—in the presence of oxygen.

Overview of Aerobic Cellular Respiration

It is a specific series of enzyme-driven reactions in which glucose is broken down to carbon dioxide and water with chemical bond energy of glucose released as ATP. The equation that explains aerobic cellular respiration is as follows:

Glucose + oxygen = carbon dioxide + water + energy

$$C_6H_{12}O_6 + 6O_2 = 6CO_2 + 6H_2O + ATP + Heat$$

Covalent bonds of glucose contain chemical potential energy. When these bonds are broken, energy of electrons is used to phosphorylate ADP and produce ATP. So, in aerobic cellular respiration, glucose is oxidized. In aerobic respiration, after removing the energy from electrons,

Experimental Insight 8.1 Redox in metabolic pathways

Bioenergetic cycle has been defined by two principal and complementary processes: photosynthesis and respiration. Carbon-containing molecules oscillate from a reduced (as glucose) to an oxidized state (as carbon dioxide) (Figure 8.8).

$$C_n(H_2O)_n + O_2 \underset{\text{photosynthesis (reduction)}}{\overset{\text{(oxidation) respiration}}{\rightleftharpoons}} CO_2 + H_2O + \Delta$$

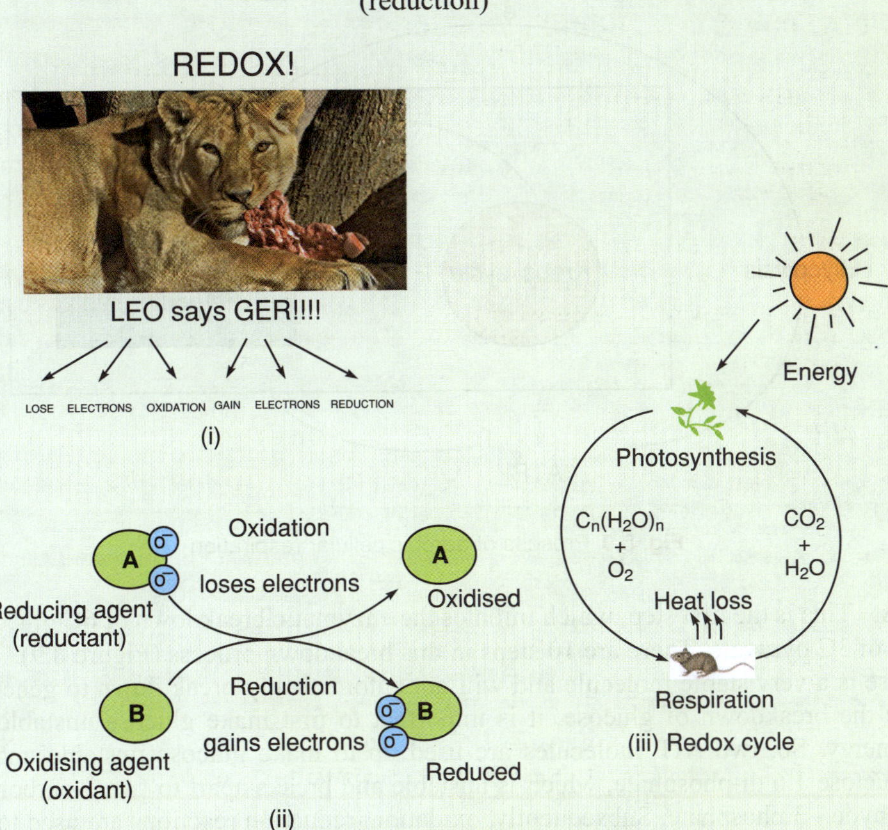

Fig. 8.8 Redox reactions in metabolic pathways

Source: Adapted from https://ib.bioninja.com.au/options/option-c-energy-and-cells/c3-cell-respiration.html https://www.intechopen.com/chapters/55282

oxygen serves as the final resting place of less energetic electrons and, thus, oxygen is reduced. After removal of electrons, the positively charged protons (H_+) released from the glucose molecule combine with the negatively charged oxygen (O^-) to form a water molecule. Once hydrogens are

stripped off, the carbon and oxygen of glucose rearrange to form individual molecules of CO_2. Thus glucose + oxygen on the reactant side produce water and CO_2 as products, with the release of energy in the form of ATP and heat.

This entire process is completed in three main stages: (i) Glycolysis (ii) Krebs' cycle or TCA cycle, also called Citric acid cycle, and (iii) Electron transport chain (ETC) or Electron transport system (ETS). Both eukaryotes and some prokaryotes carry out aerobic cellular respiration involving these three processes. In eukaryotic cells, glycolysis happens in cytoplasm, whereas Kreb's cycle and ETS happen in mitochondria. However, as prokaryotes do not contain mitochondria, all the three processes happen in cytoplasm with the support of cell membrane or plasma membrane.

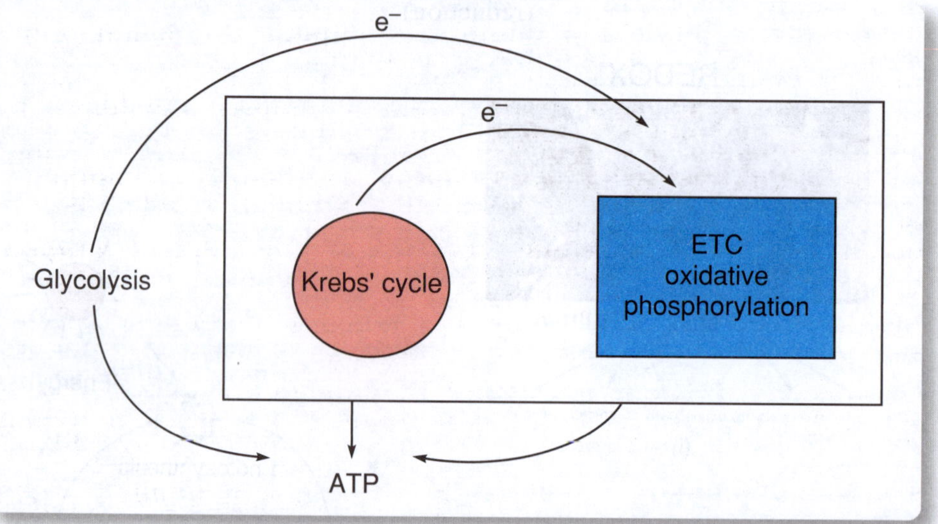

Fig. 8.9 Process of aerobic cellular respiration

Glycolysis This is the first step, which initiates the enzymatic breakdown of 6C glucose into two molecules of 3C pyruvate. There are 10 steps in this breakdown process (Figure 8.9).

Glucose is a very stable molecule and will not automatically break down to generate energy. To initiate the breakdown of glucose, it is important to first make glucose unstable by adding external energy. So, two ATP molecules are used up to make glucose unstable and convert it through fructose 1,6 di-phosphate, which is unstable and breaks apart to two 3-carbon molecules glyceraldehyde - 3 phosphate. Subsequently, oxidation–reduction reactions are used to synthesize ATP and capture hydrogen in the last five reactions. These redox reactions generate 4 ATPs and high-energy electrons released from glucose and 4 hydrogen detached from glucose are transferred to NAD^+, which acts as an electron/hydrogen carrier and gets reduced to NADH. Thus, at the end of glycolysis, the following products are formed: 2 molecules of pyruvate + 2 NADH + 4 ATP molecules. However, as 2 ATP molecules were initially used to make glucose unstable, the net gain of ATP after glycolysis is only 2 ATP.

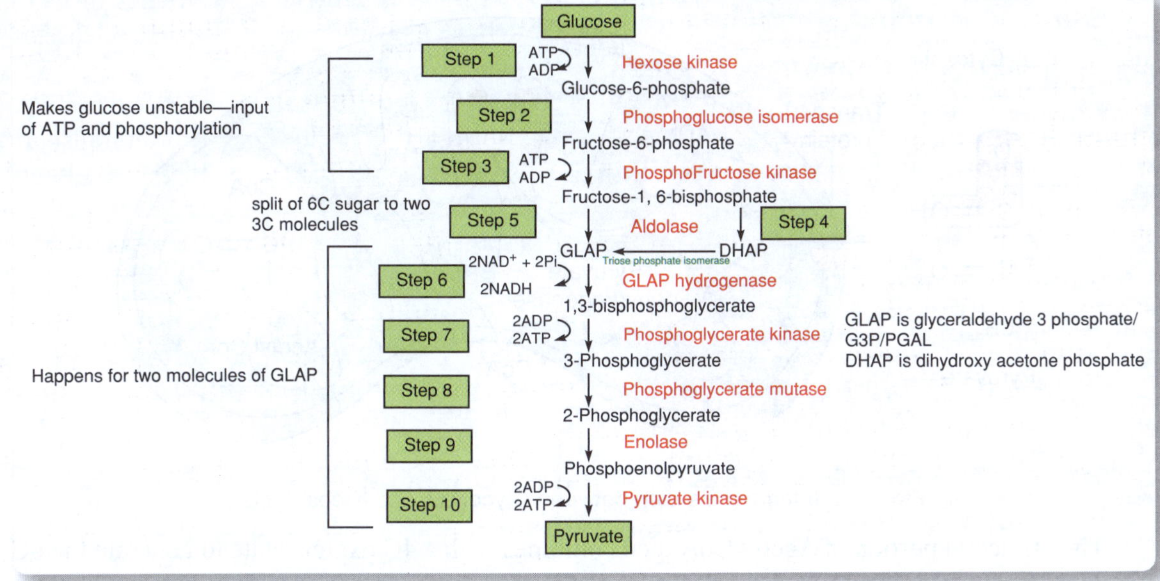

Fig. 8.10 Steps of Glycolysis

Source: Modified from Simplified Glycolysis diagram https://www.biotechfront.com/2020/11/glycolysis-metabolism-and-its-steps.html

Outcome of glycolysis: 2 pyruvate, 2 NADH, and 4 ATPs

Net gain of ATP in Glycolysis is 2ATP.

Intermediate products of steps of glycolysis are given in black while the enzymes catalyzing the steps are indicated in red. One enzyme triose phosphate isomerase between step 4 and 5 is indicated in green (Figure 8.10).

Also note that, as yet, in this first step of glycolysis, there has been no involvement of oxygen, hence glycolysis on its own is anaerobic.

Glucose + 2ATP + 2NAD+ = 4 ATP + 2 NADH + 2 pyruvic acid.

Krebs' cycle/Citric acid cycle/Tricarboxylic acid cycle: This is the second stage of aerobic cellular respiration. It is named after Hans Krebs' for his contribution in identifying the components of the citric acid cycle. In eukaryotes, this second step happens inside the mitochondrial matrix. Pyruvate, the end product of glycolysis present in cytoplasm enters mitochondria and is further broken down by Krebs' cycle with the release of carbon dioxide and ATP. As two pyruvate molecules are released in glycolysis, Krebs' cycle will happen twice for oxidation of every molecule of glucose. This cycle is thus a continuity of glycolysis (Figure 8.11).

The 3C pyruvic acid is first converted to a 2C acetyl, which then combines with Co-enzyme A to form Acetyl CoA. This reaction is catalyzed by the pyruvate dehydrogenase enzyme complex, and a molecule of CO_2 is released and one NADH molecule is formed. This intermediate step connects glycolysis and Krebs' cycle. Pyruvate enters mitochondria and is converted to acetyl Co-A with production of CO_2 and NADH in the mitochondrial matrix.

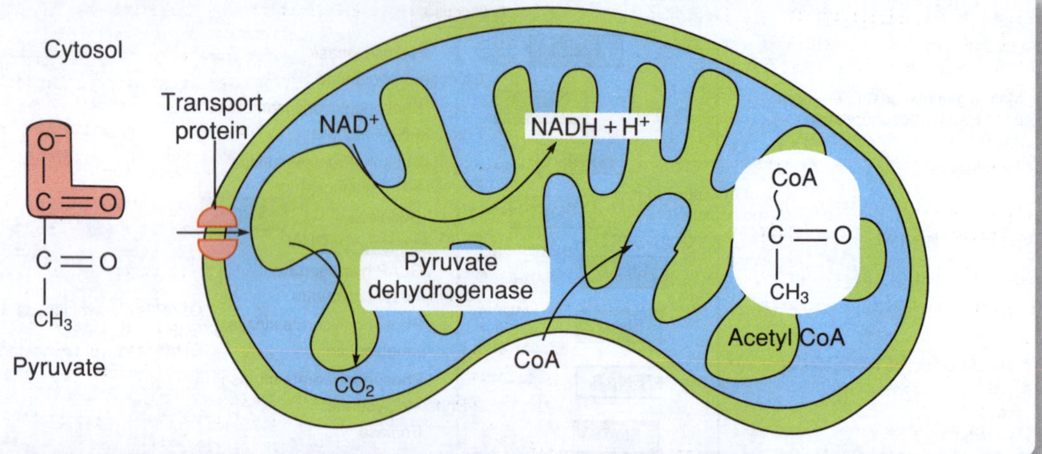

Fig. 8.11 Intermediate step between Glycolysis and Krebs' cycle

The 2C acetyl portion of Acetyl CoA then combines with a 4C oxaloacetate to generate the 6C citrate (citric acid) molecule. CoA is released to realign for another reaction with pyruvate. Citrate is then broken down by a series of enzyme-catalyzed reactions, ultimately regenerating oxaloacetate. Hence, this is named as citric acid cycle. Oxidation of acetyl will produce intermediates called keto acids. In Krebs' cycle, the electrons removed along with protons during the breakdown of acetyl are attached to two electron carriers, NAD+ and FAD, and they are reduced to NADH and $FADH_2$, respectively. One ATP molecule is produced for each pyruvate entering the Krebs' cycle as acetyl.

Thus, when both the pyruvate molecules have been processed through the cycle, all the carbon from glucose is released into the atmosphere as $6CO_2$ molecules, hydrogen from glucose are used to reduce NAD and FAD to form NADH and $FADH_2$, and a final output of 2 ATPs is also produced. A complete breakdown of glucose is achieved at the end of this second step of cellular respiration.

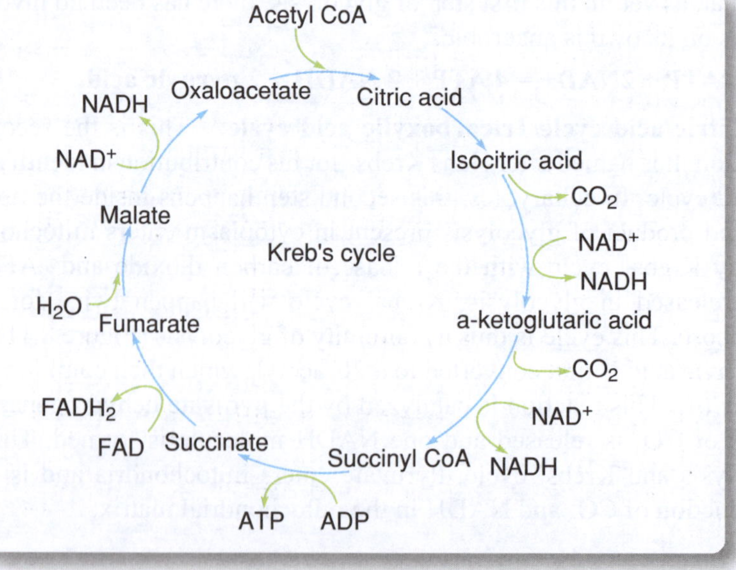

Fig. 8.12 Reactions of the Krebs' cycle

The reaction for every molecule of pyruvate entering the cycle can be given as follows:

Pyruvic acid + ADP + 4NAD+ + FAD ⟶ $3CO_2$ + ATP + 4NADH + $FADH_2$

At the end of Krebs' cycle, from one molecule of glucose, $6CO_2$ + 4ATP + 8NADH, and 2 $FADH_2$ are obtained as products.

To take stock after the second step, till now we see that in the general reaction of aerobic cellular respiration, one reactant glucose is used up; and one product CO_2 is formed; however, the role of the other reactant oxygen and production of water are not seen yet. Moreover, only 4 ATPs are produced as of now. *What will be the utility of NADH and $FADH_2$ and what will be their fate*? These questions will be answered as part of the third step, ETS.

Electron Transport System/Electron Transport Chain Energy carried by the high-energy storage molecules NADH and $FADH_2$ is harvested in a series of reactions called the electron transport system. This is the final stage of aerobic cellular respiration. The main purpose of this is to generate ATP by capturing the energy released by the electrons carried by NADH and $FADH_2$.

A series of redox reactions takes place in ETS wherein the electrons are passed from one electron carrier to another until they reach oxygen, which acts as the ultimate electron acceptor. The reactions of ETS happen in the inner membrane of mitochondria. The inner membrane has protein complexes (Complex I–IV), which contribute to the reactions that yield ATP. All these four complexes are trans-membrane complexes and their individual components are given as follows:

Complex I NADH dehydrogenase (FMN as cofactor) + Iron sulphur protein (Fe-S) + Co-enzyQ;
Complex II Succinate dehydrogenase and Fe-S protein;
Complex III Multiprotein complex of ubiquinone, cytochrome b, Fe-S protein, and cytochrome c1; and
Complex IV cyto c—cyto a—cyto a3—cytochrome oxidase.

The energy-rich electrons from NADH enter Complex I and are passed on along the chain through other complexes till they reach oxygen. Similarly, electrons from FADH2 skip Complex I and enter Complex II, from where they are deposited to oxygen after harvesting their energy through ETS. When electrons are transported, the acceptor molecules in the protein complexes alternate between reduced and oxidized states as they accept and donate electrons. When a molecule accepts electrons, it gets reduced (more electronegative), and then it becomes oxidized again by passing the electrons to its downhill neighbour.

Oxidative phosphorylation and chemiosmosis generate ATP in ETS. When electrons are transported along the ETC, the energy released pumps protons (H+) ions across the membrane in the intermembrane space between inner and outer membranes. This creates a proton gradient across the inner membrane. This is called a *proton pump*, and the pressure created by the high concentration of protons pushes them to flow down the concentration gradient through the channels in the membrane back to the side from where they were pumped. The diffusing protons drive synthesis of ATP by activating ATP synthase. ATP synthase or ATPase is a protein complex present in the inner membrane. It is a phosphorylase. It uses the energy generated to bond a phosphate to ADP (phosphorylation), thus catalysing the formation of ATP molecules (Figure 8.12).

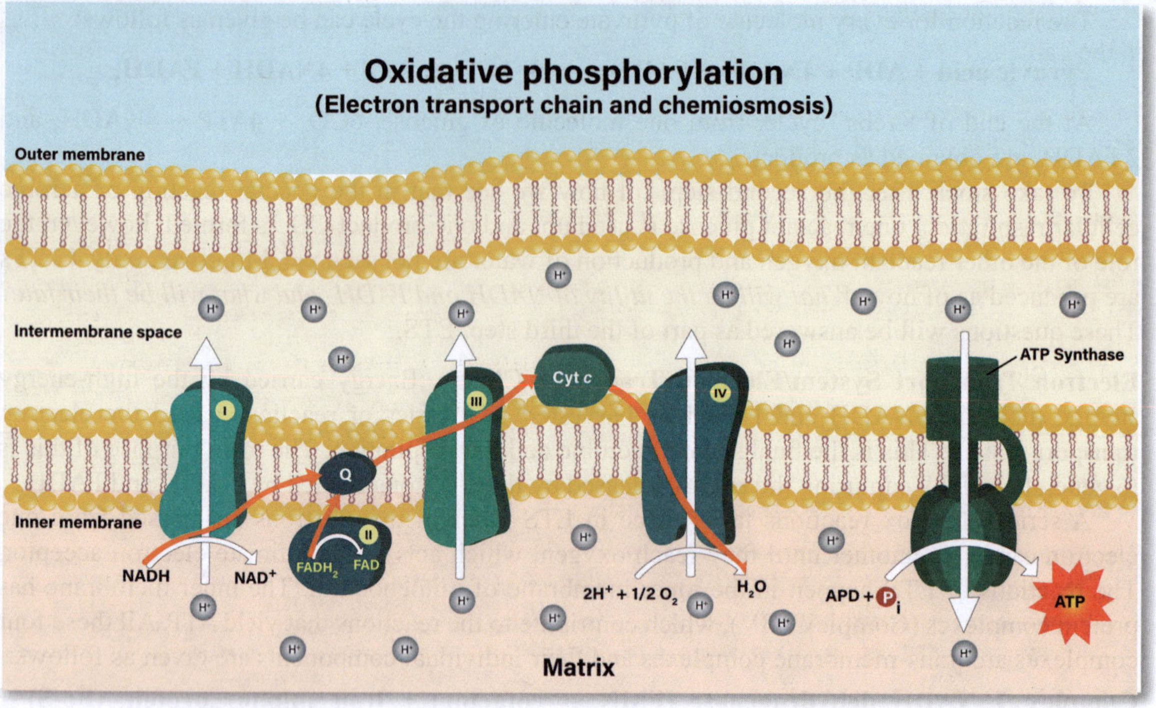

Fig. 8.13 Electron transport chain and chemiosmosis—Oxidative Phosphorylation

Finally, after all the energy is harvested from the electrons, the electrons are ultimately deposited to the electronegative oxygen (o=), which combines with the protons released to form water molecules (H_2O).

The electron energy potential decreases with each step, and the energy is used to produce ATP molecules. Each NADH produces 3 ATP from ETC (Complex I, III, and IV), while each $FADH_2$ produces 2 ATPs (Complex III and IV).

Energy Generation from Aerobic Cellular Respiration

After completing the three stages of aerobic cellular respiration, the total count of ATPs generated from the breakdown of one glucose molecule (energy gain) is 38 ATPs in prokaryotes and 36 ATPs in eukaryotes.

This difference of 2 ATPs is due to the sub-cellular locations where the three stages occur in these two kinds of organisms. In eukaryotes, glycolysis takes place in cytoplasm and Krebs' cycle and ETS occur in the mitochondria. Prokaryotes do not have mitochondria, so the membrane part for proton gradient is carried out by the folds in the plasma membrane and all the three stages of aerobic cellular respiration happen in cytoplasm. For both eukaryotes and prokaryotes, in glycolysis, there is gain of 2 ATPs and generation of 2 NADH molecules, which can produce 6 ATPs in ETS. However, eukaryotes have a challenge; NADH is unable to cross the mitochondrial membrane and enter the mitochondria for ETS to happen. Hence, eukaryotes transfer the electrons released during glycolysis carried by NADH to form $FADH_2$, which enables it to shuttle into the mitochondria. $FADH_2$ produces only 4 ATPs; hence, the eukaryotes have 2 ATPs less than in prokaryotes.

The stagewise ATP generation and comparison between prokaryotes and eukaryotes is as given in the Table 8.4:

Table 8.4 Energy Generation by Aerobic Cellular Respiration from One Molecule of Glucose Comparing Prokaryotes and Eukaryotes

S. No.	Stage of Aerobic Cellular Respiration	Energy Products Generated in Prokaryotes	Energy Products Generated in Eukaryotes	Total Direct ATP Generated in Prokaryotes	Total Direct ATP Generated in Eukaryotes
1.	Glycolysis	2 ATP + 2 NADH	2 ATP + 2NADH; however 2NADH is converted to $FADH_2$ for entering ETS in mitochondria	2 ATP	2 ATP
2.	Krebs cycle	2 ATP + 8 NADH + 2 $FADH_2$	2 ATP + 8 NADH + 2 $FADH_2$	2 ATP	2 ATP
3.	Electron transport system	10 NADH × 3 = 30 ATP 2 $FADH_2$ × 2 = 4 ATP	8 NADH × 3 = 24 ATP 4 $FADH_2$ × 2 = 8 ATP ↓ 2 from Kreb's Cycle and 2 from NADP enters in glycolysis	34 ATP	32 ATP
Total ATP from one molecule of glucose:				38 ATP	36 ATP

Anaerobic Respiration or Fermentation

All organisms need energy; hence, they have to carry out the breakdown of nutrients. Organisms that are not able to utilize oxygen are said to be *anaerobic*. It is a lifestyle for some organisms as they lack enzymes and factors to carry out oxidative phosphorylation, and anaerobic breakdown will be their only source of ATP.

Anaerobic organisms will not require oxygen as the final electron acceptor; instead, they may use other inorganic (sulphur, nitrogen, or other ions) or organic compounds (such as pyruvate) as their final electron acceptors to metabolize the nutrients. When anaerobic pathways use an organic molecule as the ultimate hydrogen acceptor to oxidize glucose, then the pathways are referred to as fermentation. So, fermentation is anaerobic respiration with an organic molecule as final electron acceptor.

Glycolysis is common between aerobic and anaerobic cellular respiration. So, glucose gets converted to pyruvate by the 10-step glycolytic reaction in both types of cellular respiration. In aerobic respiration, after glycolysis, a complete breakdown of pyruvate happens through Krebs' cycle to give CO_2 and electrons of NADH are passed on to ETS.

In anaerobic respiration, the fate of pyruvate produced after glycolysis will be different. Normally, anaerobic breakdown of glucose is incomplete and pyruvate is converted to other hydrogen-containing organic molecules, such as lactic acid, ethyl alcohol, acetone, etc., as products. This will depend on the kind of organism and the specific enzymes it possesses.

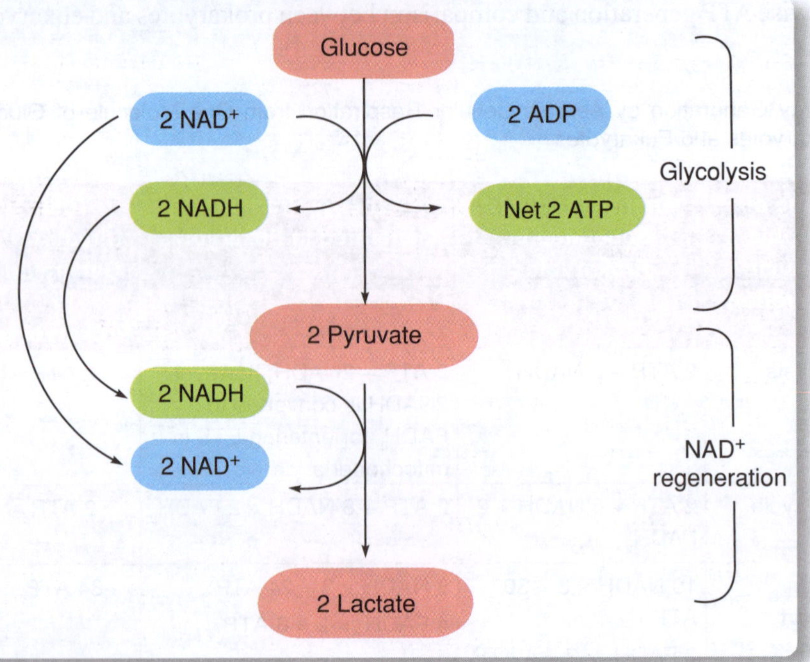

Fig. 8.14 Lactic acid fermentation

Source: Adapted from https://www.sciencefacts.net/lactic-acid-fermentation.html

Anaerobic respiration is incomplete oxidation of glucose.

Glucose + H^+/e^- acceptor = smaller hydrogen-containing molecules + energy (ATP + heat).

Two types of fermentation reactions occur, as follows:

(i) ***Lactic acid fermentation*** is a type of anaerobic fermentation process in which incomplete breakdown of glucose occurs to yield lactate as a product. The net gain of ATP in lactate fermentation is 2ATPs per glucose (whatever is formed during glycolysis). In this, pyruvate formed during glycolysis is converted to lactate. The electrons carried by 2 NADH (formed during glycolysis) are returned to pyruvate and NADH again gets oxidized to NAD^+ to participate in the next round of glycolysis. Pyruvate, an organic molecule, acts as the final electron acceptor; hence, this is a fermentation process.

Lactate fermentation is seen in some bacteria, such as *Lactobacillus*, in some plant tissues, and in some cells of human beings—the skeletal muscle cells. Build-up of lactic acid occurs. The sour taste, tart flavour, texture, and aroma of fermented dairy products, such as yogurt, buttermilk, cultured sour cream, cheese, etc., could be due to the fermentation activity of lactic acid bacteria (Figure 8.13).

Interesting Fact 8.2

Human skeletal muscle cells and lactate fermentation

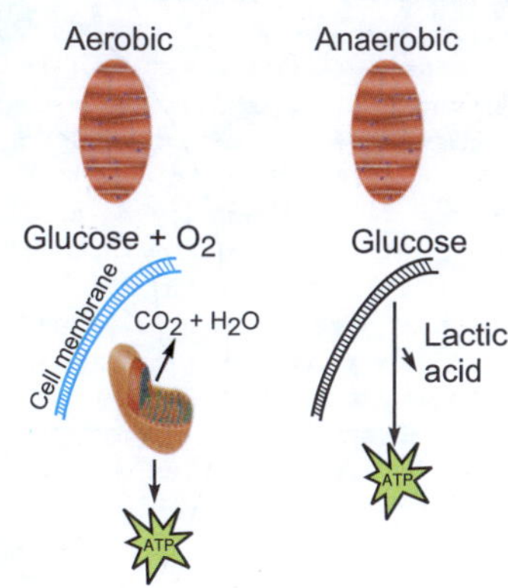

- In humans, different cells have different metabolic possibilities. Nerve cells constantly need oxygen and respire only aerobically; whereas RBCs lack mitochondria; hence, they can respire only anaerobically and rely only on lactic acid fermentation for their energy needs. The skeletal muscle cell are facultative, i.e., depending on the availability of oxygen they can switch between oxidative pathway and fermentation.
- Skeletal muscle cells constantly need ATP for their work. As long as oxygen is available, they keep metabolizing aerobically. In case of extended exercise or exertion, the ATP is spent faster than the pace at which blood supplies oxygen, so there is an oxygen debt created. Then, skeletal muscle cells begin to respire anaerobically.
- The 2 ATPs from glycolysis, though not very high but produced consistently and speedily, can still energize the muscle for short burst of activity. However, more glucose gets used up.
- The challenge from fermetation is build up of lactic acid. Recent reports suggest that it can be used as a fuel or gluconeogenic precursor. However, the excess lactic acid may be transported to liver where it converts back to pyruvic acid, or when oxygen content returns, the reaction is reversed to pyruvate (aerobic).

(ii) ***Alcoholic fermentation*** occurs in some invertebrates and protists, but the best example is the single-celled yeast—*Saccharomyces cerevisiae*. Yeast cells are facultative anaerobes. In the absence of oxygen, they follow the alcoholic fermentation pathway. This pathway also produces pyruvate from glycolysis, which is then converted to ethyl alcohol, and CO_2 in a two-step process. Pyruvate is first converted to acetaldehyde, an intermediary compound, by pyruvate decarboxylase, releasing CO_2. Acetaldehyde is then converted to ethanol and, in this step, NADH is also converted to NAD^+. The net energy generated is 2 ATPs of glycolysis (Figure 8.14).

The primary aim of yeast is not to produce ethanol and CO_2 but to obtain energy by breakdown of glucose. These products are very valuable to humans. During bread making, when yeast is added, it releases CO_2, which makes the dough rise. The alcohol produced evaporates during the baking process. On a similar note, in the brewing industry, the main product is alcohol produced by yeast cells under anaerobic conditions. If the CO_2 is trapped, then it produces sparkling wine and Champagne; if it is allowed to escape, then it produces 'still' wines.

In yeast, this is an alternative respiratory pathway performed without oxygen. When oxygen is available, they immediately shift to the aerobic path.

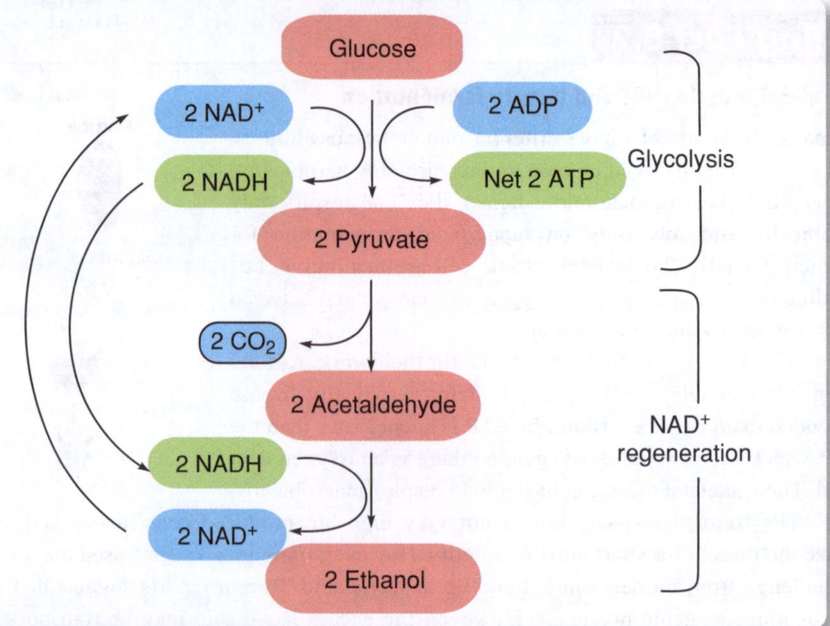

Fig. 8.15 Alcoholic fermentation—*Saccharomyces cerevisiae*

Source: Adapted from https://www.sciencefacts.net/alcoholic-fermentation.html

Metabolic Transformations of Other Nutrients

Glucose is typically considered a substrate for energy-producing biochemical pathways. Energy tied up to carbohydrates is released by the pathways studied so far. However, our diet includes other macromolecules such as fats and proteins; are they catabolized? If yes, do they provide energy? Do they have separate catabolic processes? These are some interesting questions to ponder over. Fats, proteins, and carbohydrates can all produce energy and the building blocks needed by the body.

As a first step, all these are broken down to their monomeric units—carbohydrates to simple sugars like glucose, proteins to amino acids, and fats to glycerol + fatty acid in the stomach and intestine by the process of digestion. We have already studied the glucose breakdown through the main catabolic pathway of glycolysis, Krebs' cycle, and ETS.

Fats and proteins are also broken down by this same central pathway. The monomers of proteins and fats are converted into intermediates, which then join these pathways. There are additional enzymes involved, which produce these intermediates from monomers and supply them for breakdown to the central pathway.

In case of fats, glycerol, the 3C molecule, is converted into glyceraldehyde-3 phosphate (G3P), which is one of the intermediates in glycolysis; it enters glycolysis and is broken down. Long-chain fatty acids from the lipids, such as 18C, 16C, etc., (long-chain molecules) are first broken down into 2C molecules, which are then converted to 2C acetyl, which enters Krebs' cycle. Thus, fatty acids also follow the breakdown route through Krebs' cycle in mitochondria. Based on the number of smaller lipid products that enter the catabolic pathway and the number of bonds, and the amount of potential energy stored in them, it is evident that each fat molecule produces many more ATPs. Hence, fats are excellent sources of energy and are good long-term energy storage materials. This also indicates that excess sugars can be converted to fat and vice versa as the central pathway reactions are reversible.

Our bodies first utilize the available carbohydrates; if they are not available, they move to fats; and only as a last resort break down the proteins. In case of protein breakdown, the individual amino acids are deaminated—amino group removed—in the liver. Then, the non-nitrogenous part of amino acids is rearranged to form the keto acids or organic acids, such as pyruvic acid, acetyl, or other intermediates found in Krebs' cycle and, thus, enter the central catabolic pathway through Krebs' cycle. The amino group that was removed is converted to ammonia, urea, and uric acid and is released from our body by excretion (Figure 8.16).

An important fact to note is that the breakdown of protein is not a desirable source of energy. The preferred energy molecules for the body are carbohydrates followed by fats. The primary function of protein is to provide amino acids for the synthesis of other proteins. However, proteins cannot be stored in the body. Hence, if excess proteins are there, they will be converted to carbohydrates or fats or broken down to energy. Hence, a daily supply of protein in our diet is necessary and critical for daily food requirements. Absence of protein can, hence, lead to malnutrition.

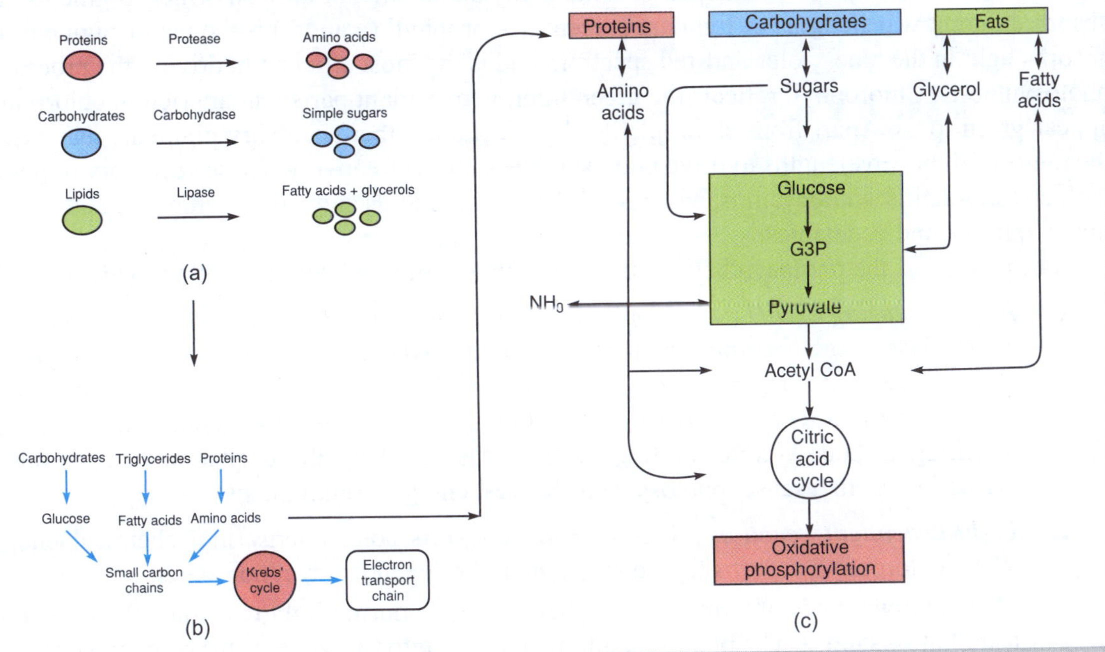

Fig. 8.16 Metabolism overview---Breakdown of proteins and lipids and interconversion of nutrient molecules: (a) Macromolecules broken down to monomers; (b) Entry into central catabolic pathway to generate ATP and building blocks; (c) Process of breakdown of nutrients through central catabolic pathway

Source: Adapted from https://studymind.co.uk/notes/protein-and-lipids-breakdown /https://healthinfo.healthengine.com.au/metabolism-and-energetics https://eweb.furman.edu/~wworthen/bio111/resp18.html

In conclusion, the central pathway ensures that as long as an individual's diet contains all the three major types of molecules, the metabolic machinery will modify them to satisfy the needs of the body. The two aspects of energy production and providing building blocks through the metabolic intermediates are balanced and regulated to ensure optimal growth and survival of the organism.

Photosynthesis—Anabolic Food Production in Autotrophs

Plants are autotrophs that produce their own food by photosynthesis. They plug into the sun's energy, convert it into ATP (chemical bond energy), and combine the carbon, hydrogen, and oxygen from nature to form carbohydrates. Plants/algae and other autotrophs are unique and amazing, as this attribute does not exist with us or other organisms.

> **The following equation summarizes the chemical reactions of photosynthesis:**
>
> **Energy from sunlight + carbon dioxide + water ⟶ glucose + oxygen**
>
> **Light energy + $6CO_2$ + $6H_2O$ ⟶ $C_6H_{12}O_6 + 6O_2$**

To carry out this task of photosynthesis, plants contain specialized organelles called chloroplasts in their leaves and stems. Chloroplasts contain thylakoid membranes on which pigments that absorb different wavelengths of light are present. Chlorophyll (a and b) is the major pigment that absorbs light in the blue-violet-and-red spectrum and is the most efficient in driving the process of photosynthesis. Chlorophyll reflects the green light; hence, plant parts that are rich in chlorophyll appear green to us. Apart from chlorophyll, plants contain other accessory pigments that extend the range of light wavelengths to drive photosynthesis more effectively. These accessory pigments include carotenoids, anthocyanins, and phycobilins. Carotenoids and anthocyanins are abundant in flowers, fruits, and vegetables.

The process of the photosynthetic pathway has three different stages as follows (Figure 8.17):

1. ***Light-capturing events*** Chlorophyll and other pigments embedded in the membranes of chloroplast absorb specific wavelengths of light. In the thylakoid membrane, the pigments are arranged in clusters called photosystems. These photosystems are the functional partners in light-dependent reactions of photosynthesis. The photosystems act as energy-capturing or energy-concentrating mechanisms that help the excitation of electrons to higher levels. In a sense, photosystems harvest energy from photons.

2. ***Light-dependent reaction*** This reaction converts solar energy into chemical energy. When chlorophyll absorbs light energy, the light-dependent reaction starts.

   There are two types of photosystems that cooperate during light reactions: Photosystem I and Photosystem II. The photosystems are divided into two zones—the reaction centre and antenna complex. The reaction centre contains the main Chlorophyll a, which is present near the membrane containing an electron acceptor. The antenna complex is a network of many chlorophyll molecules (a and b) and accessory pigments, which harvest the energy from photons and transfer it to chlorophyll a of the reaction centre.

   Photosystems I and II differ in the reaction centre proteins and the wavelength of light absorbed by chlorophyll a in the reaction centre. In Photosystem I, chlorophyll a absorbs light at wavelength 700 nm and is called P_{700} whereas in Photosystem II, chlorophyll a has a peak absorption at 680 nm and is called P_{680}. The difference is not due to the difference in chlorophyll, which is similar but due to the associated proteins that marginally change the absorption spectra. When light falls on the chlorophyll pigment in the antenna complex, light energy is gathered and collectively transferred to the chlorophyll a in the reaction centre, which becomes unstable and releases energy-rich electrons.

The outcome of these energy-rich electrons is slightly different between Photosystems I and II. Photosystem II occurs first and feeds electrons to Photosystem I. In Photosystem II, when the chlorophyll in reaction centre P_{680} is excited, the electrons pass from the primary electron acceptor to several electron acceptors like plastoquinone (Pq), a cytochrome complex, and mobile carrier protein plastocyanin (Pc). This is similar to the electron transport chain of mitochondria; the electrons, when they move from one carrier to the next, release free energy that creates a concentration gradient of H^+ ions (proton gradient), which provides the energy source for ATP synthesis (as in mitochondria) and ATP molecules are produced. The energy-depleted electrons are then deposited to P_{700} of Photosystem I. As this chemiosmotic process for ATP generation is driven by light, it is called *photophosphorylation*. Thus, the electrons released from the chlorophyll of Photosystem II moved to the chlorophyll of Photosystem I.

How are the electrons lost by P_{680} of Photosystem II replaced? Once P_{680} loses electrons, it becomes oxidized, and a special enzyme in Photosystem II extracts electrons from water and supplies it to P_{680}. The enzyme splits the water molecule into two protons and oxygen ($2H^+ + O$). The oxygen is released and the electrons from hydrogen are used to replenish the lost electrons of P_{680} in Photosystem II. Thus, when Photosystem II gives up electrons, new electrons from water replace them. Water splitting in Photosystem II is responsible for the release of oxygen as a product of photosynthesis.

In Photosystem I, light is trapped and energy is absorbed in the same manner as Photosystem II. However, there is no enzyme for water splitting, so no oxygen is released from Photosystem I. The electrons are excited in the reaction centre of Photosystem I (P_{700}), which then, again, passes through a series of redox reactions, and these electrons are picked up by $NADP^+$, which gets reduced to NADPH. So, the primary outcome of Photosystem I is production of NADPH.

So, in the light-dependent reactions, solar energy drives the release of electrons from photosystems in the thylakoid membrane. The two photosystems and the ETC that connects them transfers electrons from water to $NADP^+$, producing NADPH. In a manner similar to mitochondria, ATP synthase in chloroplast by chemiosmosis generates ATP using the proton gradient.

The products of light reaction, oxygen is released into the atmosphere, ATP and NADPH are both used in the next stage of photosynthesis, which is the dark reaction or the light-independent reaction.

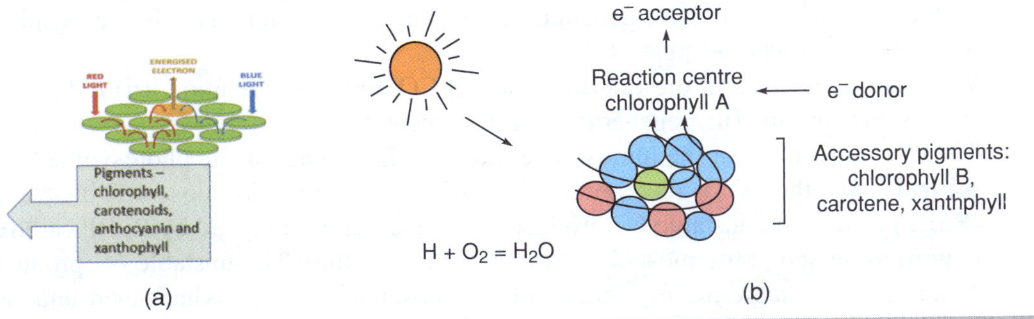

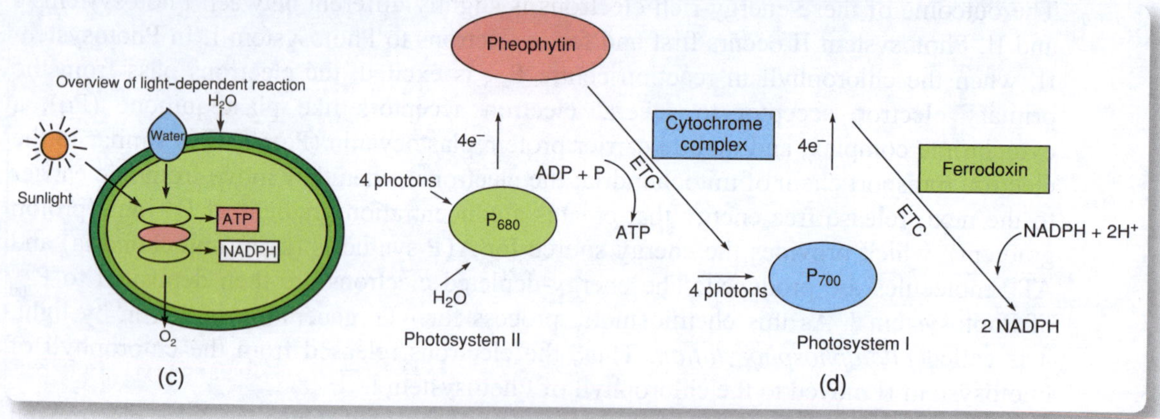

Fig. 8.17 Light-capturing and light-dependent reactions of photosynthesis: (a) Structure of chloroplast—presence of pigments that absorb light; (b) Photosystem antenna complex of photosynthesis; (c) An overview of light-dependent reaction and products; (d) Role of PS II and I in light-dependent reaction and generation of ATP and NADPH

Source: Adapted from https://flexbooks.ck12.org/cbook/ck-12-biology-flexbook-2.0/section/2.19/primary/lesson/chloroplasts-bio/
https://www.biologyexams4u.com/2017/03/what-are-photosystem-definition-and-summary-of-PS-1-and-2.html
https://www.expii.com/t/light-reaction-equation-steps-10138
https://ib.bioninja.com.au/photosystems/
https://biology4alevel.blogspot.com/2015/08/100-light-dependent-reactions.html

3. ***Light-independent reaction or dark reaction*** Dark reaction can occur with or without light. The ATP and NADPH produced during the light reaction are utilized in a cyclic series of reactions that fixes CO_2 and H_2O into glucose. The dark cycle is also called the Calvin cycle to honour the scientist for his contribution. The dark reaction happens in the stroma of the chloroplast.

Calvin cycle is the actual sugar-manufacturing machinery and is called a ‘cycle’ as the starting material is regenerated similar to Krebs’ cycle in mitochondria. The materials needed for the dark reaction are ATP, NADH, CO_2 and a 5C molecule called Ribulose. In this reaction, carbon enters as CO_2 from the atmosphere and exits in the form of sugar. Energy-rich electrons to various intermediates are added by the reducing agent NADPH with energy from ATP, both generated from the light reaction. This is the ‘synthesis or anabolic’ part of photosynthesis.

There are three steps in the Calvin cycle: (i) CO_2 entry and fixation, (ii) reduction with sugar synthesis, and (iii) regeneration of the acceptor.

The CO_2 diffuses from the atmosphere into the chloroplast of the photosynthetic cell in the leaf. First, the CO_2 is incorporated into a 5C sugar called Ribulose-1,5, bis-phosphate (RuBP) to form an unstable 6C product. This reaction is catalyzed by RuBP carboxylase (Rubisco), which is the most abundant enzyme in nature. The unstable 6C product then splits into two stable 3C molecules, PGA (phosphoglycerate), which then undergoes a series of reactions wherein each PGA gets a phosphate group from ATP and electrons

from NADPH to finally produce two molecules of glyceraldehyde-3 phosphate (G3P) or PGAL. This process of incorporating CO_2 into a stable organic compound is called carbon fixation. As a 3C compound glyceraldehyde-3 phosphate is produced, this pathway is also called the C3 pathway (Figure 8.18).

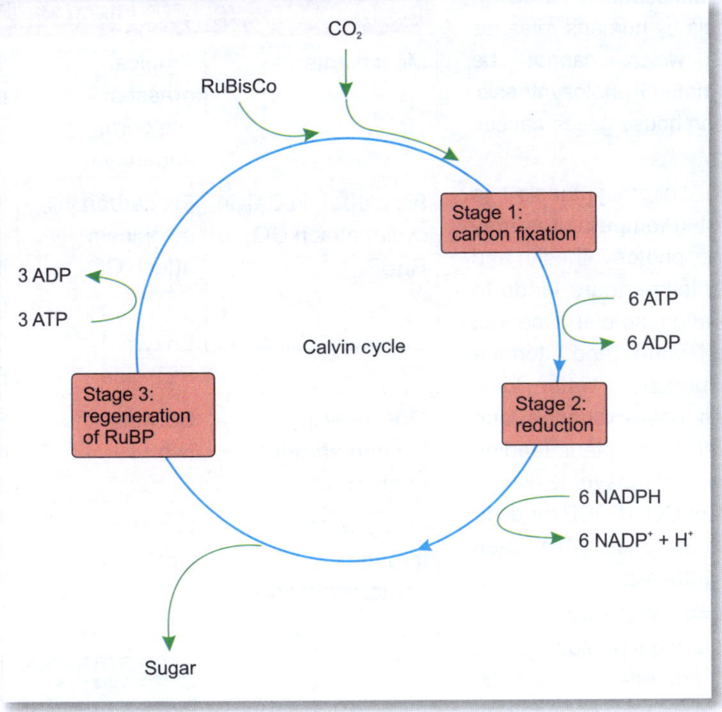

Fig. 8.18 Light-independent/dark reaction of photosynthesis; Calvin cycle

To generate one molecule of glucose, 6 CO_2 molecules have to be incorporated. Hence, six cycles, each incorporating one CO_2 is needed. After 6 molecules of CO_2 have been fixed, 12 PGAL (G3P) molecules are formed. Two of these 3C G3P /PGAL combine to form the 6C glucose molecule. The remaining 10 PGAL are phosphorylated by ATP to regenerate 5C RuBP, thus completing the cycle.

The dark reaction of photosynthesis can be overall summarized as follows.

$$CO_2 + ATP + NADH + 5C \text{ starter RuBP} \longrightarrow G3P + NADP^+ + ADP + P$$

Thus, 3C G3P is the actual product of the Calvin cycle, which is then changed into phosphorylated glucose, which is then primed to form other carbohydrates such as sucrose, cellulose, and starch. The G3P can also serve as an ingredient and lead to synthesis of lipids and amino acids forming proteins. It can also be broken down to yield energy through the glycolytic pathway. Thus, eventually, the products of photosynthesis are used as energy sources and building blocks for all the lipids, amino acids, and other organic compounds that the plants require for their survival. The ADP, NADP$^+$, and remaining phosphate diffuse through stroma, back to the light-dependent reaction.

Interesting Fact 8.3

Climate change and photosynthesis

- Photosynthesis removes green house gas CO_2 from atmosphere. However, use of fossil fuels by humans releases excess CO_2, which cannot be neutralized by natural photosynthesis. Increase in green house gases causes global warming.
- So, increased photosynthesis is needed, but rising temperatures may actually reduce photosynthesis and its effect. High temperature leads to greater evaporation, so plants cannot absorb enough water and stomata is closed to preserve water loss, which prevents gas exchange and CO_2 does not enter the plant, leading to reduction in photosynthesis. In reduced levels of CO_2, RuBP binds to oxygen to form glycolate, which then inhibits photosynthesis.
- C4 plants like corn and CAM plants like aloe and jade carry out photosynthetic pathways that are effective in such a scenario. C4 plants concentrate CO_2 by scavenging CO_2 to produce 4C sugar acetate in mesophyll cells, even when stomata are closed. These sugars then pump CO_2 to Calvin cycle. CAM plants capture CO_2 at night when temperature is low and store it as organic acid in vacuoles, which, when broken down, releases CO_2 for Calvin cycle during the day (Table 8.5).
- Evolutionarily, with climate change, C4 and CAM plants may increase and C3 plants may decline. This will overall reduce photosynthesis as C4 plants are only grasses but C3 plants are huge trees.
- Replacement of trees with grasses will add to deforestation caused by humans, leading to more deserts and snow- and ice-covered regions in future.

Table 8.5 Difference Between C3, C4, and CAM Plants

C3 plants	C4 plants	CAM plants
Most plants	Tropical grasses like corn, sugarcane	Succulents, pineapple, agave
Fix carbon in Calvin cycle; attach CO_2 to RuBP	Fix carbon in cytoplasm; attach CO_2 to PEP	Fix carbon at night only; fix it to organic molecules
Enzyme Rubisco	Enzyme PEP-ase	Enzyme PEP-ase
Most energy hyphen efficient method	Between the two	Best water conservation method
Loses water through photorespiration	Loses less water	Loses least water
CO_2 → Mesophyll: CO_2 → Calvin cycle → Glucose	CO_2 → Mesophyll: C_4 → Bundle sheath: CO_2 → Calvin cycle → Glucose	CO_2 → Night: C_4 → CO_2 → Mesophyll: Calvin cycle → Glucose

Source: Adapted from https://www.loreescience.ca/uploads/2/4/1/7/24170983/phtosynthesis___comparisons_test_review_questions_apr_2018_answers.pdf

Environmental Balance of Autotrophs and Heterotrophs

Autotrophs carry out photosynthesis and produce organic compounds from inorganic molecules. These organic molecules not only serve as food for the plants themselves but also for all the other organisms that feed on them. So, photosynthesis is done by few (plants and algae) but consumed by all. Similarly, between cellular respiration and photosynthesis, it is interesting to note that the products of photosynthesis (glucose and O_2) are the reactants in aerobic cellular respiration and the products of cellular respiration (CO_2 and H_2O) are the reactants for photosynthesis. Photosynthesis reactants are in plenty as all organisms respire and produce CO_2 and H_2O. But sufficient food and oxygen are needed for all organisms to carry out cellular respiration. As plants provide, both, nutrients and oxygen that are consumed by the majority of organisms including themselves, the right balance between producers and consumers is essential. So, in essence, it brings home the realization that 'fulfilling the demand for food for all' implies that there should be more plants; 'oxygen requirement for all' implies that there should be more plants, and 'reduction of greenhouse gases like CO_2 in the atmosphere' implies that there should be more plants. To balance the requirements of generation versus consumption, it is necessary that the plants or producers are in much higher numbers than the consumers. To sustain the environment, three things are a must: the sun should shine and there should be a right balance of autotrophs and heterotrophs with appropriate and constant cycling of resources. This will ensure that the environment is stable.

Artificial Photosynthesis Systems

Natural photosynthesis is an amazing phenomenon perfected by nature, which carries out oxidation of water to oxygen and fixation of carbon dioxide to sugars using the energy from sunlight in a multistep pathway with the support of chlorophyll pigments, as discussed above. Artificial photosynthesis uses a biomimetic approach with an aim to mimic the natural photosynthesis and capture the energy from sunlight for converting water and carbon dioxide into chemical forms of energy like hydrogen, methane, etc. Artificial photosynthesis systems (APSs) integrate various photocatalysts and biocatalysts to convert and store solar energy and use them as resource in environment, food, and energy. This gives an advantage of converting and storing solar energy for use. Hydrogen is a key solar fuel because it permits energy storage and can be used directly in combustion engines or fuel cells.

Several possibilities have been explored for this aim to be achieved by: (a) finding out templates or structural architectures that mimic chlorophyll function; (b) redox catalysts that form molecular oxygen and hydrogen in presence of oxidants and reductants; (c) possibility of decomposition of water to molecular O_2 and H_2 using sunlight; (d) direct conversion of sunlight to electricity; and (e) reduction of CO_2 to compounds that can be used as fuels and raw materials for industry.

Direct conversion of sunlight to electricity can be achieved using solar cells. Most popular and widely used solar cells are those based on semiconducting Si as single crystal, multicrystalline, or amorphous forms. Though the solar-to-electrical conversion efficiency is around 25%, the cost is enormous. Alternates made of copper–indium, gallium sulphide are low cost, but conversion efficiency drops to 15–18%. Photovoltaic solar cells with dye sensitization is emerging as a low-cost system for conversion of sunlight to electricity.

This is an exciting futuristic domain where improvements in the light energy absorption, photoexcited electron transfer, and components integration based on the confined platforms are the main aspects that will help in further boosting the performance of Artificial Photosynthesis Systems (APSs) in terms of high performance, high stability, and scalability.

SUMMARY

- Metabolism includes anabolism and catabolism.
- Metabolic reactions are carried out by enzymes.
- Enzymes are catalysts that speed up biochemical reactions without getting used up.
- The universal energy currency for most organisms is ATP. It has two high-energy phosphate bonds that are used to store and spend energy.
- Cellular respiration is a catabolic process that breaks down glucose with the release of energy.
- Cellular respiration is of two types—aerobic in the presence of oxygen and anaerobic.
- There are three stages in aerobic cellular respiration—glycolysis, Krebs' cycle, and ETS.
- Two types of fermentation are lactate fermentation and ethyl alcohol fermentation.
- Molecules other than carbohydrates, such as lipids and proteins, can also be catabolized—they enter the central catabolic pathway as intermediates to release energy.
- Photosynthesis is an anabolic process that synthesizes organic compounds from inorganic molecules using the energy of sunlight.
- Photosynthesis includes three stages: light-capturing, light-dependent, and light-independent reactions.
- Balance between autotrophs and heterotrophs is necessary to sustain the cycling of reactants and products in the environment.

EXERCISES

A. Objective-Type Questions

1. What is common between anabolism and catabolism?
   (a) Both lead to the synthesis of larger molecules from smaller ones.
   (b) Both lead to the breakdown of larger molecules to smaller ones.
   (c) Both reactions are carried out by enzymes.
   (d) Both are found in plants but not in animals.
2. Molecules acted upon by enzymes are called
   (a) Cofactors
   (b) Vitamins
   (c) Substrate
   (d) ATP
3. Which of the following terms is not directly involved in enzyme–substrate interactions?
   (a) Lock-and-key model
   (b) Induced-fit model
   (c) Binding site
   (d) Increase in activation energy
4. In eukaryotes, the sub-cellular locations of glycolysis, Krebs' cycle, followed by ETS, are ________, ________, and ________, respectively.
   (a) Chloroplast, mitochondria, endoplasmic reticulum
   (b) Cytoplasm, mitochondrial matrix, mitochondrial inner membrane

(c) Cytoplasm, nucleus, plasma membrane
(d) Mitochondrial matrix, stroma, grana

5. During aerobic cellular respiration, prokaryotes produce ________ number of ATPs from one molecule of glucose.
(a) 36
(b) 2
(c) 38
(d) 4
6. What are the products after lactate fermentation is completed?
(a) ATP, NADPH, pyruvate
(b) ATP, NAD^+, lactate
(c) ATP, NADH, lactate
(d) ATP, $FADH_2$, pyruvate
7. ______________ is the reactant that enters the Krebs' cycle and combines with oxaloacetate to form citrate.
(a) Acetyl CoA
(b) Ribulose
(c) Glucose
(d) Glyceraldehyde - 3 phosphate
8. Which of these is not a stage of photosynthesis?
(a) Light capturing
(b) Light-dependent reactions
(c) Calvin cycle
(d) Krebs' cycle
9. What does Rubisco stand for?
(a) RUBP carboxylase
(b) Ribulose dehydrogenase
(c) Pyruvate carboxylase
(d) Alcohol dehydrogenase
10. $12NADPH + 12H^+ + 18\ ATP + CO_2$ ——— $C_6H_{12}O_6 + 12NADP + 18\ ADP + 18\ Pi$
This equation represents ________________.
(a) Calvin cycle
(b) Krebs' cycle
(c) Ethanol fermentation
(d) Protein respiration
11. Photosystems are found in _____________.
(a) Stroma and stroma lamellae
(b) Thylakoids and grana
(c) Cell membrane and nuclear membrane
(d) Mitochondrial matrix and mitochondrial inner membrane
12. The molecules that link the light and dark reactions of photosynthesis are ____________.
(a) ATP and NAD
(b) RuBP and CO_2
(c) ATP and NADPH
(d) Rubisco and G3P

B. Short-Answer Questions

1. How are the reactants and products of photosynthesis and cellular respiration related?
2. What are the different pigments found in plants and how do they contribute to photosynthesis?
3. How are the C4 pathway of photosynthesis and CAM pathway different from the regular C3 pathway?
4. Explain how eukaryotes produce 2 ATPs less than prokaryotes from the breakdown of one molecule of glucose after aerobic cellular respiration.
5. What is negative feedback inhibition and enzyme competition?

C. Self-Study Questions

1. Study and analyze how our body, and a plant, as a machine and its working; from what and where energy is produced, how and where is it stored, and how is it spent. Can you draw a single model to show the same?

USEFUL WEBLINKS

1. https://www.ncbi.nlm.nih.gov/pmc/articles/PMC5354521/
2. https://www.mdpi.com/2073-4409/13/14/1177
3. https://pmc.ncbi.nlm.nih.gov/articles/PMC4030627/
4. https://www.frontiersin.org/journals/plant-science/articles/10.3389/fpls.2018.00357/full
5. https://pubs.rsc.org/en/content/articlelanding/2022/cs/d1cs01008e

VIDEO LINKS

1. https://www.youtube.com/watch?v=13h5oC4jIsk
2. https://www.youtube.com/watch?v=QTdJiG7mV40
3. https://www.youtube.com/watch?v=0IJMRsTcwcg

CHAPTER 9 Cellular Reproduction

LEARNING OBJECTIVES

After reading this chapter, you will be able to:

- Understand the necessity of cell division and its types
- Delineate the cell cycle: cycle of growth and division
- Gather a step-by-step understanding of mitotic division and its regulation
- Learn about abnormal mitosis and its implications
- Understand sexual reproduction and meiosis
- Understand meiosis—the reduction division
- Understand the differences between Mitosis and Meiosis
- Understand the genetic variability mechanisms of sexually reproducing organisms
- Understand nondisjunction due to abnormality in meiosis

CHAPTER TAKEAWAY **How cells reproduce to generate new cells (cell division)**

This chapter focuses on how cells reproduce to generate new cells (cell division), which is an integral part of our existence—cell division and regeration.

PRELUDE

Prelude: In sexually reproducing organisms as an outcome of fertilization, life begins as a single cell: zygote. This one cell then produces all the other cells to create a full-fledged organism. How does this happen? Unicellular microbes are single-celled creatures that epitomize life. They also grow and reproduce; both mean the same—increase in the number of single-celled individuals. How does this happen? How do one bacterium gives rise to many more? This increase has many implications—more microbial growth can cause disease, spoil food, ferment, etc. So, basically the question that is evident in all these instances: where do they come from whenever there is a need for more cells? The answer to all these questions is cell division. New cells come from pre-existing cells by the process of cell division. Cell division is essential for the life of both prokaryotes and eukaryotes. In larger organisms, cell division not only contributes to development of a new organism from zygote but also continues even after the organism is fully developed and is present throughout the life of an organism. One of the most important events in life is going on in your body every day, hour, or second—cells are dividing.

Cell Division

Two important life processes that define living organisms are growth and reproduction. Both processes are based on a fundamental ability of cells to divide—one cell when it becomes two new cells—this is the basic premise of cell division.

Cell division can be categorized into three types: (a) Binary fission of prokaryotes; (b) mitosis in Eukaryotic somatic or non-reproductive cells; and (c) meiosis in reproductive cells of sexually reproducing eukaryotes.

'Binary fission' is the type of cell division used by single-celled prokaryotic organisms to increase their numbers. It is a type of asexual reproduction where a cell divides into two identical daughter cells. In binary fission, the cell grows, the single loop of prokaryotic DNA replicates and then splits into two cells. The result is daughter cells that are genetic clones of the parent cell (Figure 9.1).

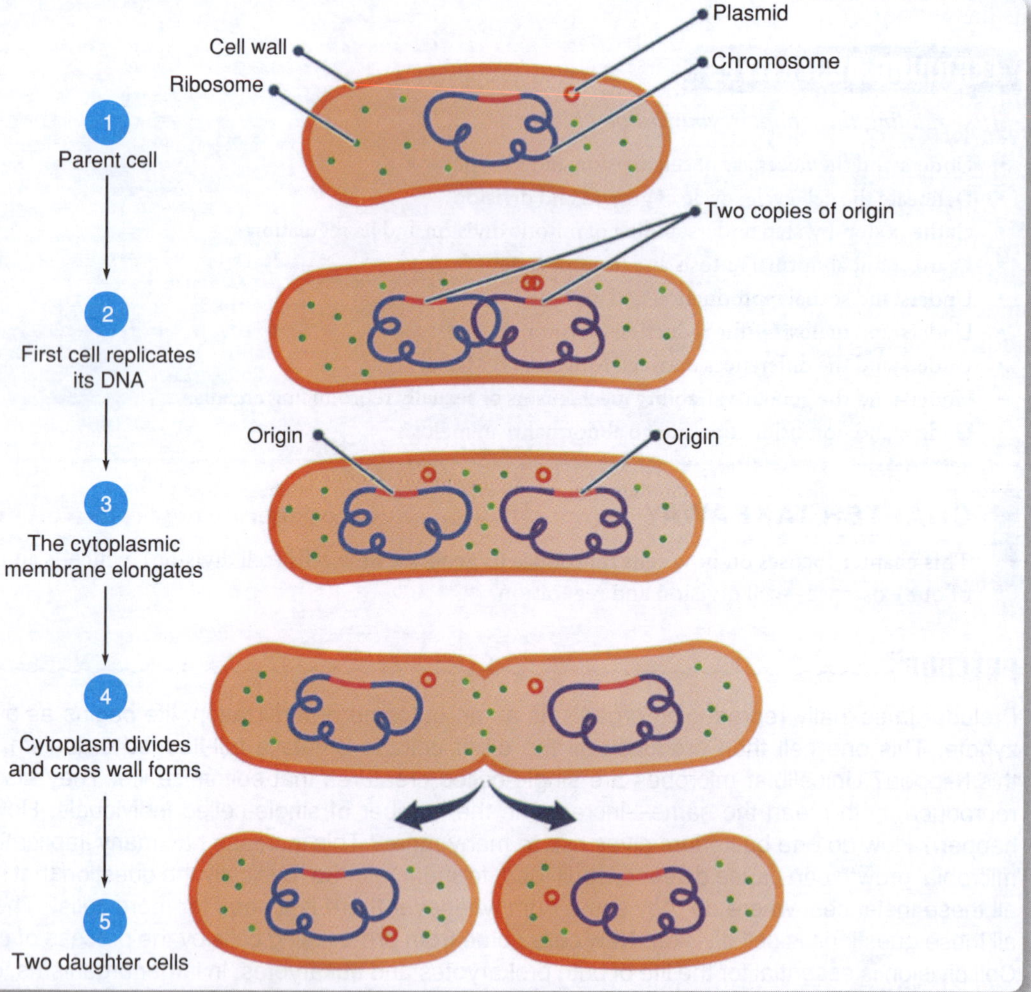

Plasmid
Cell wall
Chromosome
Ribosome
1 Parent cell
Two copies of origin
2 First cell replicates its DNA
Origin
Origin
3 The cytoplasmic membrane elongates
4 Cytoplasm divides and cross wall forms
5 Two daughter cells

Fig. 9.1 Binary fission in Prokaryotes

Multicellular organisms carry out a similar kind of cell division in their somatic or non-reproductive cells called '**Mitosis**'. Mitosis is needed for the following purposes:

Growth: We all began as a single-celled zygote; but all the other cells of our body, trillions of them, came from the process of cell division. So, essentially when we say that an individual grows from an infant, to a teen, to an adult, there is an increase in the number of cells by division, which helps

in this growth. Fixed number of cells don't become bigger during growth; instead the cell numbers are increased. So, for increase in size of an individual or growth, cell division is necessary.

Repair of tissues by replacing damaged cells: Viral infections can cause destruction of some of our cells; similarly, the cells can be damaged due to injuries and wounds. Such damaged cells need to be replaced. The repair mechanisms of damaged tissues involve replacement of damaged cells by cell division. We have all observed that whenever we have cuts in our skin, it gets filled and the skin grows back again there. In this, new cells are formed from the edges of our wound, which helps the skin to grow and cover the wound. A real example of cell division in action, each one of us has witnessed.

Replenishment of lost cells due to finite lifespan and wear and tear: Skin cells and cells lining the intestine are constantly dividing to renew themselves and eliminate the old cells due to wear and tear. RBCs lose their nucleus and have a finite life span of 120 days; therefore, the lost cells need to be continuously replenished.

In all these examples, mitosis is responsible to produce the new cells of the same kind or cells identical to parent cells. The daughter cells produced by mitosis have the same genetic information and pattern of metabolic activities similar to the parent cells. Most somatic cells of our body undergo mitosis type of cell division, which serves the purpose of replenishment in terms of addition of new cells and replacement of cells (Figure 9.2).

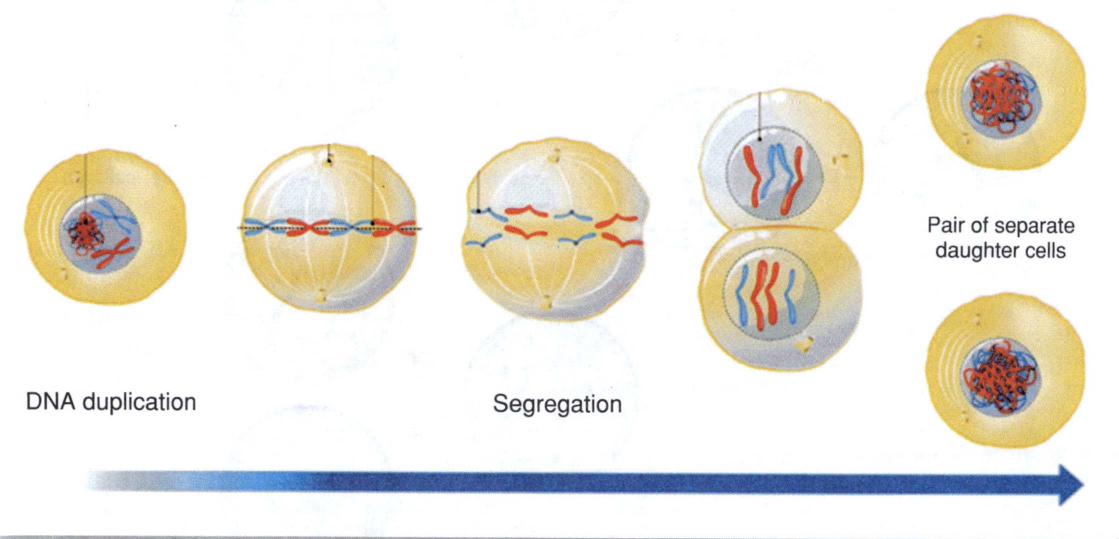

Fig. 9.2 Mitosis in Eukaryotic somatic cells

Cell division also contributes towards *generation of germ cells or reproductive cells in larger organisms. Formation of reproductive germ cells or gametes happens by a type of cell division called Meiosis.* This type of cell division occurs inside the sex organs or gonads and leads to production of gametes. In humans, the male gonads are testes, which produce the male gametes called sperm cells, and the female gonads are ovaries, where female gametes egg cells are generated. The gametes are specialized cells that contain half the number of chromosomes as parent cells. Gametes are haploid, whereas the parent cells are diploid. The gametes produced by sexually reproducing organisms by meiosis then fuse by fertilization to produce a diploid zygote, which then develops into a new genetically unique individual.

Meiosis is also called reduction division, as it reduces the chromosome number in gametes to half. Why is this reduction through meiosis necessary? In case of sexually reproducing organisms, the individuals contain diploid or two sets of chromosomes. One set is received from the mother through the egg cells or female gamete, and the other set is received from the father through the sperm cells or male gamete. If the gametes had two sets of the same chromosomes as the cell producing the gamete, then the resulting zygote produced by fusion of gametes will contain four sets of chromosomes. Hence, with each generation the chromosome number would double. Therefore, it is necessary to have a special type of cell division that reduces the chromosome number to half in gametes i.e., haploid or one copy of genetic information. Now, when the gametes fuse/fertilize, each gamete contributes to one set of chromosomes; thus, the resulting zygote will have two sets and will produce a diploid offspring similar to the parents. The chromosome number will thus be retained from generation to generation (Figure 9.3).

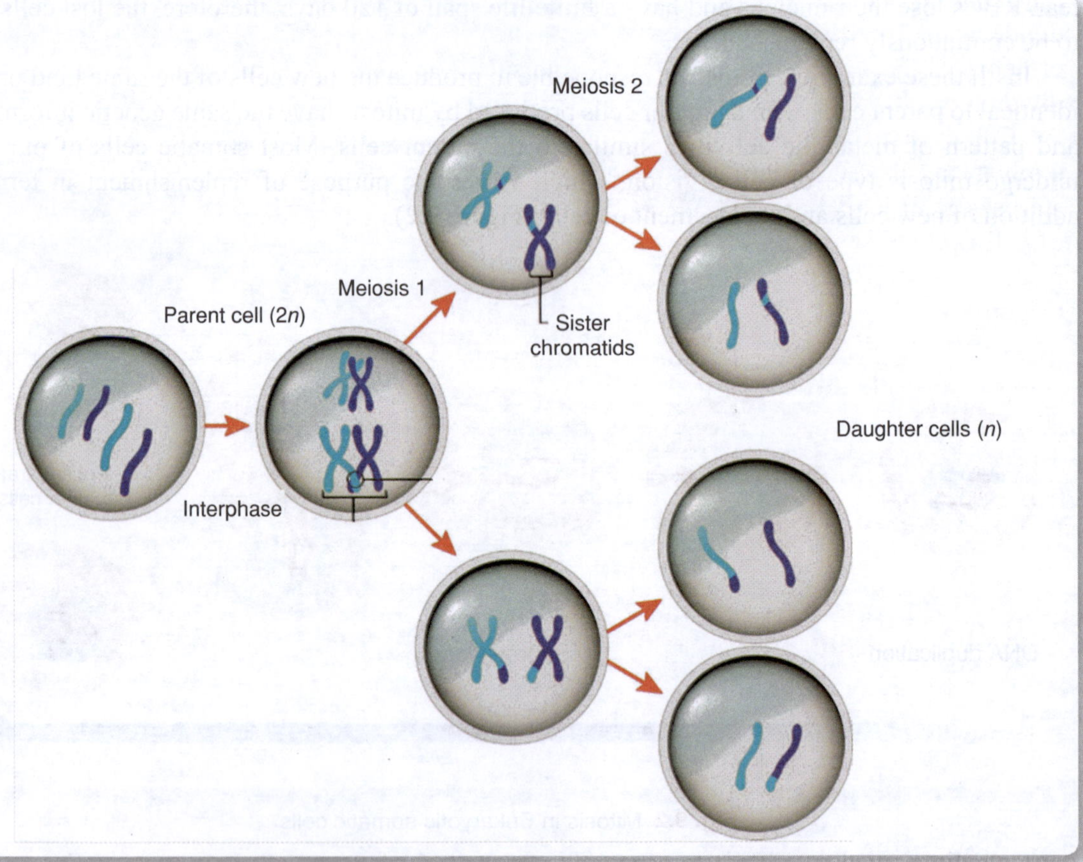

Fig. 9.3 Meiosis: Reduction division for gamete production in Eukaryotes

In humans, we know that our grandparents also had 46 ($2n$) chromosomes, our parents also have 46 ($2n$) chromosomes, and we are also born with 46 ($2n$) chromosomes. So, meiosis produces gametes in humans with 23 chromosomes (n) through reduction division, and fertilization restores the diploid state of 46 chromosomes in every generation.

23 (from egg)	+ 23 (from sperm)	= 46 (zygote) new offspring
n	$+ n$	$= 2n$
Haploid	+ Haploid	= Diploid

For any type of division some basic steps like duplication of DNA and distribution of duplicated DNA is necessary. Also, the process of cell division should happen without any errors. These are ensured by an orderly arrangements of activities called the cell cycle.

Cycle in Eukaryotes - Cycle of Growth and Division

The different stages of growth and division of a Eukaryotic cell is depicted as its cell cycle. When a new cell is formed, it is typically small. It has to undergo growth and expand to a certain size before initiating another round of division. The cell cycle includes both the time engaged in metabolism for growth and preparation for the cell to divide and the actual cell division period. The cell moves through the various stages in a sequential continuous manner.

The basic cell cycle of all cells is the same, but different cells may spend different periods of time at various stages. For example, some cells, such as nerve cells, RBCs, or muscle cells, never divide again once they mature; others, such as liver cells, only divide when stimulated by injury or damage, and some, such as skin cells, intestinal cells, and some blood cells, are continuously dividing cells.

Cell cycle of a typical Eukaryotic cell is divided into four stages: G_1, S, G_2, and M. There is a 5th stage called G_0—non-dividing stage, which is outside the cycle of division. The first three stages—G_1, S, and G_2—are collectively called Interphase; it is the growth and synthesis phase of cell cycle, whereas the final M stage or the mitotic stage is where actual cell division takes place. Each stage has an important role in the cell's journey, as detailed in the following section (Figure 9.4).

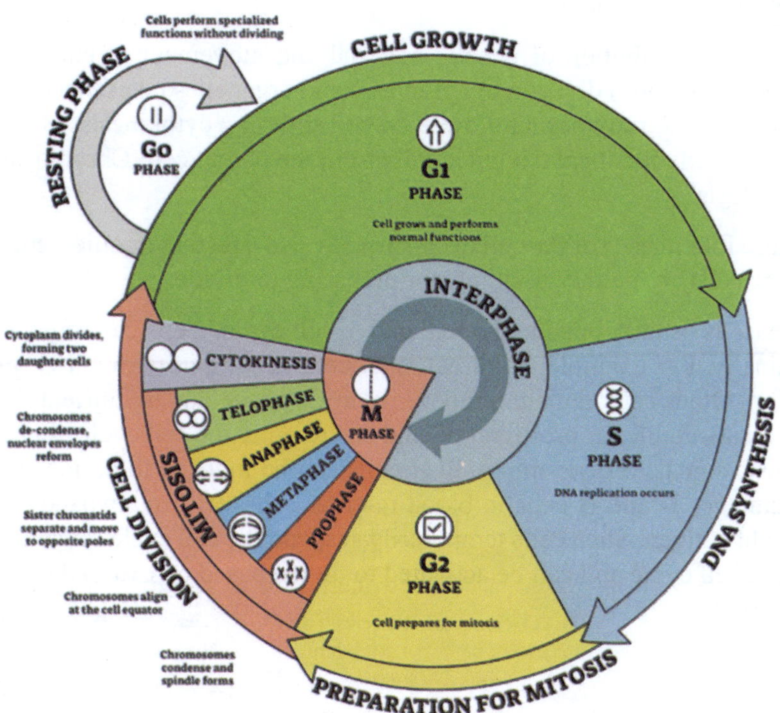

Fig. 9.4 Eukaryotic cell cycle

G_1-growth or Gap Phase I: This is the first stage of Interphase for a newly formed cell. In this stage, the cell gathers nutrients and other resources from the environment. The cell is metabolically very active at this stage, grows in volume, and synthesizes all the necessary proteins, enzymes, RNAs like tRNA, and other cell components for it to function efficiently. This is the growth stage that varies in length among different cells; the other stages are more or less uniform. So, whether a cell divides rapidly or very slowly depends on the length of this stage. In the G1 stage, the cell also makes a critical decision of whether to divide or not. If it is to divide, it will ensure that the needs of DNA replication are fulfilled, and there is no damage to DNA. In case of damage, it will arrest the cell cycle, repair the damage, or prevent cell division of the damaged cell by triggering its destruction. So, in that sense, it is a cautionary gap when preparation for the next S phase along with proper checks for a successful DNA replication is done.

S phase or Synthesis phase: Phase of DNA replication and duplication of genetic material with chromosomal proteins. This stage will produce two copies of the genetic material that can be distributed to the daughter cells as chromosomes. After replication, two replicated copies of each chromosome are called sister chromatids. Sister chromatids are the two copies of the chromosome that were produced by replication and contain the same DNA. The sister chromatids remain attached at centromere and become visible when chromosomes condense later in mitosis.

G_2 - growth or gap Phase II: This is the third phase of cell cycle and the last phase of Interphase before Mitosis. This is the preparation phase for mitosis, and the cells make all the cellular components that are required for cell division by mitosis. Any damage to DNA after S-phase is also repaired in this phase to ensure that both the daughter cells receive the accurate copy of DNA. Cell growth also continues in this phase.

M phase—Mitosis: The distribution of genetic material and movement of chromosomes from parent to daughter cells along with division of cytoplasm and organelles takes place in mitosis. So, mitosis is nuclear splitting—karyokinesis followed by cell splitting: cytokinesis. Mitosis results in production of two genetically identical daughter cells from the parent cell. Chromosome numbers are also retained in mitosis.

In addition to these four phases of the cell cycle, there is also a resting or quiescent phase called the G_0 phase when the cells are neither dividing nor preparing to divide.

G_0 phase: This phase is usually outside the replicative cell cycle. Cells can remain in G_0 phase temporarily or terminally. For example, adult neuronal cells remain in terminal G_0 phase as part of their differentiation. Other cells remain temporarily in G_0 phase due to nutrient deprivation or limitations in resources available for proliferation and can re-enter the cell cycle after activation or in response to extrinsic signals. For example, liver cells can be induced to divide and regenerate whenever there is damage; T- and B-cells in blood re-enter cell cycle and expand clonally when presented with the right antigen. Such cells temporarily suspend the cell cycle for some time but do not fully terminate the cell cycle and can be activated to divide based on external signals.

M-phase Decoded: Steps in Mitotic Cell Division

In mitosis, primarily two major events occur: splitting of the replicated chromosomes for distribution of genetic information to the two daughter cells (karyokinesis or nuclear division) followed by cell splitting—division of cell cytoplasm and its organelles (cytokinesis). So, mitosis is karyokinesis followed by cytokinesis.

These two events happen through a seamless transition across different individual stages of mitosis. The precise order of events as they progress one after the other provide characteristic features that define each stage of mitosis. The activities of mitotic stages are continuous, and there is no distinct starting and end point for each stage. However, the different stages of mitosis are defined based on the characteristics features that appear.

There are four major stages of mitosis: Prophase, Metaphase, Anaphase, and Telophase.

Prophase: First stage of mitosis. The thin extended chromatin that was replicated during interphase begins to gradually coil, thicken, and condense into compact separate chromosomes, each chromosome consisting of two sister chromatids. This packs the long DNA molecule into smaller rod-like units that can be separated successfully in mitosis. At the same time, the nuclear membrane disappears and the nucleolus shrinks, shutting down its activities. Spindle formation begins between the two centrosomes as they move towards the opposite ends of the cell. Spindle is a structure made of microtubules that radiates from the two spindle poles in each end of the cell, and the individual spindle fibres interact with the centromere of each chromosome to facilitate their separation in the later stages of mitosis. So, the prophase stage is defined by three major events: dismantling of the nuclear membrane and nucleolus, chromosome condensation, and spindle formation (Figure 9.5).

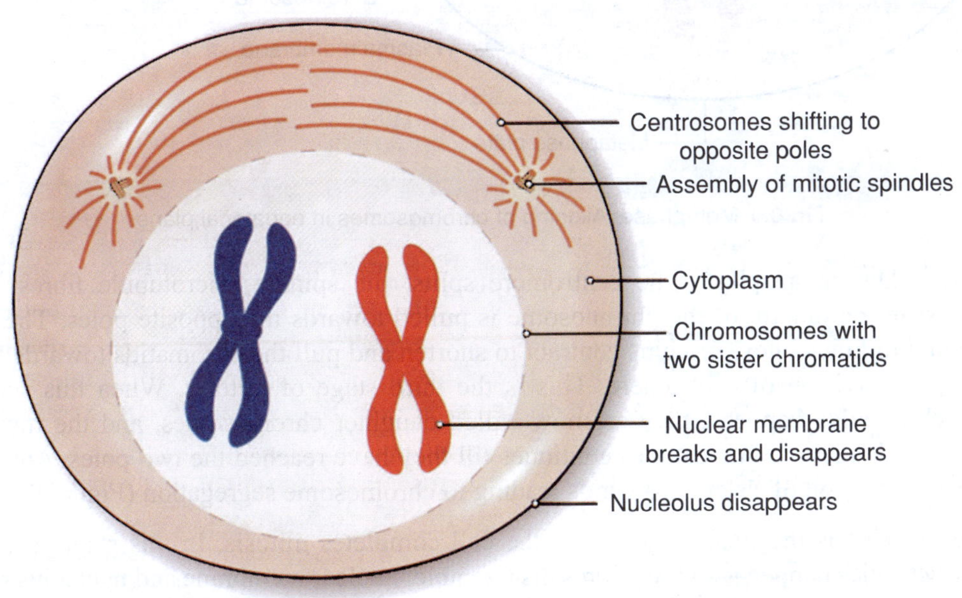

Fig. 9.5 Prophase: Activities of the first stage of mitotic cell division

Metaphase: During this second phase of mitosis, the completed spindle from prophase bound to the centromere of chromosomes aligns the replicated chromosomes across the middle or in the equatorial plane. Centromere is the region of the chromosome that has a complex of several proteins assembled on it and forms a structure called kinetochore. Kinetochore is a complex of proteins associated with the centromere of a chromosome during cell division to which the microtubules of the spindle attach. So, the metaphase stage is defined by the lineup of chromosomes as a single file in the spindle midpoint across the middle of the cell. In this stage the chromosomes are still as two sister chromatids, each one attached to a spindle from the opposite pole, ensuring their separation in the next stage (Figure 9.6).

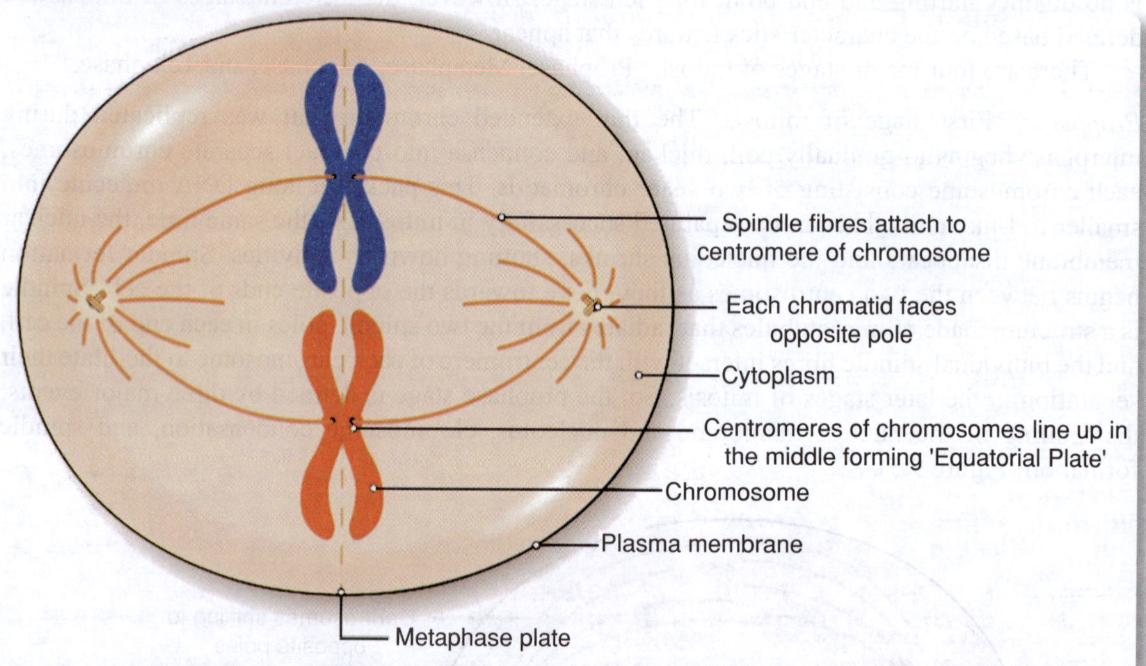

Fig 9.6 Metaphase: Aligning of chromosomes in equatorial plane

Anaphase: During anaphase, the centromere splits, the spindle microtubule fibres shorten, and each sister chromatid of the chromosome is pulled towards the opposite poles. The spindle fibres bound to kinetochore proteins contract to shorten and pull the chromatids towards the pole following the division of centromere. This is the third stage of mitosis. When this separation of sister chromatids happens, they are now called daughter chromosomes, and the movement/separation of daughter chromosomes continues till they have reached the two poles. Anaphase is defined by separation of sister chromatids leading to chromosome segregation (Figure 9.7).

Telophase: This is the final stage where the cell completes mitosis. In this stage, reversal of prophase activities happen—spindle fibres disassemble, nuclear membrane and nucleolus reforms, and chromosomes uncoil. This brings back the genetic material to its extended state as found during interphase. A new nuclear envelope is formed around the daughter chromosomes at the two poles, and at this point the cell has completed the nuclear division or karyokinesis. Two daughter

Contraction of spindle fibres pulls sister chromatids apart
Centrioles
Centromeres also divide one for each sister chromatid
Chromosome
Cytoplasm
Plasma membrane

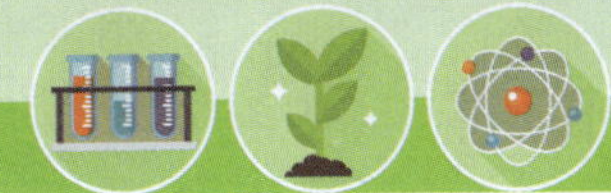

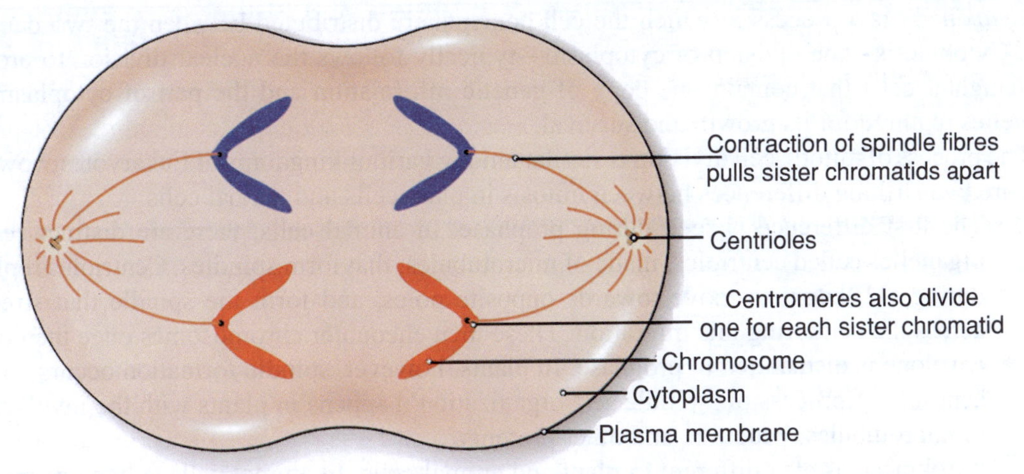

Fig. 9.7 Anaphase stage of mitosis

nuclei having each with identical genetic information to the parent cell is formed, which needs to be separated into two daughter cells by cytokinesis or cell splitting (Figure 9.8).

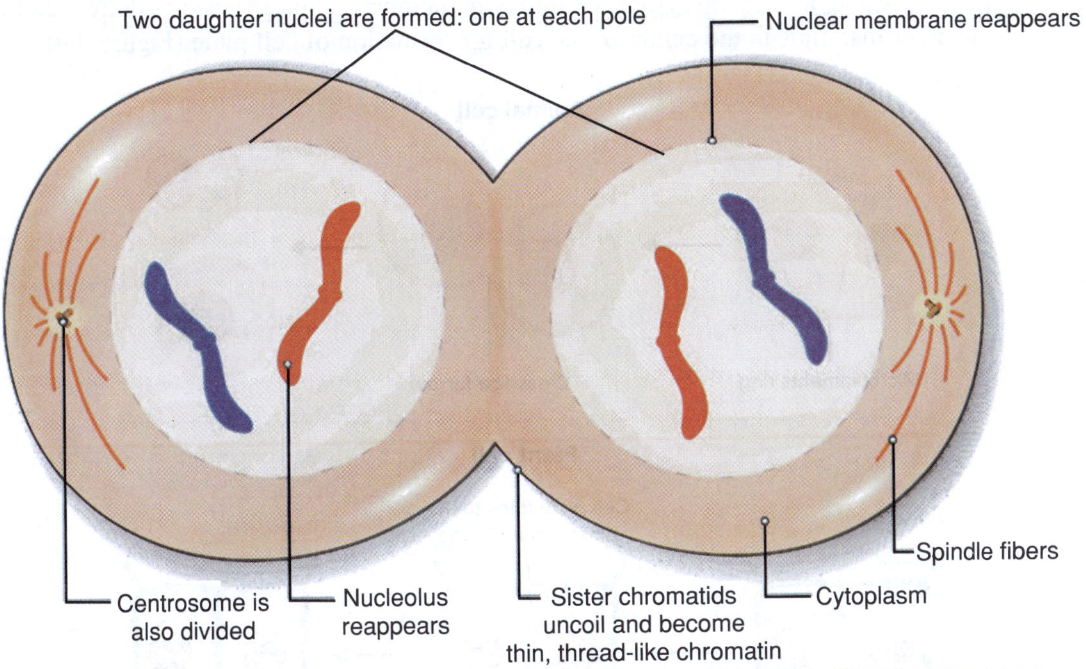

Two daughter nuclei are formed: one at each pole
Nuclear membrane reappears
Spindle fibers
Centrosome is also divided
Nucleolus reappears
Sister chromatids uncoil and become thin, thread-like chromatin
Cytoplasm

Fig. 9.8 Telophase of mitosis

Cytokinesis is a process in which the cell contents are distributed between the two daughter cells. Cytokinesis—the division of cytoplasm—typically follows the nuclear division to produce two daughter cells that contain one copy of genetic information and the part of cytoplasm and organelles required for its growth and survival.

The process of mitotic cell division is similar among various kingdoms of Eukaryotes; however, there are two striking differences between mitosis in plant cells and animal cells.

1. The first difference is seen during prophase. In animal cells, there are distinct cellular organelles called centrioles, made of microtubules, that form spindles. Centrioles replicate during G2, begin to move towards opposite poles, and form the spindle that stretches across the cellduring this migration. These then encounter chromosomes once the nuclear envelope is dismantled in prophase. In plants, however, spindle formation occurs without centriole. A process of 'spindle self-organization' happens in plants with the involvement of microtubules, γ tubulin, and other proteins.
2. Cytokinesis is also different in plant and animal cells. In animal cells, a band of proteins encircles the cell at equator to form a cleavage furrow by indentation of the plasma membrane that pinches in towards the centre of the cell and thus splits cytoplasm into two. This can be visualized as a rubber band put in the middle of an inflated balloon and the way it constricts and splits the balloon into two smaller ones. So, cytokinesis in animal cells starts at the plasma membrane and moves to the centre. In plant cells, though a new cell wall called cell plate forms at the centre of the cell that grows laterally to the plasma membrane and divides the cytoplasm into two, membrane-bound vesicles deliver cellulose and other materials to the centre of the cell for formation of cell plate (Figure 9.9).

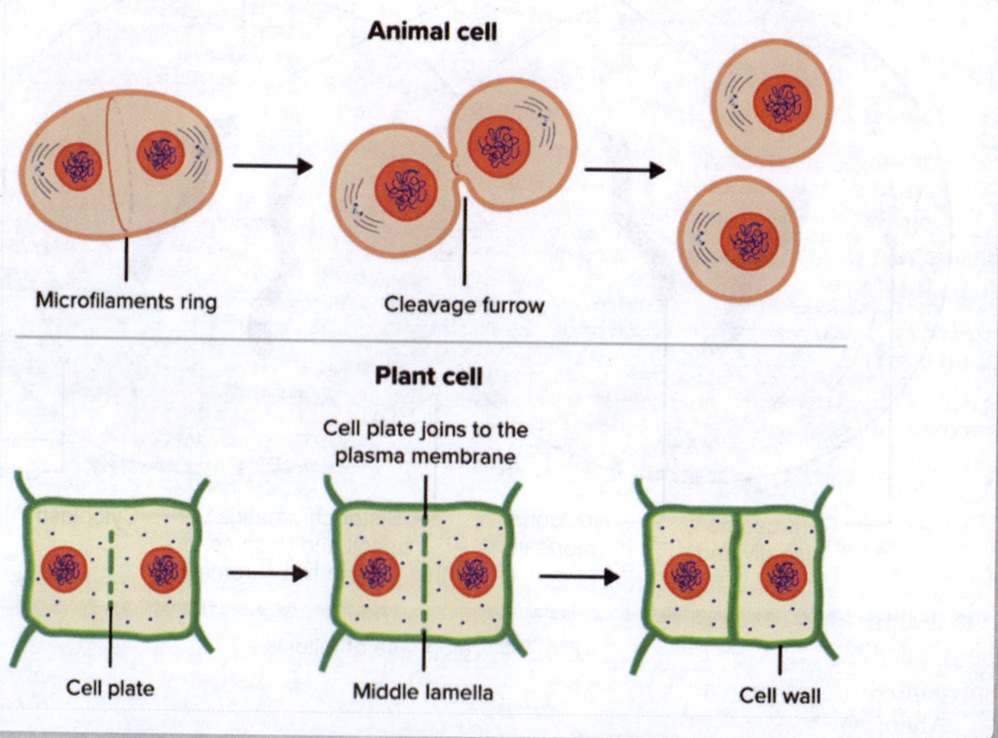

Fig 9.9 Cytokinesis in animal cell versus plant cell

Experimental Insight 9.1 Mitosis Cell Division

Where can we see mitosis in action? We have all observed that when we cut our skin, new skin grows over a period and the wound is healed. The mitotic division of the cells at the periphery of the wound leads to new cells that cover the wound. Skin cells are one of the most mitotically active cells of the body.

Similarly, in plants, roots continuously grow in search of water and nutrients. Onion (*Allium cepa*) root tip experiment is a common hands-on exercise for students to study cell cycle and observe mitosis under a microscope. The cells in the tip of onion root can be studied by taking a thin slice and staining the chromosomes with acetocarmine stain. All four stages of mitosis—Prophase, Metaphase, Anaphase, and Telophase can be identified in these cells under a light microscope.

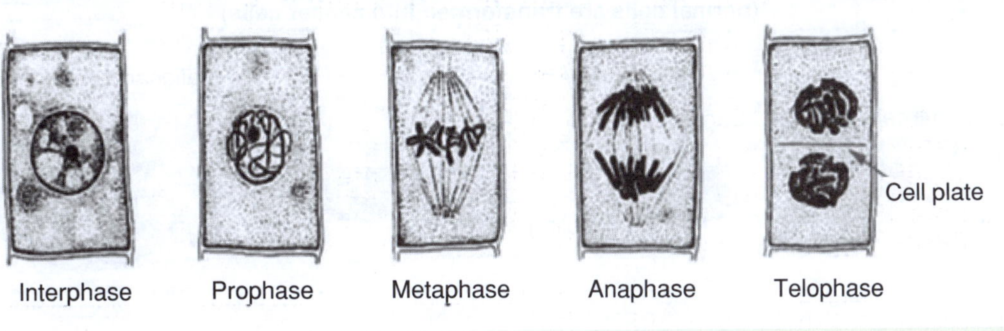

Regulation of Mitosis

Regulatory signals trigger each phase of the cell cycle, ensuring accuracy and order and that one phase is completed successfully before the next one begins. Cell cycle regulation is a critical component that also controls the rate of division and helps to maintains cell numbers in the body.

Cell division in Eukaryotes is well-regulated. Every stage of cell cycle has mechanisms to ensure that proper sequence of events is completed accurately before the cell moves to the next stage. Cell cycle control proteins monitor the activities of cell cycle at specific points called cell cycle checkpoints. Several factors such as poor genetic health or damage to DNA, crowded cell numbers, and wrong environment signal the cell from proceeding with division.

The cell cycle control proteins fall into two categories—Proto-oncogenes that provide signals encouraging cell division and tumour suppressor genes that discourage cell division. A healthy cell receives signals from both these groups of proteins to make a controlled decision on division.

Abnormal Mitosis and its Implications

Uncontrolled abnormal mitosis leads to immortal cells that cause cancer. Excess cells produced due to this result in tumours that compete for space and resources and interfere with the normal body functions. Benign tumours are localized in nature, while malignant ones can metastasize and spread to different regions.

Abnormal mitosis results due to mutations in cell cycle control genes. Mutations can be inherited or caused by external agents called mutagens. Mutagenic agents that cause cancer are called carcinogens.

Gain-of-function mutations in proto-oncogenes result in 'oncogenes' that produce hyperactive or overexpressed proteins, and these abnormal high levels of proteins stimulate abnormal and uncontrolled cell division. For example, oncogenes - *fos*, *myc*, *jun*, and growth factors. Gain-of-function mutations are impactful even if the mutation is seen in one copy of the gene. On the other side, loss-of-function mutations in tumour suppressor genes fail to prevent and discourage cell division. For example, *p53*, *pRB*-retinoblastoma, etc. Loss-of-function mutations can be effective only if both copies of the gene are mutated, which makes the protein non-functional. When both these cell cycle control genes are impacted, the mitotic cell division becomes uncontrolled, leading to increased division and cancer. Cancer is an outcome of excess activity of oncogene and a loss or lack of activity of tumour suppressor gene (Figure 9.10).

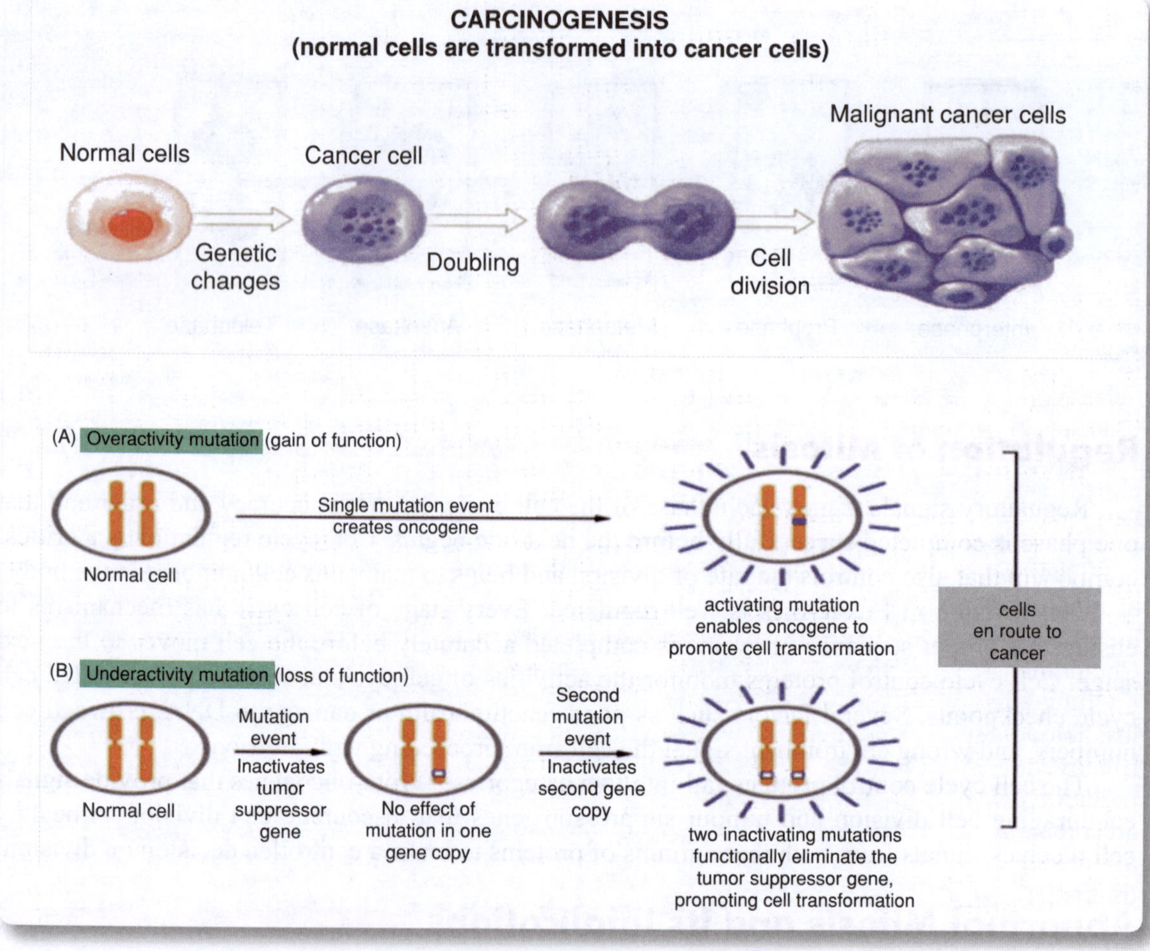

Fig. 9.10 Cancer-causing mutations and carcinogenesis

Some viruses can also cause cancer by inserting their DNA into the host genome. If the insertion happens in a cell cycle regulating the gene, it disrupts the cell's ability to control division and can also result in cancer. For example, herpes simplex virus (HSV-1); human papillomavirus (HPV), etc.

Interesting Fact 9.1

Policing by tumor-suppressor protein—*p53*

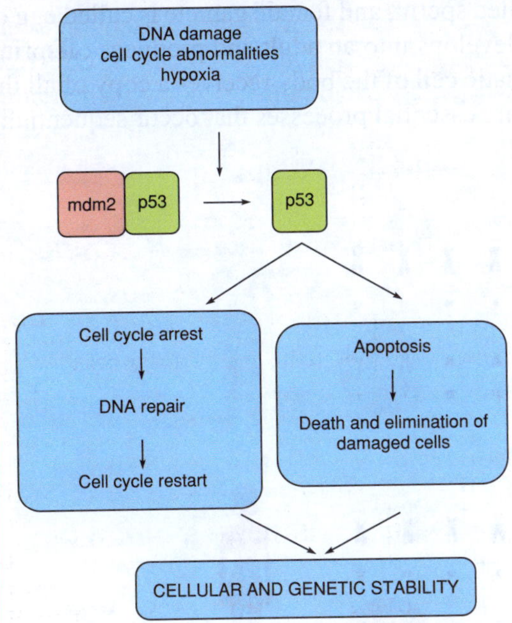

p53 prevents cancer by preventing the genetically abnormal cells from dividing. Hence tumor suppressor.

- Responsible for monitoring DNA damage during mitosis in G1/G2 stage.
- If DNA is healthy, allows the cell to proceed with division.
- If DNA damage is detected, it immediately does the following three things:
  - causes a cell cycle arrest
  - triggers DNA repair proteins and initiates repair of the DNA
  - initiates self-destruction of the cell by Apoptosis or Programmed cell death, if DNA is beyond repair.
- Then why do cancers happen?
- *p53* is also a protein coded by a gene. If mutations are created in *p53* gene by carcinogens, and the *p53* becomes non functional, the tumor suppressive protection is lost.
- The solace is cells have two copies of *p53*. Loss of function mutations of *p53* have to inactivate both the copies for the protection to be lost.
- Combination of *p53* mutations along with oncogenic mutations are a cause of most cancers.

Treatment Strategies

Traditionally, surgery, chemotherapy, and radiation are usually the three treatment options available for cancer. If cancer is localized, it can be removed surgically; however, invasive and metastasised tumours are difficult to be removed by surgery without impacting normal tissues. Chemotherapy uses chemicals to kill cancer cells, but some have side effects because normal dividing cells of the body, such as intestinal cells, skin cells, etc., are also harmed, causing hair loss, blood vomits, etc., and weaken the immune system. Radiation kills cancer cells by inflicting irrecoverable damage to its DNA by enhancing the existing genetic instability. Harmful effects of radiation on healthy cells can be minimized by targeting only cancer cells, and the minor damages are repaired by the DNA repair mechanisms of the body even if the DNA of normal cells are affected. New strategies based on emerging molecular knowledge has also opened up alteration to trait and therapies. Cancer research is one of the most important part of the health research.

Sexual Reproduction and Meiosis

Meiosis is a critical component in sexually reproducing organisms and is responsible for production of gametes. Humans and many large eukaryotes are diploid, and they contain two sets of chromosomes inherited from each parent. This diploid nature of the genome is a key factor in sexually reproducing organisms. All body cells contain pairs of homologous chromosomes—one inherited from mother and the other from the father. Every chromosome has a matching set of diploid genomes. In humans, there are 23 pairs of chromosomes, of which 22 pairs are autosomes and 1 pair is sex chromosome (Figure 9.11).

In multicellular sexually reproducing organisms, continuity of life happens, wherein adults of one generation reproduce and give birth to the next generation of offspring. In each adult, diploid germ cells help in producing haploid gametes by a special type of cell division called meiosis or reduction division. These gametes fuse by fertilization to produce the diploid zygote, which results in offspring in the next generation. Male gamete is called sperm, and female gamete is called egg or ovum. The life cycle is continued when this zygote develops into an adult and produces offspring in the next generation. Mitosis ensures that every somatic cell of the body receives a copy of all the zygote's 46 chromosomes. Meiosis and mitosis thus are essential processes that occur sequentially in the life cycle of organisms.

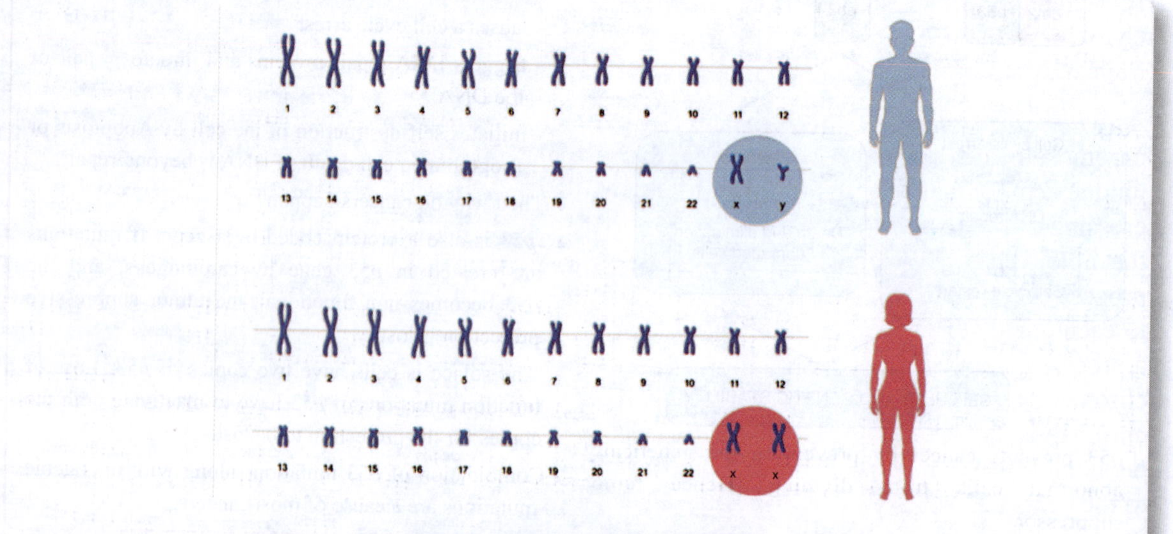

Fig. 9.11 Human karyotype* of a female showing 22 pairs of autosomes and 1 pair of sex chromosomes (XX)

*A karyotype refers to a laboratory-produced image of a person's chromosomes isolated from an individual cell and arranged in numerical order—an individual's complete set of chromosomes.

Stages of Meiosis

Meiosis occurs in two parts: Meiosis I and Meiosis II. It has a premeiotic interphase similar to pre-mitotic interphase that goes through G1, S, and G2 stages where DNA replicates in S phase. After S phase, each chromosome has two identical sister chromatids before meiosis begins subsequently after G2.

After interphase, the cells enter meiosis, which begins with meiosis I. Both the phases of meiosis have the same four stages like mitosis for division. So, Meiosis I has Prophase I, Metaphase I, Anaphase I and Telophase I, which produces two cells that then undergo Meiosis II: Prophase II, Metaphase II, Anaphase II, and Telophase II to produce four cells after completion of meiosis. Cytokinesis happens both after Meiosis I and Meiosis II. Thus, by the two-part division process of Meiosis, one parent cell produces four daughter cells or gametes.

Activities in Meiosis I: In Prophase I, the same events as mitosis occur, which includes chromosome condensation, spindle formation, and nuclear membrane disassembly. However, one unique activity that is different from Prophase of mitosis is in Prophase I of meiosis synapsis and crossing over occurs, which was absent in Prophase of mitosis. The homologous chromosomes of the diploid pair with their sister chromatids recognize each other and line upsideside by sidethrough process called synapsis, where the four chromatids (2 from each chromosome) of tetrad lie close to each other. When they are paired at synapsis, the chromatids physically exchange segments by a process called 'crossing over'. In this process, the enzymes break and rejoin DNA segments of chromatids with equivalent information. This exchange of genetic material between non-sister chromatids leads to mixing of characteristics between the maternal and paternal chromosomes of the homologous pair. Crossing over is important as it helps in mixing of genes from one generation to the next (Figure 9.12).

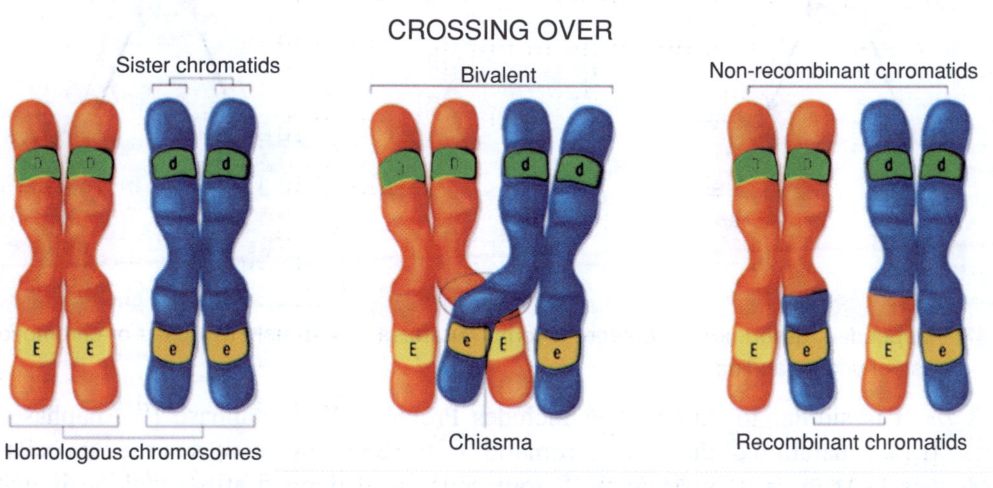

Fig. 9.12 Synapsis and crossing over in Prophase I

Prophase I is followed by Metaphase I—the homologous chromosomes are aligned at the equatorial plane. Each chromosome of the diploid pair is next to each other; therefore, there will be two rows containing homologous chromosomes bound to the spindle in Metaphase I. In humans,

the diploid number of 46 chromosomes is arranged in two rows, and, hence, there will be 23 chromosomes in each row—23 pairs that lie next to each other. Now the attached spindle will separate these homologous chromosomes to opposite spindle poles in Anaphase I. This will result in 23 chromosomes moving to one spindle pole and 23 to the other. Segregation of these homologous pairs will deliver one-half of the diploid number of chromosomes to each spindle pole. Remember each chromosome is still with the sister chromatids, and they have not yet been separated. Unlike mitosis, the centromeres do not divide yet and keep the sister chromatids bound together.

When the 23 pairs are separated, one homologous pair does not influence the other chromosomal pairs; therefore, they are said to assort independently in Anaphase I. Depending on how the pairs were positioned in Metaphase I, the maternal chromosome may go to one pole and paternal to other in one chromosome pair, whereas it may be reversed in others. This is called the independent assortment of chromosomes. Telophase I is replay of mitosis, wherein the spindle fibres disassemble, chromosomes uncoil, nuclear membrane reforms, and cytokinesis occurs to form two daughter cells.

Meiosis I results in reduction division, wherein each daughter cell receives one-half of the genetic information of the parent cell. In humans, after Meiosis I, each daughter cell will have 23 chromosomes (Figure 9.13).

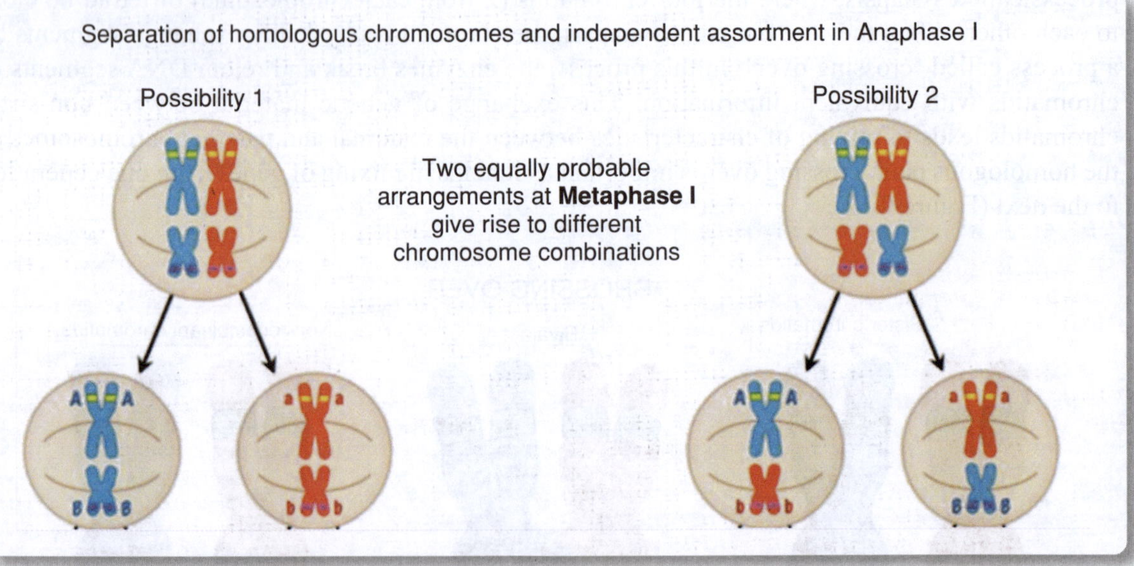

Fig. 9.13 Independent assortment and segregation in Anaphase I with only two pairs of chromosomes for ease of understanding

Meiosis II: It is similar to mitosis and includes Prophase II, Metaphase II, Anaphase II, and telophase II for separation of the sister chromatids of the 23 chromosomes in the two cells formed after Meiosis I. At the end of Meiosis II, four cells are generated after cytokinesis, with each

daughter cell containing 23 chromosomes. Key difference in Anaphase I and Anaphase II is that Anaphase I separates homologous chromosomes to reduce it from 46 to 23 chromosomes, while Anaphase II separates the sister chromatids of these 23 chromosomes in the newly formed four cells (Figure 9.14).

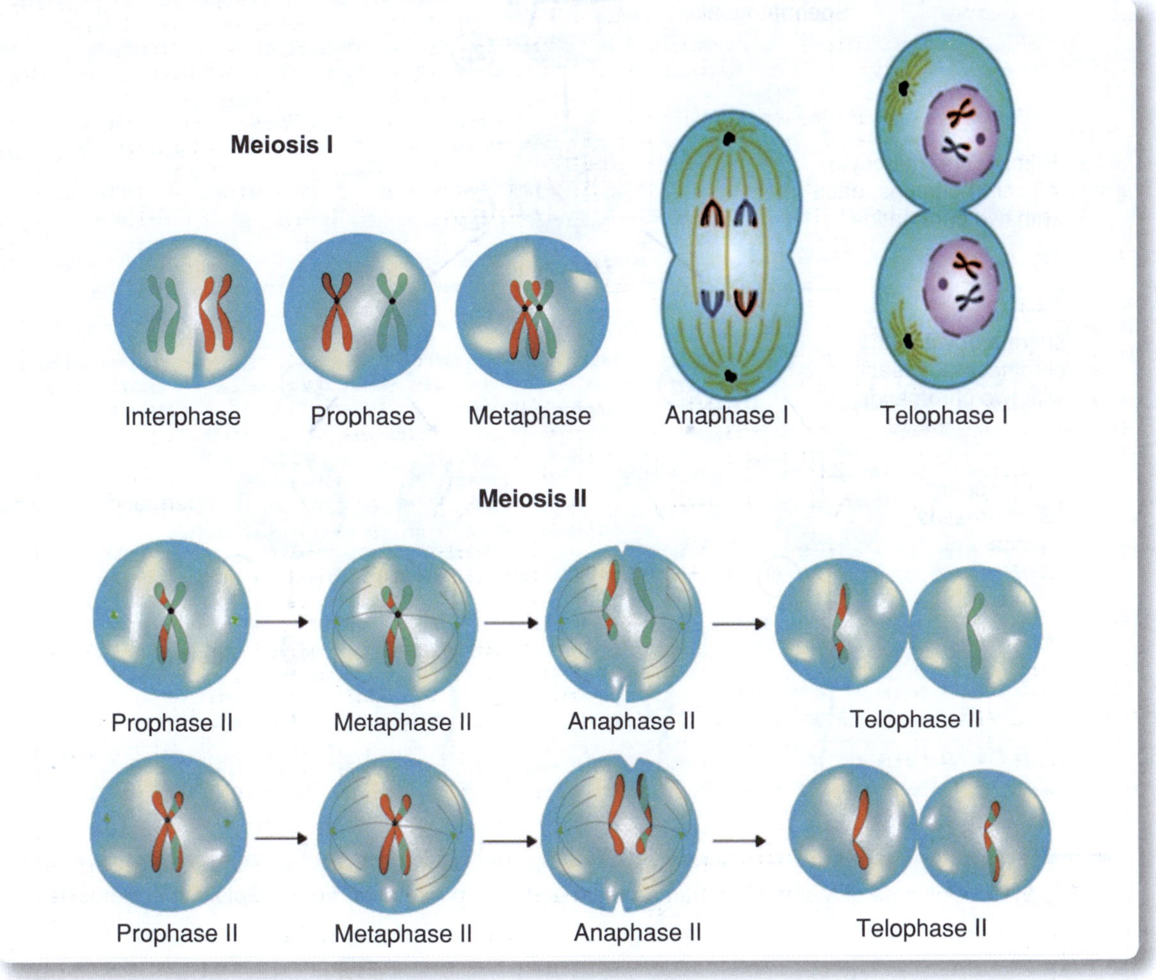

Fig. 9.14 Different stages of Meiosis: Meiosis I + Meiosis II. The steps are explained using two pairs of homologous chromosomes

In humans, after Meiosis, there is reduction from the diploid (46) to haploid (23) chromosomes in gametes. Only one complete set or copy of genetic material is present in the four cells produced after Meiosis. These four cells thus result in formation of haploid gametes completing the process of Meiosis. The formation of haploid sperm from the diploid spermatocytes in humans is depicted below (Figure 9.15).

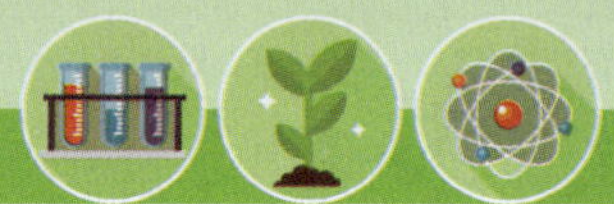

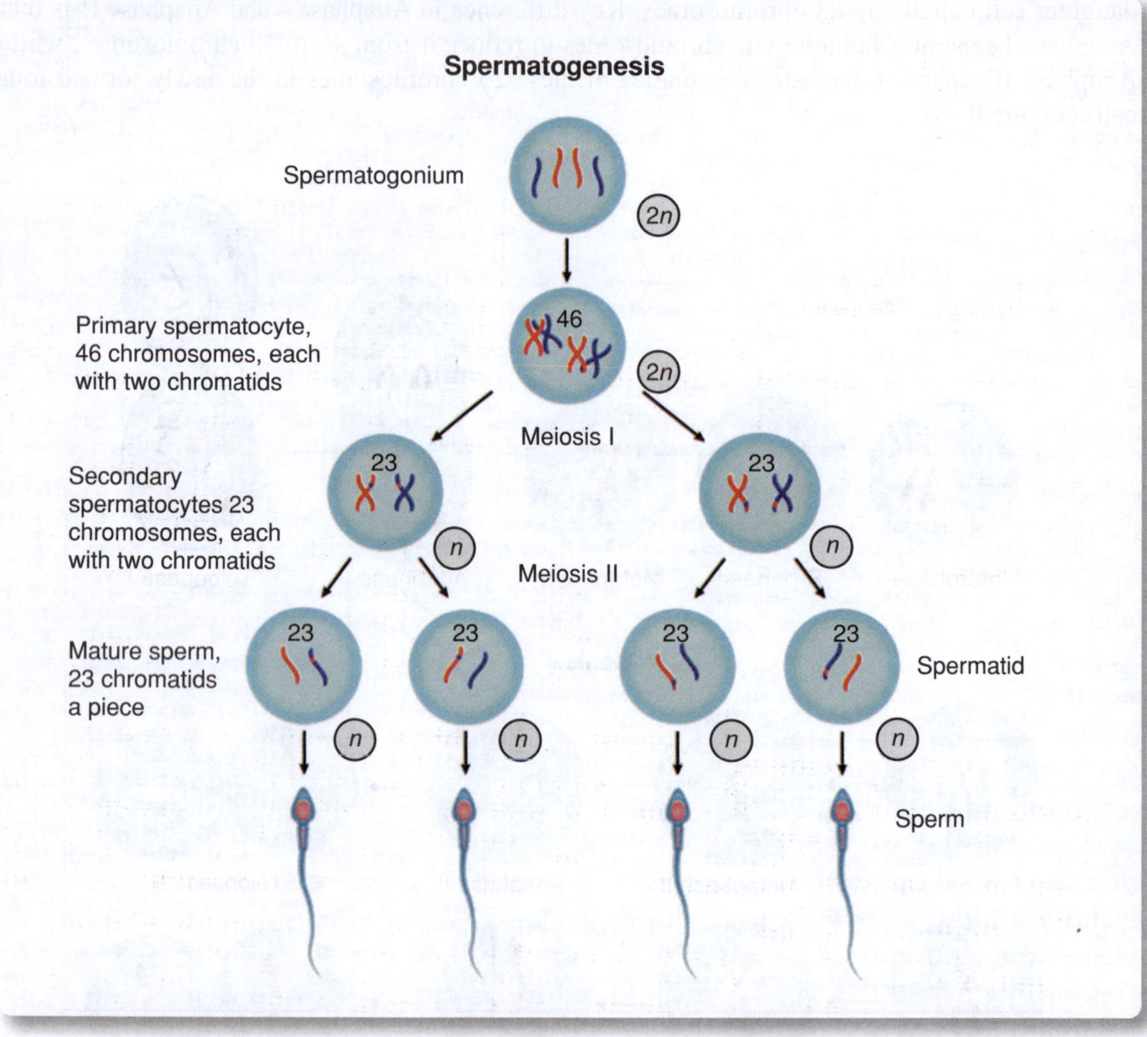

Spermatogenesis
Spermatogonium
2n
Primary spermatocyte, 46 chromosomes, each with two chromatids
46
2n
Meiosis I
Secondary spermatocytes 23 chromosomes, each with two chromatids
23
23
n
n
Meiosis II
Mature sperm, 23 chromatids a piece
23
23
23
23
Spermatid
n
n
n
n
Sperm

Fig. 9.15 Outcome of Meiosis in a human male and sperm formation from a diploid spermatocyte

2n
2n
mitosis
2n
2n
Two diploid daughter cells
2n
2n
meiosis I
1n
1n
meiosis II
1n
1n
1n
1n
Four haploid daughter cells

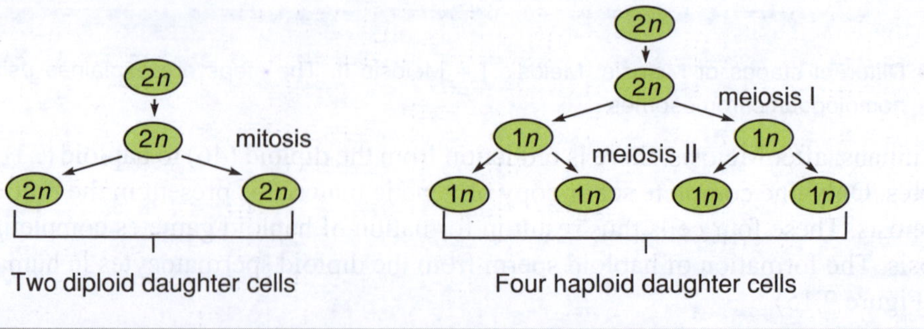

Mitosis	Meiosis
One-step division process	Two-step division process—Meiosis I and Meiosis II
No synapsis and crossing over	Homologous chromosomes synapse and crossing over occurs in Prophase I
Centromeres divide in anaphase	Centromeres divide in Anaphase II, not in Anaphase I
Chromosome number is retained daughter cells have same chromosome number as parent	Reduction division chromosome number reduced to half in daughter cells
Occurs in somatic cells for growth, repair, and replacement of cells	Occurs in reproductive cells to generate gametes

Mechanisms for Genetic Variation in Sexually Reproducing Organisms

Production of gametes by meiosis followed by fertilization creates genetic variation in sexually reproducing populations. Haploid gametes combine to form new, unique combinations of genetic information in the offspring. While the genes for the same basic traits are present in all gametes, creation of new alleles or new combinations of alleles is responsible for genetic diversity. Five factors contribute to this: mutations, crossing over, segregation, independent assortment, and fertilization.

Mutations can either change nucleotide sequences producing different proteins or can rearrange genes at chromosomal level, thus creating new alleles and increasing genetic diversity.

Crossing over is another source of variation that allows new combinations of genetic information to be generated. Crossed over chromosomes in gametes contain a mix of father's and mother's genes, hence a combination of traits is inherited in one piece of DNA. The closer genes will be linked, and the farther genes have a higher chance of getting separated during crossing over. Crossing over can occur in all chromosomes at multiple loci; hence great variation is possible.

Segregation and independent assortment of chromosomes: It differentiates gametes in terms of their characteristics and helps in producing offspring that have combinations different from their parents. The differing combinations of maternal and paternal chromosomes segregated to poles during anaphase I is another source of variation. Independent assortment greatly increases the diversity of gametes, because the number of possible types of gamete cells will increase based on chromosome combinations. For example, with two chromosomes, four possible combinations can be seen; with three chromosomes, eight combinations can be seen; and so on, with 23 pairs of chromosomes in humans: $2^n = 2^{23} = 8{,}388{,}608$, i.e., nearly 8 million different types of sperm or egg cells are possible by independent assortment alone. The number of unique gametes is huge

when incorporating mutation and crossing over. Humans produce millions of genetically different gametes; any of these can fuse with each other during *fertilization*, so the combination of zygote and hence the possible off springs is almost infinite. Hence, each individual born is genetically unique, except for monozygotic identical twins. Thus, meiotic cell division contributes significantly along with sexual reproduction towards genetlc variation and population diversity.

Abnormal Meiosis and its Implications

Accidents during meiosis can alter chromosome number in the off spring produced. Normally, during meiosis, every gamete should have one copy of all chromosomes (n) or haploid. However, if mishap occurs and the homologous chromosomes during reduction division do not segregate properly during meiosis, and if both chromosomes end up in the same gamete, it results in **nondisjunction** of chromosomes. In such a case of non-disjunction during Meiosis I, 50% of gametes will have two copies of the chromosome ($n = 24$, both in one gamete) and 50% of gametes will not have the chromosome at all ($n = 22$). Instead of homologous chromosomes of a pair segregating as one in each daughter cell, both have gone to one daughter cell and the other daughter cell is devoid of the chromosome from that pair. If such an abnormal gamete unites with a normal gamete, then the resulting zygote and the offspring produced from it will have abnormal number of chromosomes. Normally, zygote will have 46 chromosomes: 23 from male gamete and 23 from the female gamete.

Due to nondisjunction, two conditions can result in the zygote and the resulting offspring: monosomy or trisomy. Monosomic zygotes have only one copy of chromosome. This happens when the gamete without a chromosome ($n = 22$) fertilizes with a normal gamete ($n = 23$) with one copy. All the cells produced by mitosis for development of this individual will have one copy of that chromosome, so the abnormal chromosome number of those diploid individuals will be $22 + 23 = 45$ ($2n - 1$), instead of the normal 46 ($2n$). In trisomy, the zygote carries three copies of a specific chromosome. This happens when the gamete with two copies ($n = 24$) fertilize with a normal haploid gamete ($n = 23$) with one copy. Trisomic individuals have an abnormal diploid chromosome number of $24 + 23 = 47$ ($2n + 1$), instead of normal 46 ($2n$).

Most of such abnormal zygotes with atypical number of chromosomes are usually aborted. However, nondisjunction in chromosome number 21 produces a condition called Down syndrome in offspring. Individuals with Down syndrome have an extra chromosome 21 in them, are trisomic with three copies of 21 chromosomes present in them, and their diploid chromosome number is 47 or $2n + 1$ (Figure 9.16(a)). Characteristic symptoms of Down syndrome include thickened eyelids, large tongue, flattened facial features, mental impairments, with short stature and faulty speech. Frequency of Down syndrome was higher in child births due to late pregnancies in older women in their late 30s. Nondisjunction can also result in sex chromosomes leading to abnormal numbers of X and Y chromosomes in individuals. Klinefelter syndrome (XXY, male) and Turner syndrome (X0, female) are examples of non-disjunction in sex chromosomes (Figures 9.16(a,b)). These individuals can have some abnormalities in their reproductive characteristics and appearance but otherwise survive normally.

Down syndrome

(a)

Turner syndrome

(b)

Klinefelter syndrome

(c)

Fig. 9.16 Abnormal chromosome numbers due to nondisjunction. a–c Karyotypes of Down syndrome, Turner syndrome, and Klinefelter syndrome

Experimental Insight 9.2 Karyotyping of Chromosomes

A karyotype refers to a laboratory-produced image of a person's chromosomes isolated from an individual cell and are arranged in numerical order.

A human cell is broken open in the metaphase of mitosis, stained with dyes, a picture taken with the aid of a microscope, and chromosomes are arranged in matching pairs by size, banding pattern, and centromere positions as guides to generate a karyotype (Figure 9.17).

Karyotypes are used for cytogenetic studies: Prenatal diagnosis, tumour studies, study of genetic aberrations, capturing chromosomal nondisjunction events, etc.

Karyotyping can be performed on a variety of tissues, including blood, bone marrow, amniotic fluid, or placental tissue.

Human male and female karyotypes can be easily distinguished by the sex chromosomes: XX for females and XY for males.

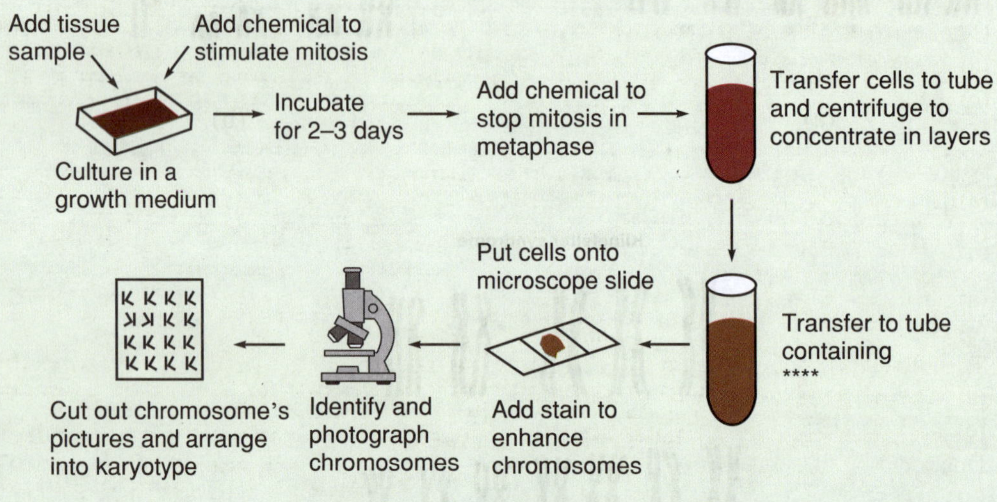

Fig. 9.17 The method of Karyotyping

SUMMARY

- Cell division is necessary for maintaining cell numbers in body for growth, repair, and production of gametes.
- There are two types of cell division: Mitosis and Meiosis.
- The cell cycle includes the three stages of Interphase: G1, S, G2, and M or Mitosis stage. G1 and G2 stages are growth or Gap stages, whereas DNA replicates in S phase.
- Mitotic cell division involves karyokinesis and cytokinesis. It has four stages: prophase, metaphase, anaphase, and telophase. Cytokinesis is the division of cytoplasm.
- Abnormal mitosis or uncontrolled division leads to cancer.
- Meiosis forms gametes in reproductive cells.

- Meiosis has two steps: Meiosis I and Meiosis II. These include Prophase I, Metaphase I, Anaphase
- I and Telophase I + Prophase II, Metaphase II, Anaphase II, and Telophase II. Cytokinesis happens in both Meiosis I and II.
- Meiosis generates genetic variation in sexually reproducing organisms.
- Abnormality in meiosis causes nondisjunction of chromosomes during gamete formation and alters chromosome numbers in the offsprings of next generation.
- Down syndrome is an example of an autosomal nondisjunction event, whereas Turner syndrome and Klinefelter syndrome have nondisjunction in sex chromosomes.

EXERCISES

A. Objective-Type Questions

1. Which of these is not an activity of Prophase?
   (a) Nuclear envelope breakdown
   (b) Spindle assembly
   (c) Condensation of chromosomes
   (d) Separation of daughter chromosomes to opposite spindle poles.
2. Cytokinesis in plant cell occurs by:
   (a) cleavage furrow. (b) formation of cell plate.
   (c) both cleavage furrow and cell plate. (d) cytokinesis does not happen in plant cells.
3. If a diploid cell ($2n = 78$) undergoes mitosis, the diploid chromosome number in the daughter cells will be:
   (a) 78. (b) 39.
   (c) 156. (d) None of the above.
4. Number of chromatids in Metaphase of a mitotically dividing human cell is:
   (a) 46. (b) 92.
   (c) 23. (d) 0.
5. Trisomic individuals have ________________ chromosomes in number
   (a) 47 (b) 46
   (c) 45 (d) $2n - 1$
6. Down syndrome is due to chromosome non-disjunction in chromosome number______.
   (a) 21 (b) Xchromosome
   (c) ychromosome (d) 3
7. __________________ is responsible for mixing of characteristics in Prophase I of Meiosis.
   (a) Condensation of chromosomes (b) Crossing over of chromosomes
   (c) Chromosomal segregation (d) Independent assortment of chromosomes
8. Number of cells formed at the end of meiosis from one parent cell is:
   (a) 2 cells. (b) 4 cells.
   (c) 0 cells. (d) 23 cells.
9. X0 is a monosomic condition called as ________________
   (a) Down syndrome. (b) Turner syndrome.
   (c) Klinefelter syndrome. (d) Cancer.
10. Human sperm cells are ___________ and have _______ chromosomes.
   (a) Diploid, 23 (b) Diploid, 46
   (c) Haploid, 23 (d) Haploid, 46

11. Which of the following is true for the G_0 phase?
    (a) It is a non-dividing quiescent phase.
    (b) DNA replication happens in this phase.
    (c) It is part of mitosis.
    (d) Reduction division is hallmark of this phase.
12. Chromosomes appear to be lining up near the middle of the cell during which phase of mitosis?
    (a) Prophase
    (b) Metaphase
    (c) Anaphase
    (d) Telophase

B. Short-Answer Questions

1. How does *p53* prevent cancer?
2. What is the difference in cell division between skin cell and germ cell?
3. What is the fundamental difference between Anaphase I and Anaphase II of Meiosis?
4. A cell has 44 chromosomes. After mitosis and cell division, one daughter cell has 45 chromosomes and the other has 43 chromosomes. Which stage of mitosis was faulty to produce this result?
5. Can agents that disrupt spindle fibres be a good chemotherapeutic agent for cancer? Explain your answer.

FURTHER READINGS

1. https://www.news-medical.net/life-sciences/The-Role-of-Cell-Division-in-Tumor-Formation.aspx
2. https://www.bbc.co.uk/bitesize/guides/zws8h39/revision/4
3. https://www.hopkinsmedicine.org/news/newsroom/news-releases/2024/05/scientists-track-doubling-in-origin-of-cancer-cells
4. https://www.yourgenome.org/theme/mitosis-versus-meiosis/
5. https://pmc.ncbi.nlm.nih.gov/articles/PMC2732749/

USEFUL WEBLINKS

1. https://www.youtube.com/watch?v=QVCjdNxJreE
2. https://www.youtube.com/watch?v=VzDMG7ke69g
3. https://www.youtube.com/watch?v=zrKdz93WlVk
4. https://www.youtube.com/watch?v=4bzY9e-YQqI

PART II HUMAN BIOLOGY

CHAPTER 10

Structural Organization and Internal Control in Humans

LEARNING OBJECTIVES

After reading this chapter, you will be able to:

- Understand the levels of structural complexity
- Get an understanding of tissues, organs, and organ systems
- Study concepts of internal control or homeostasis
- Learn about mechanisms for maintaining homeostasis

CHAPTER TAKEAWAY **Introduction to human systems and their control**

This chapter is the introductory chapter for understanding the structure and function of human systems and their control mechanisms

PRELUDE

Humans fall under the domain Animalia within Eukaryotes. Animals are extremely diverse in terms of their structure and function. While some common features exist, such as they are eukaryotic, multicellular, heterotrophic, and mobile with well-differentiated organ systems, the diversity seen in different categories of animals is huge. However, one common trait as a part of evolution is that their unique structures and adaptations reflect and support their survival in their niches. Just as for a particular task at hand, we may pick the right tool—between a screwdriver and a hammer; nature picks the right design and adaptations for organisms to function. Thus, there is a clear correlation between structure and function. Among animals, mammals are complex and have evolved advanced systems and characteristics that define them. The mimicking of design seen in biological systems to specific products like robots, submarines, and airplanes is an important area of bionics. Hence, understanding animal structure and function is a useful endeavour. Within the animal kingdom, our primary focus will be on human systems—their structure, functions, and mechanisms of internal control.

Levels of Structural Complexity

At the fundamental level, in terms of structure and function, all cells have the same requirements and characteristics, as we have seen in the previous chapters, with variations. Animal cells also have the same requirements as any other cell. They transport ions, molecules, and organic compounds from their external aqueous environment, maintain osmotic balance, carry out oxidative reactions in presence of oxygen for energy needs, release waste products, and maintain the internal cellular environment within certain tolerable limits for them to survive.

Multicellularity in larger animals generated a need to create an internal environment that supplies, ensures, and facilitates the individual needs of cells like nutrient supply, osmotic balance, and waste elimination, etc. This internal stabilized environment facilitates the animal to adapt and survive irrespective of their external environment, which otherwise would have been harsh to their individual cells. Multicellularity also helped animals to become large, though the cells remain small for their effective functioning. Multicellularity also allowed the categorization of major life functions and subdivided them to create specialized groups of cells performing individual functions. Specialized group of cells evolved—one handling movement, other digestion of food, some for reproduction, etc.

In large multicellular organisms, specialized cells are grouped into tissues, various tissues combine to form functional organs, and several organs together form organ systems. A tissue is a group of cells with the same structure and function, whereas an organ integrates two or more different tissues to execute one specific function. In organ systems, individual organs specialized for a particular task collaborate as an integrated unit. To understand these complex systems, it is important to cover two aspects: structural design and functional outcomes. Anatomy is the study of structures of an organism, while physiology is the study of their functions and physico-chemical processes of the organisms.

Tissues, Organs, and Organ Systems

Tissues

Though complex animals have several different cell types forming tissues, they can be classified into one of the four basic tissue types: epithelial tissue, connective tissue, muscle tissue, and nerve tissue. The properties of the individual cells in these tissues determine their structure and function. The structural integrity of the tissue is also dependent on organization of cytoskeleton within the cell, the extracellular matrix surrounding the cell, and the junctions that hold the cells together. Architectural differences among these tissues contribute to structural variations and effective functional outcomes.

Epithelial tissue

Also known as 'epithelium', it refers to the surface layer of cells that covers the surface of the body and lines organs. If the epithelium is made up of a single layer of cells, it is called simple epithelium; and if the epithelium is made of multiple cell layers, it is called stratified. Cells within the epithelium can be squamous (flattened), cuboidal (like dice or cubes), or columnar (brick shaped). The cells of epithelium are closely aligned and tightly packed together, which contributes to its function of protection, absorption, and secretion. They protect the body from abrasions, invasion by pathogens, and aid in secreting and absorbing substances. Epithelial tissue inside our body lines the mouth, nasal cavity, respiratory system, reproductive tract, stomach, and digestive and excretory track. In addition, the outermost layer of skin—the epidermis—is made of epithelial tissue; the blood vessels and secretive glands are also lined by epithelium, which helps in secretion. The body continually renews the epithelial layers, and these are one of the most actively dividing cells of the body. As they are exposed to the environment and divide actively, exposure to carcinogens leads to high frequency of cancers in these cells. Nearly 80% of cancers have an epithelial origin and are called as carcinomas (Figure 10.1).

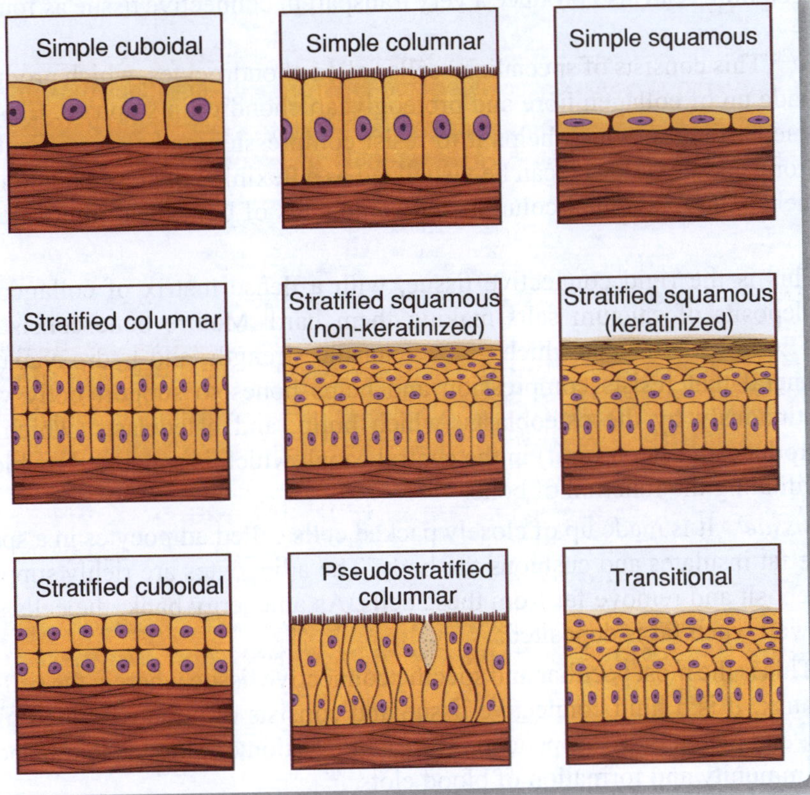

Fig. 10.1 Types of epithelial tissues

Connective tissue

Connective tissue is found throughout the body and has a role in protection of other tissues, transmit mechanical forces, act as filters, and in transportation of cells and molecules. Connective tissue is organized and primarily consists of well-separated cells that are embedded in a non-living matrix called extracellular matrix (ECM) secreted by the cells of the tissue. The matrix material is higher in terms of weight and volume as compared to the cellular counterparts. The properties of the connective tissue are based on the type and quantity of ECM. The ECM is mainly made of proteoglycans and protein-based fibres. The protein fibres of ECM are of different types such as collagen, fibronectin, elastin, etc. Collagen is the most abundant protein in animals. The structure and composition of ECM varies and matches the requirements of each tissue. There are six major types of connective tissues in vertebrates.

Loose connective tissue This is the most widespread connective tissue in the body made of sparsely distributed fibroblasts cells in a network of collagen and other glycoprotein fibres. The main function is to bind and support epithelia to underlying tissues. It appears as a corset-like interwoven band around blood vessels, nerves, and some internal organs and provides great strength and elasticity.

Fibrous connective tissue This includes tendons, which attach muscles to bones, and ligaments, which connect bones to one another. In this connective tissue, collagen and elastin form highly ordered parallel bundles in which the few fibroblasts cells are distributed. The highly

ordered collagen fibres can also produce a very transparent connective tissue as found in the cornea of the eye.

Cartilage This consists of specialized cells called chondrocytes, which produce the cartilage in a matrix made up of collagen fibre and proteoglycan chondroitin sulphate. Elastin may also be present in some cartilages, which helps it to resist compression and stay resilient. The matrix of cartilage is strong and flexible and can be visualized as a flexible boneless skeleton. Human organs like lungs, trachea, larynx, spinal column, and some ends of bones and joints are also supported by cartilage.

Bone This is the rigid connective tissue, with a dense matrix of collagen fibres and the presence of deposits of calcium salts making them hard. Mature bone consists of osteocytes embedded in ECM of collagen, which gives it tensile strength, and hydroxyapatite—a calcium phosphate mineral that resists compression and helps bones to support body weight. Bone is reshaped continuously by the osteoblasts, which build, and osteoclasts, which degrade bone. Osteoblasts produce an osteon (unit) in the central canal, which is supported by blood vessels and nerves for facilitating the function of bone.

Adipose tissue It is made up of closely packed cells called adipocytes in a sparse matrix and stores fat. The fat insulates and cushions the body. The adipocytes are richly supplied with blood vessels that deposit and remove fat from these cells. As an energy bank, the cells shrink when fat is spent and swell when fat is deposited.

Blood This is the most peculiar and specific connective tissue wherein the cells are suspended in a liquid matrix. It is a fluid connective tissue and consists of a matrix called plasma in which RBCs, WBCs, and platelets are suspended. The main function of blood is to transport material and contribute to immunity and formation of blood clots.

From the different types, it is clear that connective tissue shows great diversity from a transparent cornea to a fluid blood to a hard bone (Figure 10.2).

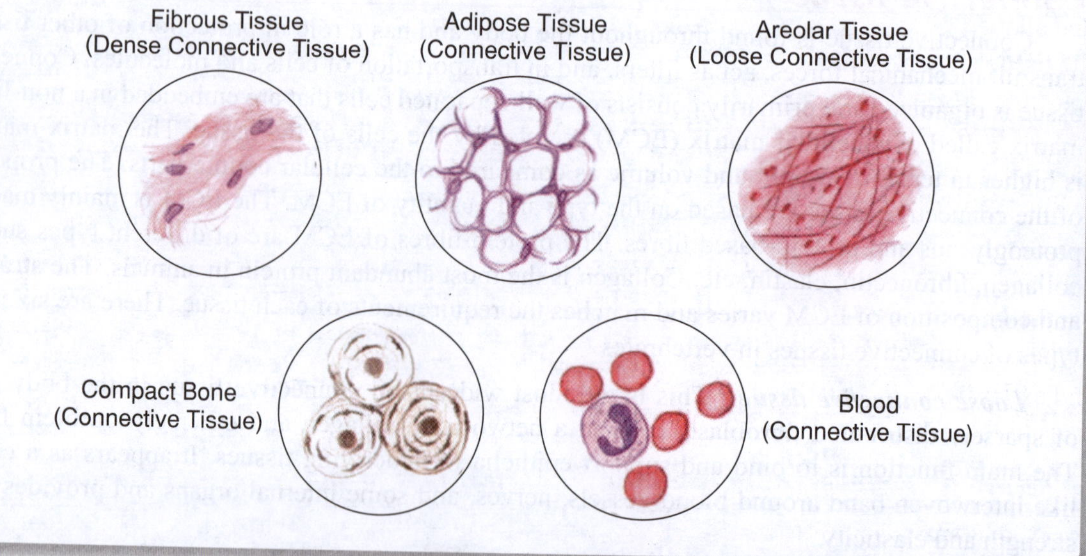

Fig. 10.2 Different types of connective tissues

Muscle tissue

This tissue produces the force for body movements. It is the most abundant tissue in most of the animals and is made up of bundles of long, thin cylindrical cells called muscle fibres aligned in parallel. Within each muscle tissue are myofibrils made up of contractile proteins: actin and myosin. Vertebrate muscles are classified into three types: skeletal muscle, smooth muscle, and cardiac muscle.

Skeletal muscle is attached to your bones by tendons and is responsible for voluntary movements like walking, talking, throwing, etc. Skeletal muscles contract in response to signals from the nervous system. These are also called striated muscles due to the banded appearance because of the perpendicular arrangement of myofibrils to the direction of muscle fibres. The number of muscle fibres of this type are finite for life, but exercise can expand myofibrils and enlarge individual muscle.

Smooth muscle is found in walls of tubes and cavities of the body such as blood vessels, stomach, intestine, bladder, and uterus. It contracts slowly as a unit, producing a squeezy motion, and can sustain for a long period of time. They give a smooth appearance due to the loose network of actin and myosin rather than a banded pattern.

Cardiac muscle is the contractile tissue of the heart. The contractions of this muscle do not need any external stimulus, and are rhythmic and intrinsic, thus producing a coordinated heartbeat. It has a striated appearance similar to skeletal muscles but contracts by squeezing action similar to smooth muscle. It has cells that are branched and allow flow of electric current throughout the tissue (Figure 10.3).

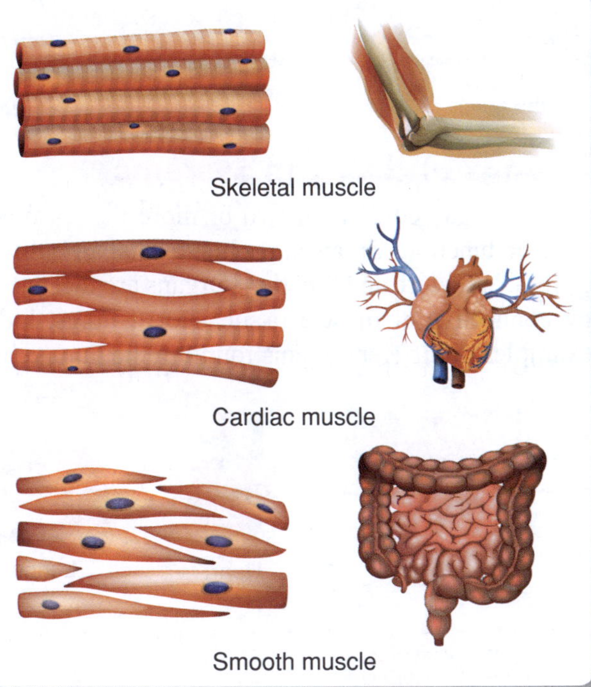

Fig. 10.3 Types of muscle tissues in humans

Nerve tissue

It contains cells called neuron, which is the major contributor to sensing and transmitting signals that act as mechanisms for adaptation and response to the environment. Neurons contain a cell body with two types of extensions called dendrites and axons. Dendrites receive signals that are transmitted through axons. Nervous tissue is found in the brain and spinal cord, as well as in the nerves that connect these organs to other parts of the body. Nervous tissue also contains glial cells that physically support and provide nutrients to neurons, provide insulation, and scavenge cellular debris and foreign matter (Figure 10.4).

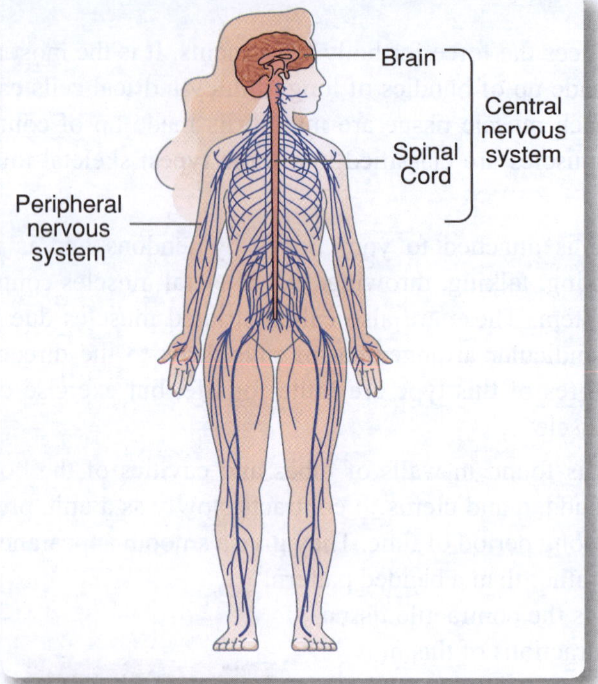

Fig. 10.4 Nervous tissue

Organs and Organ Systems

An organ consists of two or more tissues functioning together as a single unit to perform a specific function. Organ can achieve what an individual tissue component cannot. Different organs differ in structure and function. Organs are located at specified places in the body and are supported by connective or muscle tissue. Heart, lungs, kidney, brain, small intestine, stomach are some examples of different organs found in the human body (Figure 10.5).

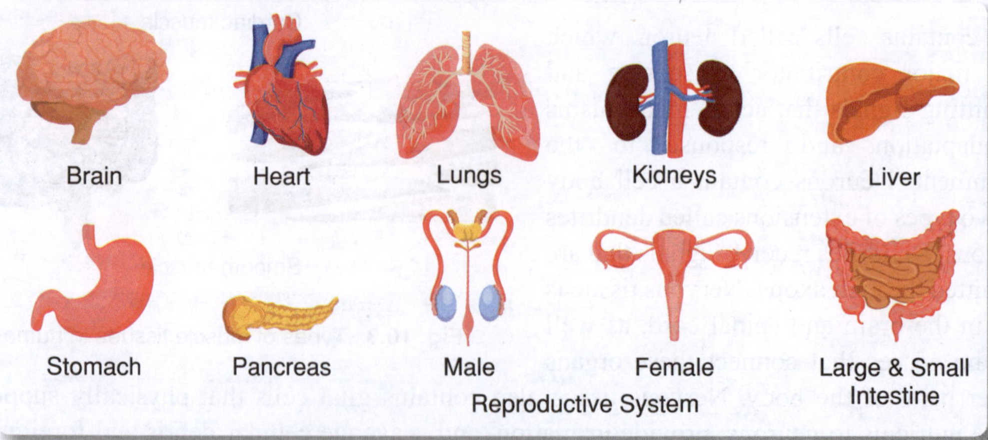

Fig. 10.5 Human organs

Different organs working together as a team to perform vital functions of the body is called organ system. The component organs can either be physically connected or can be dispersed throughout the body. These organ systems are highly interdependent in coordinating various bodily functions. There are 11 major organ systems in humans: circulatory system, respiratory system, reproductive system, lymphatic and immune system, nervous system, endocrine system, urinary system, digestive system, muscular system, and integumentary system. In some cases, the same organ may be part of multiple organ systems (Figure 10.6).

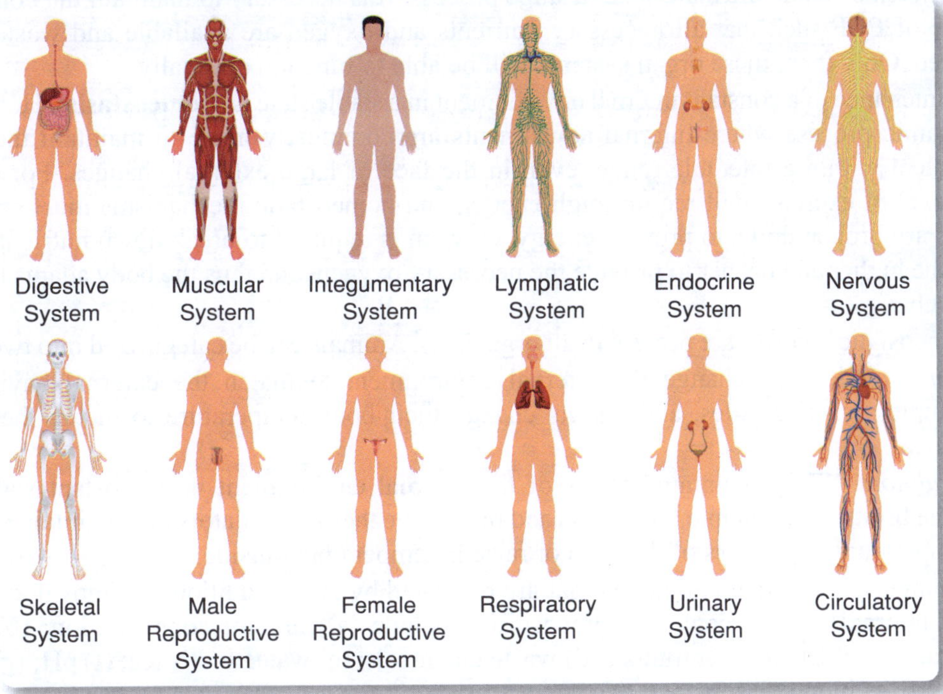

Fig. 10.6 The human organ systems

Vital functions of the body, critical for all organisms, such as acquiring nutrients, coordinating their processing, obtaining oxygen, distributing nutrients, elimination of wastes, synthesizing macromolecules for body structures, responding to changes inside the body like pH, temperature and ion concentrations, protecting body from pathogens, reproducing, and nourishing young ones for their growth and development, are accomplished by the coordinated interplay of various organ systems. When all organ systems function effectively and all the vital tasks are carried out with high level and order of efficiency, the internal environment of the body called homeostasis is preserved.

Homeostasis or Internal Control

Most animals have a protective layer that separates the internal environment from the world outside. However, to live, organisms continuously need to exchange chemicals and energy with the external environment. For example, nutrients are taken in and wastes are eliminated out, oxygen is taken in and carbon dioxide is released out. Unicellular organisms are in direct contact with their environment, which facilitates exchange; however, in multicellular organisms, as some cells could

be isolated from the external environment, additional mechanisms need to be evolved to facilitate effective material and energy exchange.

An aqueous extracellular fluid (ECF), which includes the fluid portion of blood called plasma and the interstitial fluid that surrounds the cells, provides the internal mechanisms for material exchange. These circulatory fluids (ECF) enable that the needs of all cells are delivered from the external environment and substances dissolved in water cross through the plasma membrane of cells where exchange can occur. Different organ systems such as circulatory, digestive, respiratory, and urinary systems enable the material exchange process. It is necessary to maintain the composition and state of ECF such that all necessary nutrients and oxygen are available and wastes can be eliminated. Only then, these organ systems will be able to function optimally.

Maintenance of a constant internal environment in a stable state is **'homeostasis'**. Homeostasis is a dynamic process where internal adjustments are continually made to maintain the internal environment within a tolerable range, even in the face of large external changes. For example, a high level of exercise will require high energy, and homeostatic mechanisms must ensure that nutrients are broken down to provide energy. Oxygen is required to break down nutrients; hence, we breathe higher and deeper to provide the necessary oxygen, and thus the body adapts to sudden external change.

Homeostatic controls are present in all organisms. Animals can be categorized into two groups.

Conformers They change the internal environment to match the external environment. For example, fishes, reptiles, and insects change their body temperature to match the external environment.

Regulators They maintain factors of the internal environment in a constant state within acceptable limits. For example, mammals and birds have thermoregulators that maintain a constant body temperature, regardless of the drop or raise in temperature outside.

In humans, major internal factors that are regulated by either addition or removal to maintain constant internal environment or 'homeostasis' include (a) nutrient concentration; (b) oxygen concentration; (c) CO_2 concentration; (d) waste chemicals; (e) water and NaCl; (f) pH; (g) volume and pressure of blood plasma; and (h) temperature.

To achieve these, homeostatic controls operate locally within the tissue, organ by intrinsic regulation, or systemically by other external systems by extrinsic regulation.

Intrinsic Local self-regulating mechanisms within a tissue or organ contributing to maintaining internal balance and control is called as intrinsic homeostatic regulation. For example, dilation of blood vessels to provide more oxygen to exercising skeletal muscles is a local control affecting only one organ.

Extrinsic Systemic control is mainly handled by the nervous and endocrine systems of the body and regulates multiple organs and organ systems. Here, bodily functions are extrinsically controlled by an outside system such as nervous or endocrine. For example, when body temperature drops, the hypothalamus (in the brain) sends signals to initiate shivering—a response controlled by the nervous system. In similar ways, blood pressure, blood pH, and temperature are regulated by nervous and endocrine systems together to obtain a systemic effect.

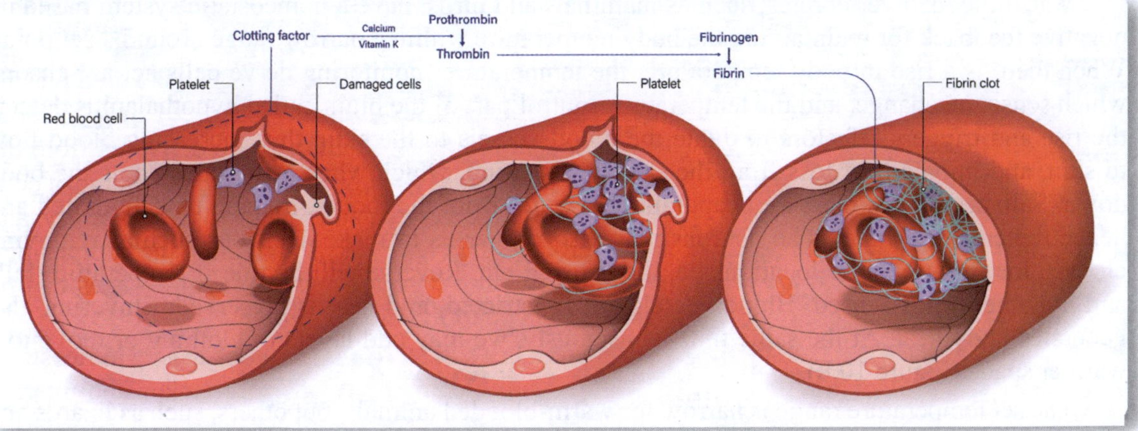

Blood clotting as an event for homeostasis

Mechanisms for Maintaining Homeostasis

The primary mechanism of homeostasis is ***negative feedback***. In this, when an environmental change acts as a *stimulus*, the body component acts as a sensor and detects this change; a control centre then integrates this information from the sensor and decodes the deviation from the acceptable limit and sends out commands for correction, which the effector executes and brings the condition under control to the set point. Integrators are control centres such as brain and endocrine systems, whereas the effector would include parts of any body tissue or organ (Figure 10.7).

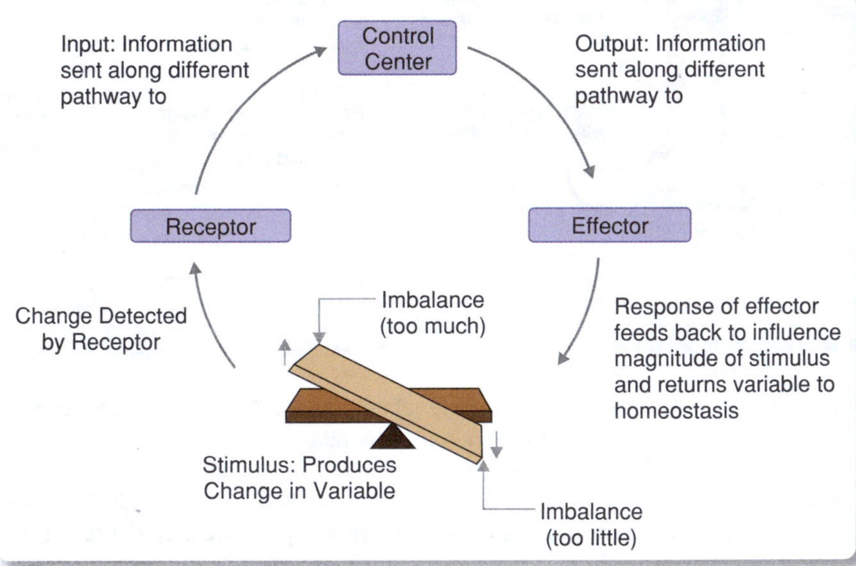

Fig. 10.7 Negative feedback response for homeostatic control

Warm-blooded vertebrates, such as mammals and birds, have a homeostatic system based on negative feedback for maintaining the body temperature within a narrow range around a set point. When there is a rise in body temperature, the temperature-monitoring nerve cells act as sensors, which sense the change, and the temperature control part of the brain called hypothalamus detects the rise and triggers effectors to dilate the blood vessels to the skin, thus increasing blood flow to skin, and initiate heat loss from the skin by sweating, which when evaporates cools the body down. Salivary glands increase evaporation of water from the mouth, making us dehydrated and hence replenishing water. The neurons may also trigger us to make decisions such as removing clothes, moving to cooler locations, taking shower, etc. Reverse will be true if temperature falls and hypothermia is sensed. Blood vessels are constricted, and heat is preserved. Shivering also generates body heat. At the same time, consciously, we may add layers of clothing or move to a warmer space (Figure 10.8).

The set temperature range is narrow for warm-blooded animals, but others, such as lizards and snakes, regulate temperature over a broader range and do not have such precise negative feedback homeostasis mechanisms. They respond behaviourally by moving to sunny rocks or cooler shaded spots.

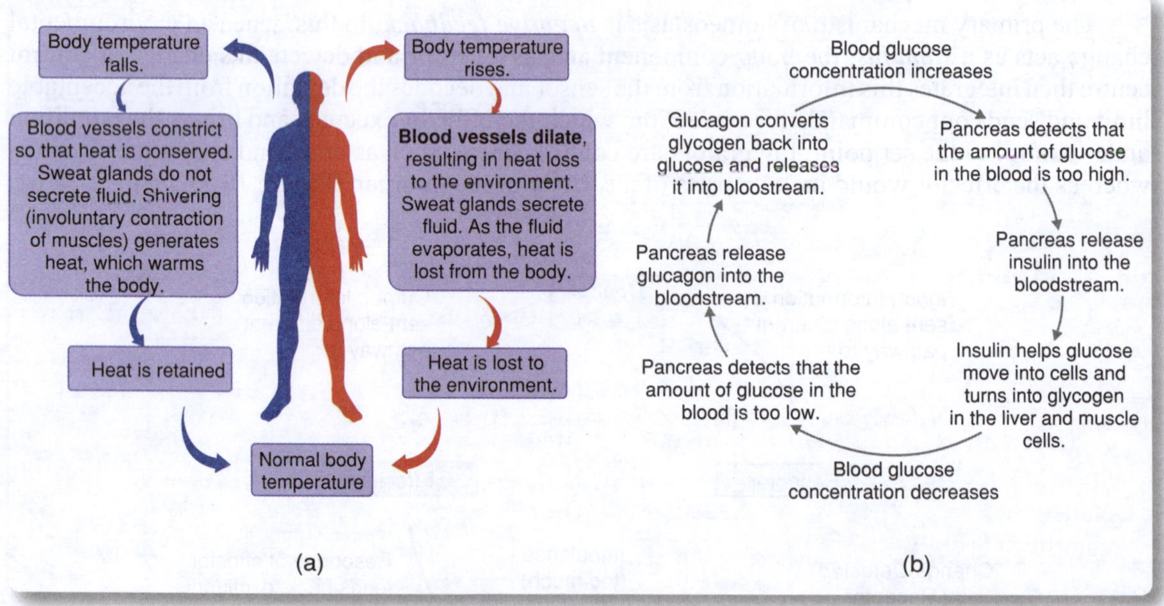

Fig. 10.8 (a) Thermoregulation through negative feedback homeostasis; (b) Blood glucose homeostasis through negative feedback

Similar negative feedback mechanisms also exist to maintain glucose levels in blood. Too much insulin sends a signal to stop producing insulin by pancreas, and the amount decreases. Insulin production is activated when it becomes low again. A certain concentration of insulin is required in the body to break down blood glucose. This regulation maintains the level of glucose in the body within set limits (Figure 10.8(b)).

Positive feedback mechanisms that exist in animals intensify the process and do not result in homeostasis. Positive feedback happens when a response to a stimulus causes further increase in stimulus, which further intensifies the response. Positive feedback is rarer than negative feedback. Positive feedback is mainly seen during life-threatening situations and as part of reproductive control. For example, during childbirth in humans, contractions of the uterus push the child towards the cervix. Sensors in the cervix detect the signal, and the hypothalamus directs the pituitary to release oxytocin. Oxytocin increases the uterine contractions, causing a positive feedback circuit, and the increased squeezing pressure eventually helps in pushing the foetus out of the uterus (Figure 10.9).

The nervous, endocrine, and immune systems are major key systems contributing to maintaining homeostasis in the body. Thermoregulation and osmoregulation work through negative feedback, while uterine contractions during childbirth follow a positive feedback loop.

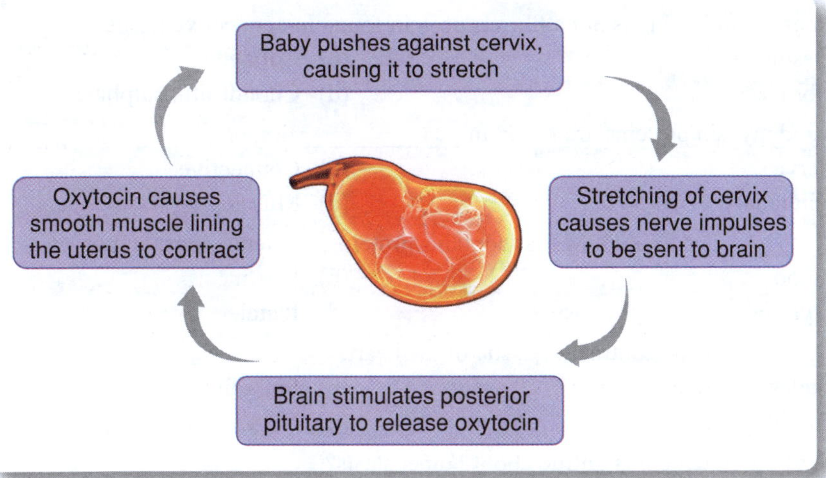

Fig. 10.9 Positive feedback in birth process

Thus, understanding of tissues, organs, and organ systems along with homeostatic mechanisms help in maintaining and coordinating the vital functions of multicellular organisms efficiently. Detailed understanding of all the individual organ systems in subsequent chapters will provide greater insights to the functioning of human biology.

SUMMARY

- The benefits of multicellularity in animals and the hierarchical levels of structural complexity were discussed.
- Insights into the organization of animal bodies were obtained.
- Various tissues, organs, and organ systems along with their defining features were detailed.
- The need for homeostasis and internal mechanisms for the same were studied.
- Thermoregulation, glucose regulation through negative feedback control, and contribution of positive feedback for the birth process in humans was elaborated.

EXERCISES

A. Objective-Type Questions

1. Which of these is not a kind of epithelial tissue?
   (a) Squamous (b) Columnar
   (c) Cuboidal (d) Fibrous
2. ____________ is a connective tissue that stores fat.
   (a) Loose connective tissue (b) Cartilage
   (c) Blood (d) Adipose tissue
3. Cells that make the bone are called
   (a) Osteoclasts. (b) Osteoblasts.
   (c) Osteon. (d) Osteocytes.
4. Which of the following is not a fibrous protein found in connective tissue?
   (a) Elastin (b) Collagen
   (c) Fibronectin (d) Chondroitin sulphate
5. Actin and myosin proteins are found in ____________ tissue.
   (a) Nerve (b) Connective
   (c) Epithelial (d) Muscle
6. ____________ is a fluid ECM that contains proteins, nutrients, ions, and gases.
   (a) Blood (b) Plasma
   (c) Erythrocyte (d) Platelet
7. ____________ is the contractile tissue of the heart.
   (a) Skeletal muscle (b) Smooth muscle
   (c) Cardiac muscle (d) None of the above.
8. Which of the statement is untrue about homeostasis?
   (a) Constant internal environment regardless of outside
   (b) It is controlled by negative feedback.
   (c) It is a dynamic mechanism than maintains levels around set point.
   (d) Positive feedback brings homeostasis.
9. Which of this is not an organ system?
   (a) Immune system (b) Respiratory system
   (c) Digestive system (d) Pancreas
10. Which muscle type appears straited under a microscope?
   (a) Skeletal muscle only (b) Cardiac muscle only
   (c) Smooth and skeletal muscle (d) Skeletal and cardiac muscle.

B. Short-Answer Questions

1. Bones of elderly women break easily. Identify the cell type that diminishes in activity.
2. In a contest, a student eats an entire chocolate bar but his blood glucose does not increase dramatically. How is this regulated?
3. Oxytocin is released during childbirth. What triggers the release, and what kind of regulation is this?
4. Snakes and lizards move to shaded areas during peak summer. Explain this adaptation in the context of homeostasis.

FURTHER READINGS

1. https://www.frontiersin.org/journals/physiology/articles/10.3389/fphys.2020.00200/full
2. https://biocyclopedia.com/index/general_zoology/the_hierarchical_organization_of_animal_complexity.php#google_vignette
3. https://bio.libretexts.org/Courses/Cosumnes_River_College/Introductory_Anatomy_and_Physiology_(Aptekar)/01%3A_Introduction_to_the_Human_Body/1.03%3A_Structural_Organization_of_the_Human_Body
4. https://www.bbc.co.uk/bitesize/articles/zrp3ydm

USEFUL WEBLINKS

1. https://www.youtube.com/watch?app=desktop&v=Iz0Q9nTZCw4
2. https://www.youtube.com/watch?v=1i_iJBcB51I
3. https://www.youtube.com/watch?v=O0ZvbPak4ck
4. https://www.britannica.com/video/discussion-organ-systems-human-body-another-influence/-189649

CHAPTER 12

Physiological Systems of Humans: Response Control

LEARNING OBJECTIVES

After reading this chapter, you will be able to:

- To study the components and functioning of the nervous system
- To understand how the nerve impulse is generated
- Understand the working of the brain and its complexities
- Understanding endocrine system—the different glands and their hormones
- To establish the integration between the nervous and the endocrine systems.

CHAPTER TAKEAWAY

Mechanisms of control and protection in the human body

Ability to respond to stimuli is one of the characteristic features of life. This chapter deals with mechanisms of control and protection in the human body. Fine tuning responses and coordinating the right level of response depending on the intensity of stimuli to provide varied responses is one of the hallmarks of response control.

PRELUDE

The human body is designed to respond to threats in a quick and meaningful way. Multiple systems of the body function in an integrated way to provide the response. Imagine you are enjoying a walk in the neighbourhood woods and suddenly you spot a snake. In an instant, your body goes into high alert, and two systems—the nervous and endocrine—step up to help you survive. The nervous system will be the first responder as it will process the information and trigger you to take action and run. It directs your muscles, and you are ready to flee in an instant. When you are running, the nervous system co-opts the endocrine system, and the brain guides the adrenal glands to secrete epinephrine or adrenaline into the bloodstream. This results in increase in your heart rate, rise in blood pressure, dilation of pupils for better vision, and better flow of oxygen to muscles; all these lead to an intense surge of energy and focus helping you to run. So, the endocrine system ensures that your body is at its peak performance. Once the danger subsides and the threat is gone, the two systems gradually return to normal. This shows that the nervous and the endocrine systems are working in harmony, constantly communicating and adjusting to keep you alive and thriving. Thus, this is an interesting combination that serves as a survival toolkit inside the body.

Nervous System

The nervous system is an intricately organized data processing system of the human body. The nervous system of the body functions like a computer, which obtains information through input devices; the CPU processes the information; and the output devices provide the outcomes. In a similar manner, our nervous system takes the inputs about the stimuli and the environment from the sensory organs, integrates the input through a central organ like the brain, processes the information, and sends out an appropriate response. The nervous system is at the forefront for managing rapid adjustments that must take place within the body. The message is sent very rapidly along a specific route from the starting point to the destination end point.

The basic unit of the nervous system is the nerve cell or neuron. Neurons generate neural signals and transmit them from one part of the body to another. The neuron consists of a cell body with a nucleus and several long extensions called nerve fibres. The nerve fibres in neurons are of two types: dendrites and axon. Dendrites carry information towards the cell body, and axon carries the information away from the cell body. Every neuron has one axon and many dendrites. The nerve signal is relayed along the neuron, and the speed of nerve signal is attributed to the myelin sheath—a chain of bead-like supporting cells that wrap around the insulate nerves (Figure 12.1(a)). Nerve signals are only generated at the gaps in myelin sheath and hence transmitted rapidly from one gap to the next. Neurons are connected to each other through synapse (Figure 12.1(b)).

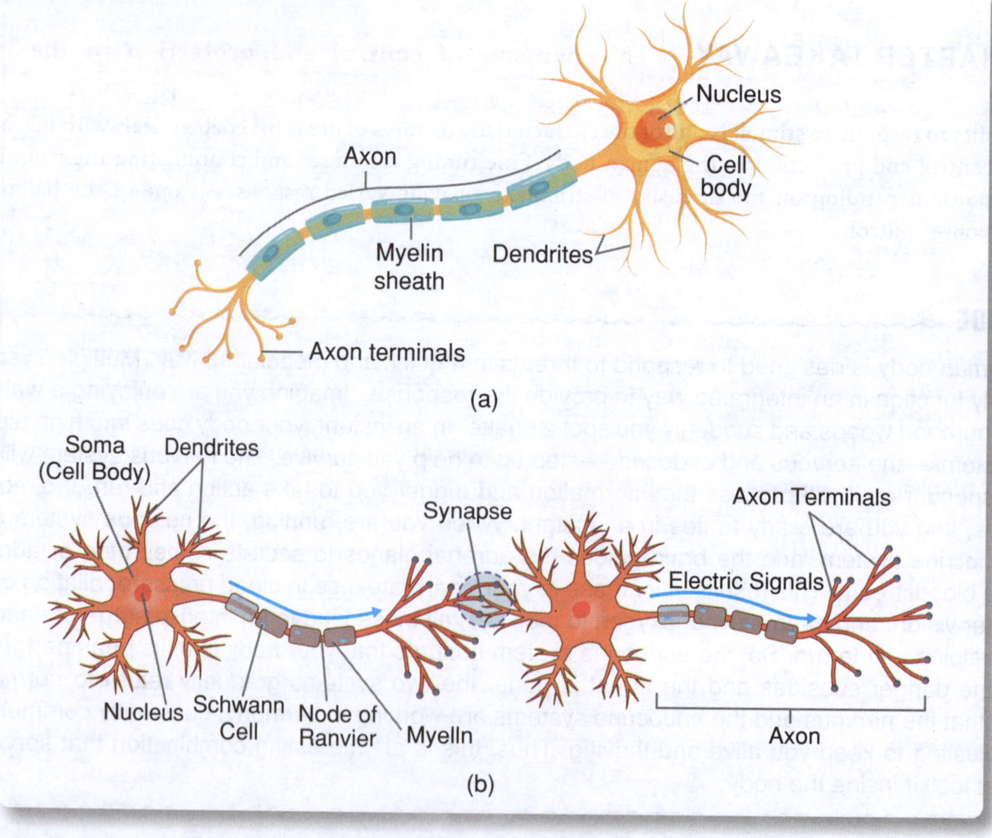

(a)

(b)

Fig. 12.1 Structure of a neuron: (a) Neuron; (b) Synapse

Organization of the Nervous System

The nervous system has two main anatomical divisions:

(a) Central nervous system: It includes the brain and the spinal cord. Brain is surrounded by the skull and the spinal cord by vertebrae. The CNS is the mainstay of intelligence, learning, memory, and emotion.

(b) Peripheral nervous system: It consists of bundles of long axons and dendrites called nerve fibres. These nerve fibres radiate out of the brain and spinal cord, gathering information and relaying it to CNS as well as back to the organs. The cranial nerve fibres connect to the brain, and the spinal nerves connect to the spinal cord. Spinal nerves carry signals between the spinal cord and the body trunk and limbs, while the head, neck, and body trunk are connected to the brain by the cranial nerves (Figure 12.2(a)).

The peripheral nervous system functions with two sets of neurons: sensory neurons and motor neurons. ***Sensory neurons*** (also called afferent systems) transmit information from the sense organs (eye, ear, nose, etc.) to the brain, whereas ***motor neurons*** (called efferent systems) carry the output from the brain to different parts of the body—mainly to muscles and glands. Interneurons are located between sensory and motor neurons within the brain and the spinal cord (Figure 12.2(b)).

Within the motor nerves or efferent system, some control the skeletal muscles (voluntary muscles)—This is called ***somatic nervous system***. It controls body movements that are conscious and voluntary. Others control the smooth (involuntary) muscles, heart, and glands, and these are called ***autonomous nervous systems***. The autonomic nervous system controls involuntary processes like digestion, secretion, circulation, and reproductive and excretory functions. The two sides of the autonomic nervous system—***sympathetic and parasympathetic*** are always active and have opposing effects on organs they control, thereby enabling precise control.

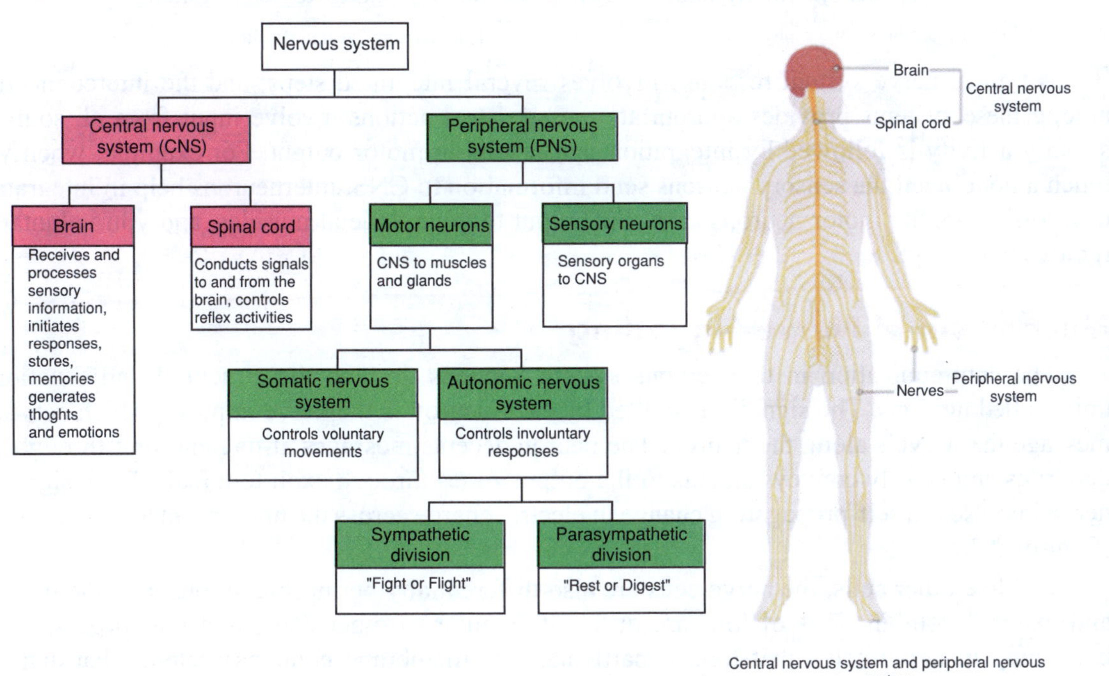

(a)

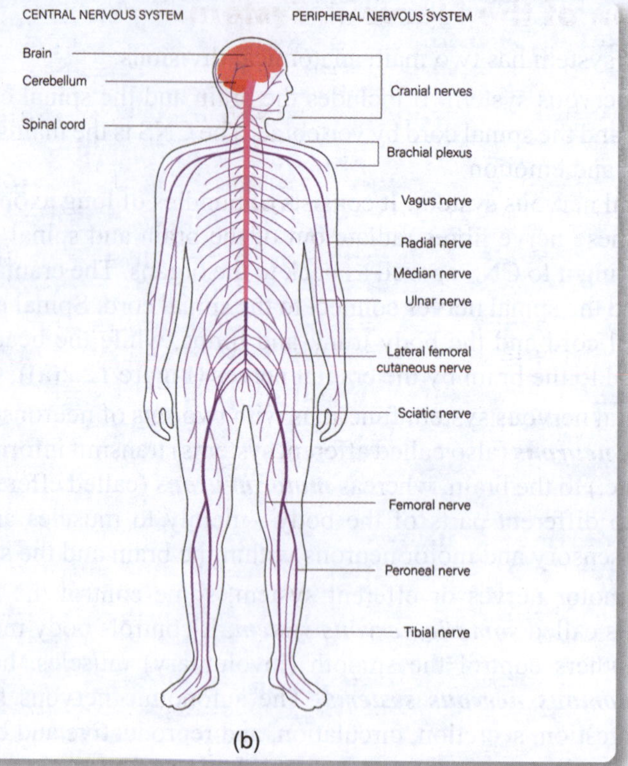

Fig. 12.2 (a) Nervous system and its functions; (b) Organization of the nervous system: Central Nervous System and Peripheral Nervous System

Source: Adapted from https://www.physio-pedia.com/Introduction_to_Neuroanatomy?veaction=edit

A typical nerve system response involves several integrated steps, and the interconnection among these systems provides appropriate output. Most actions involve input from all sources; sensory activity is followed by integration that results in motor output. For example, when you touch a hot object, the sensory neurons send information to CNS, interneurons help in integration of information, the motor neurons carry the output to your skeletal muscles, and you voluntarily remove your hand.

Mechanism of Nerve Signalling

The communication in the nervous system happens through the structural and functional units called neurons. The signal transmitted by each neuron is the nerve impulse, which acts as a message that travels along the neuron. The neuron accepts messages at its input unit through the dendrites and swiftly conveys signals to the output zones through axon terminals. Each signal or nerve impulse is a self-propagating change in electric charge across the neuron's plasma membrane (Figure 12.3(a)).

Just like other cells, the nerve cells are also differentially permeable to ions, and the internal and external combinations of ions are different. Similar to other cells, transport processes are happening in nerve cells, that too, in particular, the membrane contains proteins that actively transport ions from one side of the cell to other. Active transport requires ATP; this kind of transport is usually seen only in living cells, and they lose this ability once the cell dies.

Na^{2+}–K^+ pump is an example of active transport that is operating in nerve cells too—Na^{2+} ions are transported out of the cell and K^+ ions are transported inside the cell against the concentration gradient.

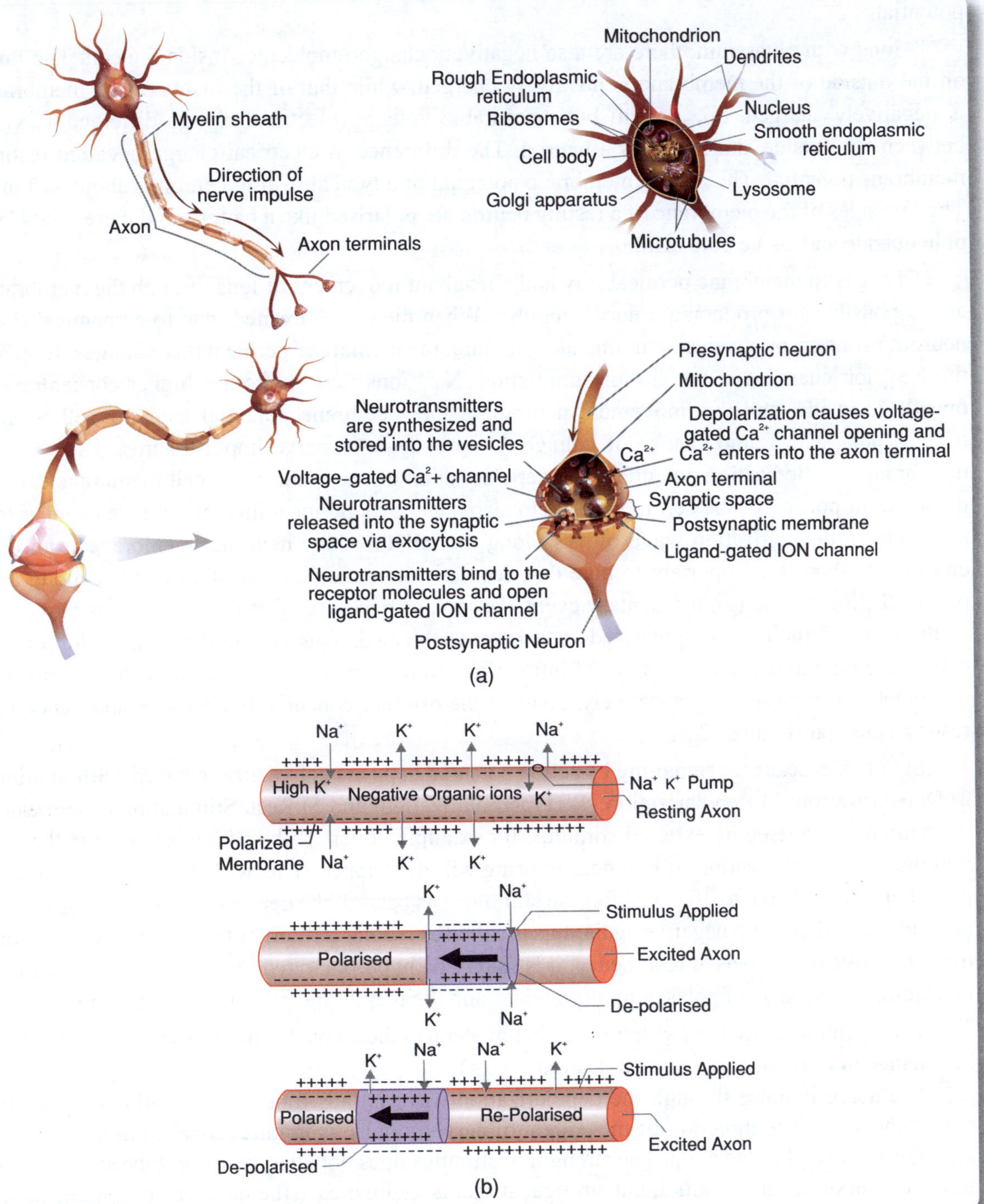

Fig. 12.3 (a) Signal transmission in neuron; (b) Changes to membrane permeability and a resultant movement of ions in nerve impulse

Normal resting neuron has more positively charged Na^{2+} on the outside of the cell. More sodium ions are transported out than the potassium ions transported in. Each ATP moves $3Na^{2+}$ and $2K^{+}$ ions in, thus the Na^{2+}–K^{+} pumps establish and maintain ion concentrations and membrane potential.

Along with potassium, there are also negatively charged molecules inside the cell. The fluid on the outside of the membrane is positively charged, while that of the inside of the membrane is negatively charged. So, a small but measurable voltage (electrical charge difference) exists between the outside and inside of a neuron. The difference in electrical charge is called resting membrane potential. The resting membrane potential of a typical resting neuron is about –70 mV. The two sides of the membrane of a resting neuron are polarised like a battery, and there is a '+' ve pole outside and '–' ve pole inside.

Changes to membrane permeability and a resultant movement of ions through the membrane are responsible for producing a nerve impulse. When the cell is excited, due to a chemical (like neurotransmitter) or physical stimulus like pressure, the membrane permeability changes. It opens the Na^{2+} ion channels at the site of stimulation. Na^{2+} ions flow in from a higher concentration outside to inside, and this movement neutralizes the membrane potential, and the cell is said to be depolarized. Rapid change of voltage that propagates a nerve impulse across a nerve cell membrane is called action potential. The depolarization at one point in the cell membrane causes the adjacent portion of the cell membrane to also change its permeability, and it gets depolarized too. Thus, a depolarization wave moves along the length of the neuronal membrane from one end to the other. It is important to note that depolarization and passage of an impulse along any part of the membrane is a momentary event. As soon as depolarization occurs, K^{+} ions move out of the cell and the original polarized state is re-established. This is called repolarization of the cell. The continuous active transport of influx (inward movement) of sodium and outflux (inward movement) of potassium, respectively, restores the original concentration of ions and hence the resting potential (Figure 12.3).

So, in essence, nerve signalling by action potential happens in five different steps: Stimulation; 2) Depolarization; 3) Repolarization; 4) Hyperpolarization; and 5) Rest. Stimulation occurs when the neuron is triggered by external stimulus, for example, touch. If the stimulus is strong, the cell will undergo depolarization; if not, neuron firing will not happen. This will change the membrane potential from –70 mV to 30 mV. After the stimulus has passed, the depolarization wave will drop, and the cell will become negative again; this is called as repolarization. At this stage, it will become more negative than it was at rest; hence, it is said to be hyperpolarized. After some time, the ionic concentration, such as of potassium, normalizes and the membrane potential returns to rest.

The membrane potential graph given below depicts the steps discussed above and illustrates the change in the cell action potential (Figure 12.4).

The nerve impulse through the depolarization wave is thus transmitted continuously to the end of the axon. The direction of movement of nerve impulse is unidirectional from the point of stimulus to the end of the axon. The strength of stimulus does not alter the magnitude and speed of transmission of action potential, but stronger stimulus will increase the number of impulses along the axon. The nerve impulses in human neurons jump directly between the node of Ranvier, thus speeding up the signal transmission.

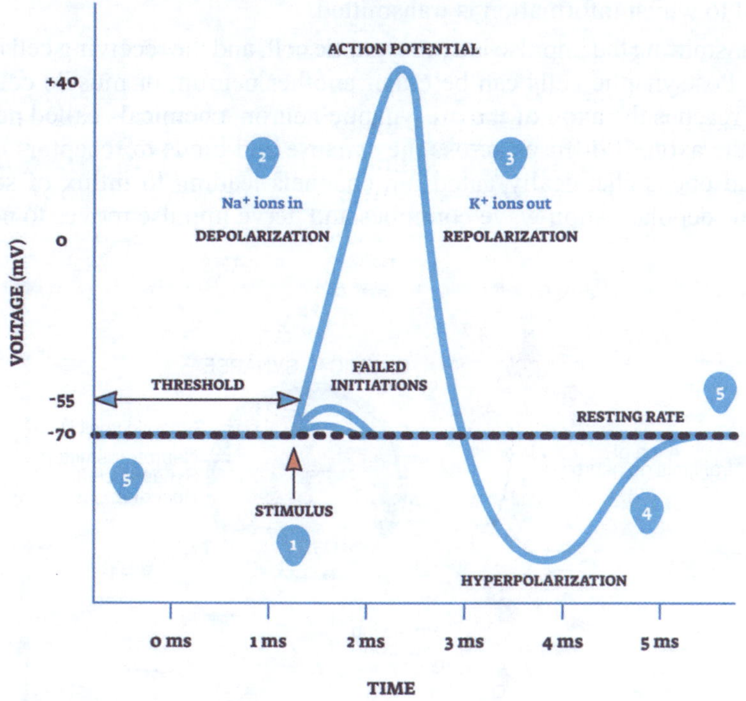

Fig. 12.4 Action potential graph depicting steps and characteristic rise and fall of action potential across neuron

Cell–Cell Signalling

Once it reaches the tip of the axon, how does the nerve impulse transmit to the next neuron? Neurons are arranged in a head-to-tail orientation. So, the axon of one neuron will face the dendrite

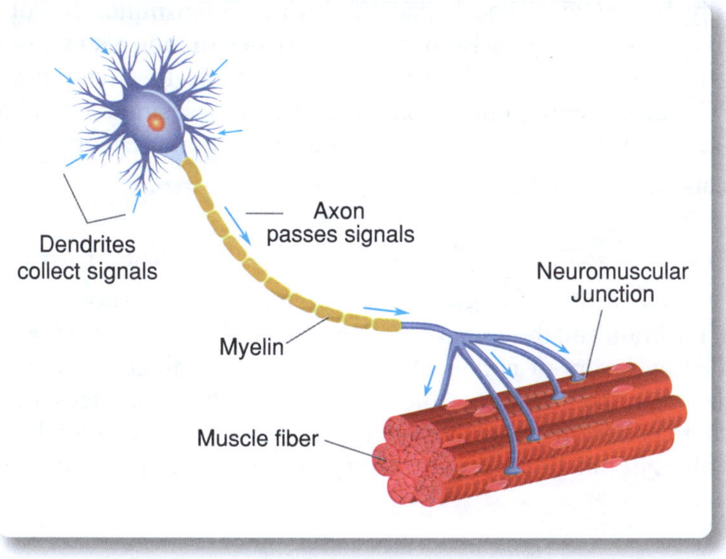

of another neuron. The junction between the axon of one cell and the dendrite of the other is called synapse. Synapses are not only formed between two neurons but are also found between neurons and their target cell to which information is transmitted.

The neuron transmitting the impulse is a presynaptic cell, and the receiving cell is a postsynaptic cell (Figure 12.5). Postsynaptic cells can be either another neuron, or muscle cell, or gland cell. When the impulse reaches the axon of the presynaptic neuron, chemicals called neurotransmitters are released from the axon. It diffuses across the synapse and binds to receptors on the dendrites of the next cell and opens chemically gated ion channels leading to influx of sodium ions and depolarization. Thus depolarization wave continues and nerve impulse moves to next neurons.

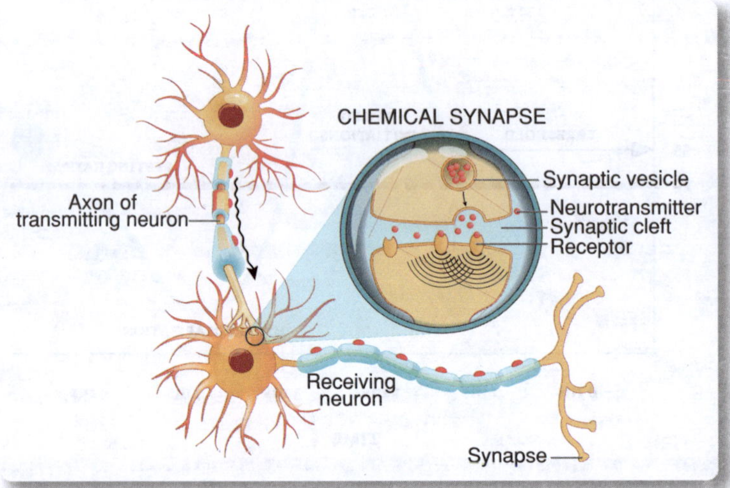

Fig. 12.5 Neurotransmission across neurons—presynaptic and postsynaptic cells

About 50 neurotransmitters are known, and examples include acetylcholine, dopamine, serotonin, GABA, epinephrine, and norepinephrine, etc. The neurotransmitters change the resting potential of the postsynaptic cell, and thus transmission of nerve impulse continues. Once this happens, the neurotransmitters are inactivated. For example, neurotransmitter acetylcholine is degraded rapidly by acetylcholinesterase. If the neurotransmitter is not broken down, the postsynaptic cell would be continuously stimulated. The destruction also allows the cell to become ready to receive the next burst of neurotransmitter from the first neuron a while later. Nerve impulses go only in one direction; only axons release neurotransmitters, and only dendrites have receptors for the same. This also explains why there are separate sensory and motor neurons that carry impulses to and from the CNS.

Central Nervous System—Brain and Spinal Cord

The central nervous system integrates sensory inputs into responses. The central nervous system consists of the brain and the spinal cord. Inside the skull and vertebrae, the brain and spinal cord are surrounded by layers of connective tissue called meninges and by cerebrospinal fluid, which circulates between the meninges. The cerebrospinal fluid nourishes, cushions, and protects the brain and spinal cord from toxic substances. A highly selective blood–brain barrier (BBB), sealed together with tight junctions, prevents harmful substances like viruses, bacteria, and toxic substances from entering the cerebrospinal fluid. Ions like oxygen, carbon dioxide, alcohol, and anaesthetics can cross the BBB by diffusion, while selective transport proteins help in movement of glucose for energy.

Interesting Fact 12.1

Glial Cells in Nervous System

1. Glial cells act as 'glue' of the nervous system. These are most abundant cells in CNS and help in maintaining homeostasis, supporting neurons, and forming myelin.
2. There are four primary types of glial cells in CNS.
   (a) Astrocytes: Star shaped, regulate neurotransmitters, form blood–brain barrier, control blood flow to brain, synchronize axon activity, fuel neurons, and also clear dead neurons.
   (b) Oligodendrocytes: Mainly responsible for creating myelin around neurons.
   (c) Microglia: Star shaped, brain's resident immune cells, and protect the brain against pathogens and disease-causing agents.
   (d) Ependymal cells: Provide lining to fluid-filled spaces and help in formation and transport of cerebrospinal fluid.
3. Glial cells of PNS ensure proper communication between PNS and CNS. Two types of glial cells are there in PNS
   (a) Schwann cells – similar to oligodendrocytes, create myelin sheath around nerves, and repair damaged axons and regenerate them.
   (b) Satellite cells – maintain health and activity of neurons in PNS.

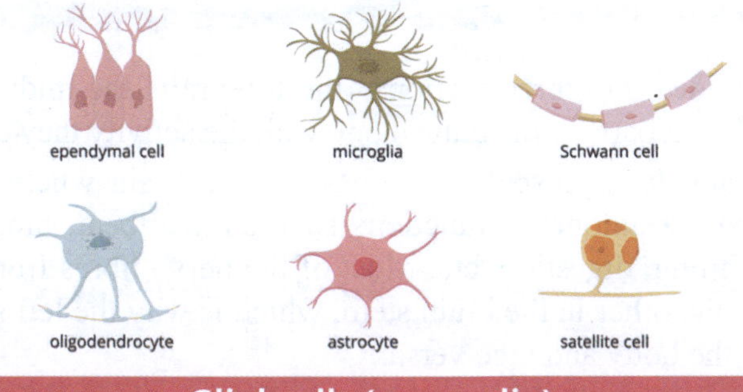

Glial cells (neuroglia)

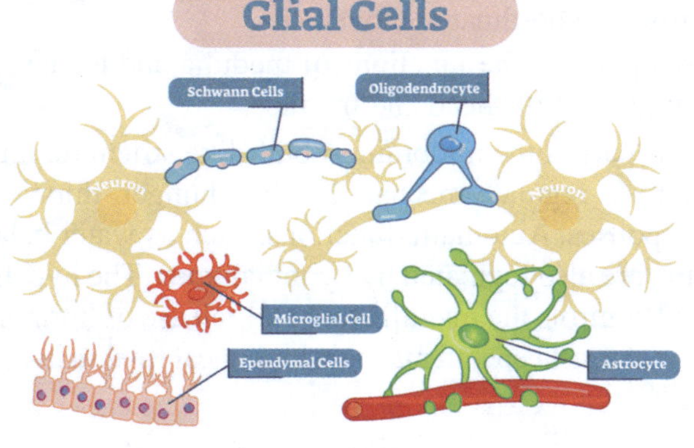

The spinal cord acts as a link and carries impulses between the brain and the peripheral nervous system. Brain functions as the master control centre that receives, integrates, stores and retrieves information, and generates responses for movement, consciousness, behaviour, emotions, learning, reasoning, language, memory, and other complex functions. Brain as an organ is a combination of billions of neurons, supporting cells, and fluid-filled cavities.

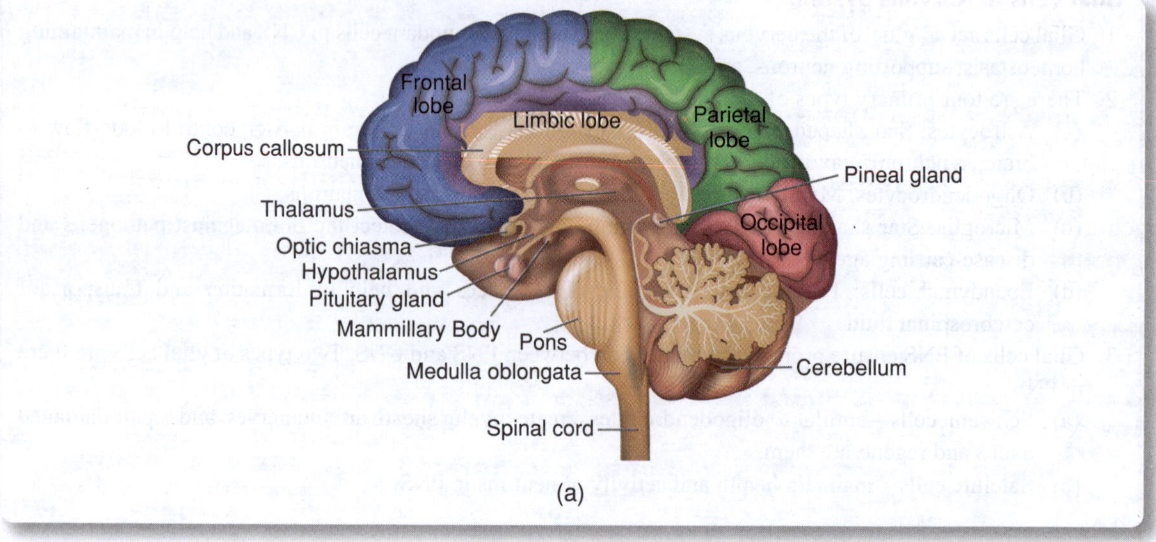

(a)

The brain can be divided into three sections—the forebrain, the midbrain, and the hindbrain (Figure 12.6(a)). The different parts of the brain along with the activity they control are given below:

1. Medulla oblongata: It is present at the base of the brain where the spinal cord enters the brain. It controls automatic functions such as heartbeat, blood pressure, breathing, swallowing, vomiting, digestion, etc. Most of the nerve fibres from the spinal cord cross from one side to the other in the brain stem, which is why the left side of the brain affects the right side of the body and vice versa.
2. Cerebellum: It is a large bulge at the base of the brain that is connected to medulla oblongata. Cerebellum coordinates balance and monitors cognitive and motor functions. It uses the information from sensory organs, such as eyes, ears, etc., and adjusts the muscular contractions to ensure a coordinated movement.
3. Pons: It provides support to the functions of medulla and is mainly involved in functions pertaining to the organs of the head and the face.
4. Hypothalamus: It is part of the midbrain. A bulb-like structure, it is our body's thermostat and acts as a centre to regulate mechanisms like hunger, thirst, sleep, arousal, pleasure, fear, and anger responses. Automatic functions, such as water balance, body temperature, and blood pressure, are also regulated by hypothalamus. The hypothalamus connects to the pituitary gland and controls the manufacture and release of hormones from the pituitary. It also acts as an internal timekeeper to maintain biological rhythm in cycles of sleep and hunger.

5. Thalamus: It is also a bulb-like structure that connects to cerebrum. It acts as the main input for the sensory information, which is sorted and sent to the appropriate part of the cerebrum, and as an output centre for the motor information of responses to be directed out from the cerebrum. At some level, it also controls emotions and arousal.
6. Cerebrum: It is the largest, highly evolved, and logical part of the forebrain. It is divided into two hemispheres: the right and the left hemispheres. Each hemisphere is composed of outer gray matter called the cerebral cortex and internal white matter. Communication between the right and left hemispheres is maintained by a band of neural fibres called corpus callosum. The cerebral cortex has four lobes: frontal, parietal, occipital, and temporal lobes. Cerebral cortex is mapped to several functions in learning, thinking, analysing, decision-making, imagination, dreaming, consciousness, etc. (Figure 12.6(b)).

The function and performance of the brain is not determined by its structure alone; the intricate neural circuitry comprised of specialized neurons secreting neurotransmitters are all involved in appropriate functioning of the brain. It is believed that the left hemisphere of the brain plays a dominant role in language, mathematics, and logical reasoning, whereas the right hemisphere functionally contributes in arts, music, spatial sense, perception, and imagination (Figure 12.6).

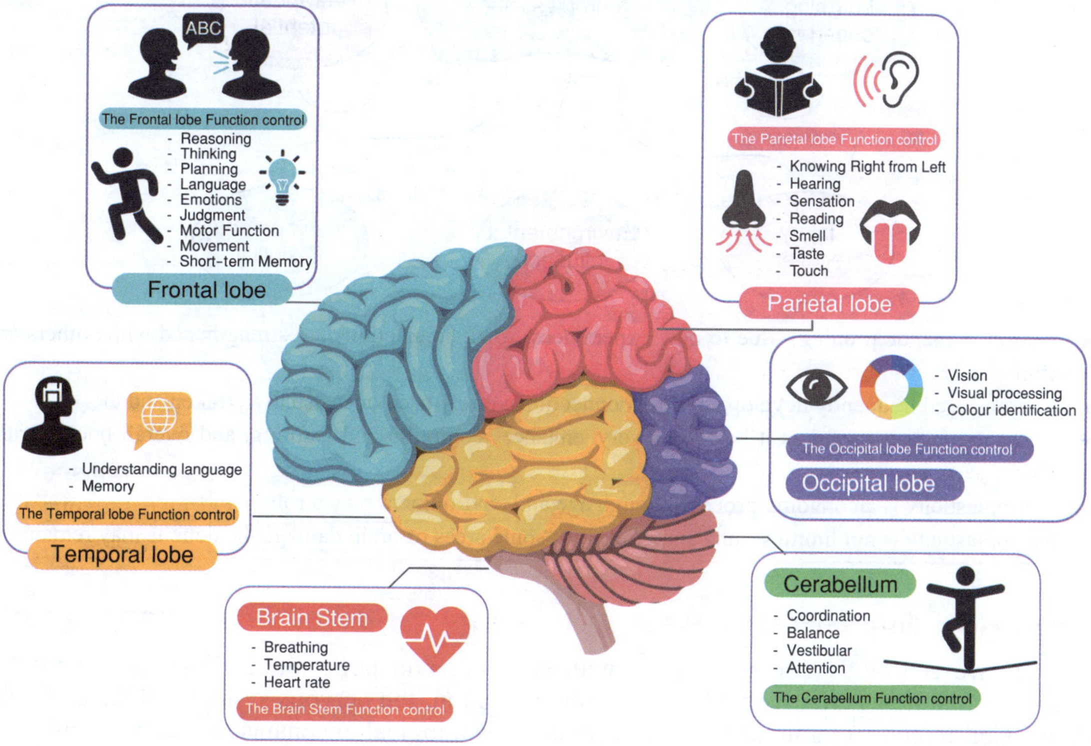

ABC

The Frontal lobe Function control
- Reasoning
- Thinking
- Planning
- Language
- Emotions
- Judgment
- Motor Function
- Movement
- Short-term Memory

Frontal lobe

The Parietal lobe Function control
- Knowing Right from Left
- Hearing
- Sensation
- Reading
- Smell
- Taste
- Touch

Parietal lobe

- Vision
- Visual processing
- Colour identification

The Occipital lobe Function control

Occipital lobe

- Understanding language
- Memory

The Temporal lobe Function control

Temporal lobe

Brain Stem
- Breathing
- Temperature
- Heart rate

The Brain Stem Function control

Cerabellum
- Coordination
- Balance
- Vestibular
- Attention

The Cerabellum Function control

(b)

Fig. 12.6 (a) Anatomy of brain, spinal cord, and their functions; (b) functions of various parts of brain and spinal cord

Interesting Fact 12.2

How experiences change the brain!

Neuroplasticity

- Brain has the ability to change and adapt due to experiences. It can change, recognize, and grow new networks. This ability of brain is called neuroplasticity. Plasticity—brain's malleability or ability to change; neuro—neuron.
- Neuroplasticity is of two types:
  - ✓ Functional neuroplasticity: When brain moves its functions from damaged parts to undamaged parts.
  - ✓ Structural neuroplasticity: Change in physical structures as a result of learning.

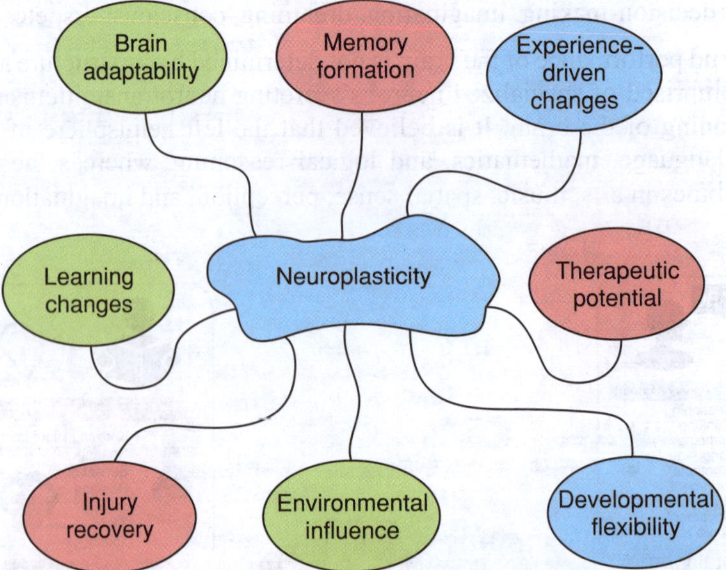

- Reason: Synaptic pruning—due to new experiences, some connections are strengthened while others are eliminated.
- Neurons used frequently develop stronger connections; rarely used ones usually weaken and die.
- Neuroplasticity is useful as it helps recovery, enhances cognition and learning, and overall boosts brain fitness.
- Neuroplasticity is an ongoing process, though age and environment play a role.
- Neuroplasticity is not limitless, and sometimes for some areas of brain damage the deficits may remain.

Endocrine System

Cells are constantly communicating with each other in large multicellular organisms like humans. Two systems are involved in this communication: the nervous system and the endocrine system. Cell-to-cell communication to regulate physiological mechanisms happens either by depolarization waves of action potential like in the nervous system or through chemical signals as seen in the endocrine system. The endocrine system secretes hormones to regulate the target cell functions. Hormones are chemical messengers that are released by one organ and are transported to another organ, where it triggers a change in activity. In humans, apart from the nervous system, the endocrine system facilitates systemic cell–cell communication and signalling through long distance by way of chemicals. The endocrine system secretes hormones to regulate the target cell

Experimental Insight 12.1

(a) Electroencephalogram (EEG)

Electroencephalogram (EEG) helps record brain activity. It is a painless test where small metal discs or electrodes that can act as sensors are attached to the scalp to pick up naturally occurring electrical activity produced by the brain. Brain cells communicate via electrical impulses, and this activity shows up as wavy lines on an EEG recording. Brain cells are active all the time, even during sleep. Recording EEG helps to understand the changes in brain activity, which aids in diagnosing brain conditions in epilepsy, identifying brain tumours, and brain damage due to injury, inflammation, stroke, etc. (Figure 12.7).

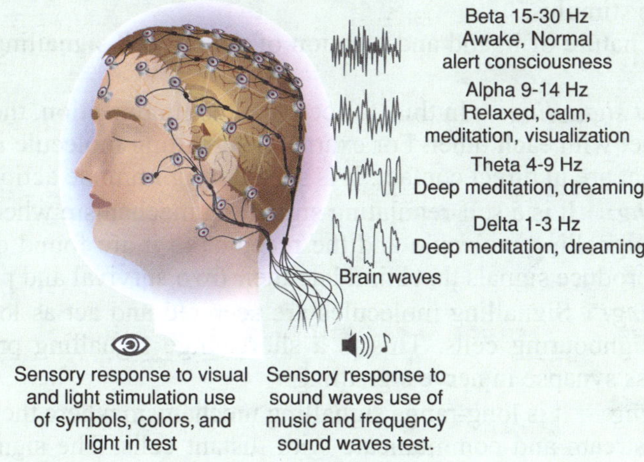

Fig. 12.7 Electroencephalogram (EEG)

(b) Computer-Based Neural Networks

These are computer-based machine learning models that use a network of interconnected nodes to process data. These models are inspired by the structure and functions of biological neural networks in the human brain. Thus, the manner in which the computers process data is inspired by the human brain. These models enable tasks related to pattern recognition and decision-making by training programs to learn and recognize patterns and process data for solving common problems in artificial intelligence, machine learning, and deep learning.

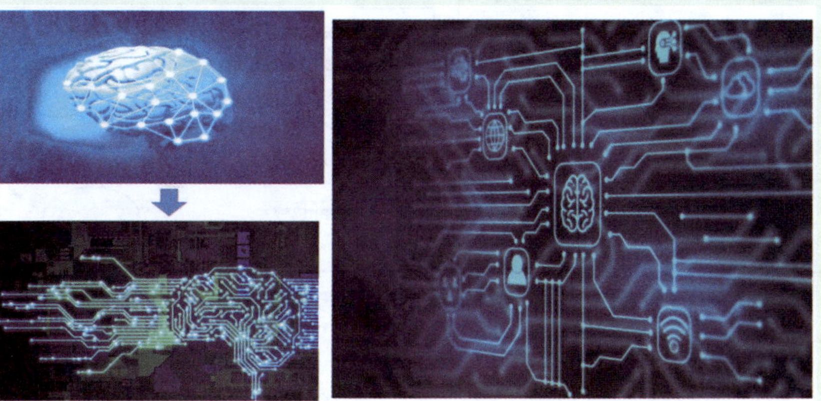

Computer-generated neural networks inspired by biological neurons and networks

functions. Hormones are chemical messengers that are released by one organ and are transported to another organ where it triggers a change in activity.

Cell signalling through regulatory proteins that produce chemical signals help in cellular interactions with itself, other cells, and the environment. Signal transduction through chemicals involves:

(a) Signalling molecule that is either found inside or on the surface of the cell
(b) Cell surface or intracellular protein receptors to which these signalling molecules bind and activate
(c) A signalling cascade from one molecule to another, which finally culminates in cells' response to the stimulus.

Depending on the nature of ligand and location of action, cell signalling can be divided into the following types:

Contact-dependent signalling: In this, for cell–cell communication, the membranes of cells must be in direct contact with each other. For example, signalling molecule is bound to a cell and influences only cells that are in direct contact. It is seen during immune action and development.

Autocrine signalling: It is a self-regulating signalling mechanism where cells communicate with themselves. The signalling molecule and the receptor to it are found on the same cell. For example, cancer cells produce signals that stimulate their own survival and proliferation.

Paracrine signalling: Signalling molecules are secreted and act as local mediators within a short distance or neighbouring cells. This is a short-range signalling process. For example, neurotransmitters across synapse in nerve signalling.

Endocrine signalling: It is long-range signalling mechanism where the signalling molecules are secreted in blood stream and communicate with distant cells. The signalling molecules are called hormones, and distant cells having the specific receptor to the hormone will respond to it (Figure 12.8).

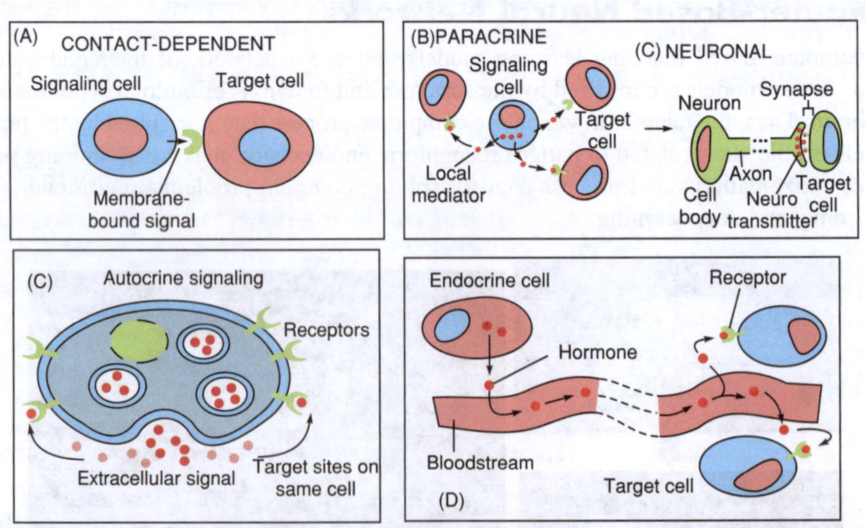

Fig. 12.8 Different types of cell signalling

https://x.com/CbeHx/status/1735689952493510935?mx=2

https://www.linkedin.com/pulse/types-signaling-paracrine-autocrine-endocrine-neuronal-ibrar-khan/

Endocrine Glands

There are two types of glands in the body. Glands are organs that manufacture molecules and are either secreted into the bloodstream or secreted through specific ducts into the target organ or tissue. Endocrine glands have no ducts; they secrete their hormones directly into the bloodstream, whereas exocrine glands, such as digestive glands, sweat glands, etc., empty their products through ducts.

There are nine endocrine glands in the human body that secrete hormones and are part of generating responses. They include hypothalamus, pineal gland, pituitary gland (anterior and posterior), thyroid gland, parathyroid gland, thymus, adrenal glands (adrenal cortex and adrenal medulla), pancreas (Islets of Langerhans), and sex glands (ovaries in females and testes in males) (Figure 12.9).

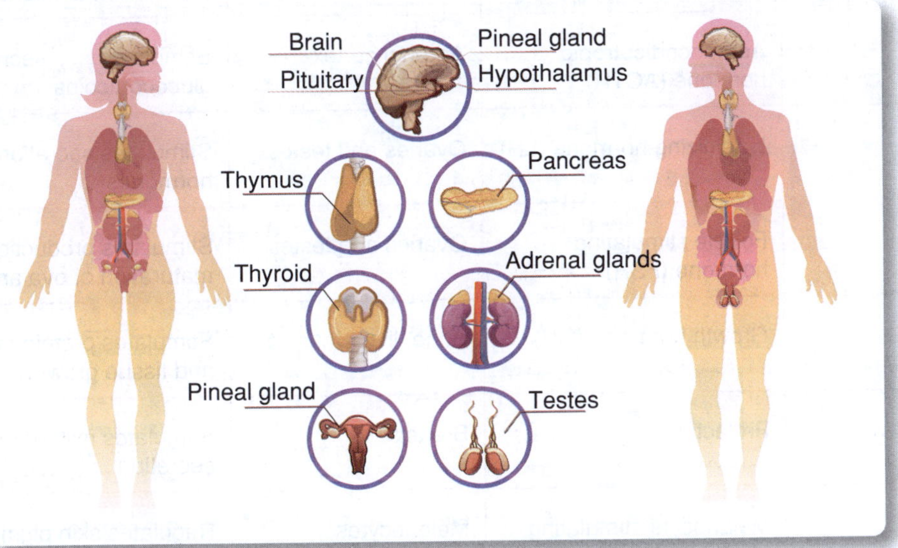

Fig. 12.9 Different glands of the endocrine system in humans

The endocrine system functions like a radio broadcast system just as radio broadcast systems send their signals in all directions and radio instruments tuned to a specific frequency receive the signals. On similar lines, hormones are distributed throughout the body by the circulatory system; but only cells that have the appropriate receptors will be able to respond to them. Hormones have two specific features: they exist in low concentration in blood (0.01–0.1 mg/ml), and they trigger specific target cells to either stimulate or suppress a particular reaction. The target cells upon being stimulated by hormones can respond in one of the three ways:

- some cells grow and divide
- others start manufacturing products and initiate metabolic activities
- and some release pre-formed molecules from the cells by secretion.

Table 12.1 Endocrine Glands, Their Secretions, Target Cells, and Their Functions

Gland	Hormone Secreted	Target Organ	Major Functions
Hypothalamus	Multiple peptide neurohormones, called releasing hormones (RH) and inhibiting hormones (IH). RH and IH regulate hormone secretion by other endocrine glands	Pituitary gland	Main control centre of endocrine system
Anterior pituitary	Thyroid stimulating hormone (TSH)	Thyroid gland	Stimulates secretion of thyroxine and triiodothyronine
	Adrenocorticotropic hormone (ACTH)	Adrenal cortex	Stimulates secretion of glucocorticoids
	Luteinizing hormone (LH)	Ovaries and testes	Stimulates secretion of sex hormones
	Follicle stimulating hormone (FSH)	Ovaries and testes	Stimulates production and maturation of ova and sperm
	Growth hormone (GH)	Bone, liver, muscle	Stimulates protein synthesis and tissue growth
	Prolactin	Breasts	Stimulates milk production and secretion
	Melanocyte stimulating hormone (MSH)	Melanocytes	Regulates skin pigmentation.
	Endorphins	Peripheral nervous system - pain pathway	Inhibits pain sensation and perception
Posterior pituitary	Anti-diuretic hormone (ADH)	Kidney	Involved in water balance
	Oxytocin	Uterus	Stimulates uterine contractions
Thyroid gland	Thyroxine and Triiodothyronine	Multiple tissues	Stimulates metabolism and necessary for normal body growth
	Calcitonin	Bone	Lower blood calcium level and promote bone formation

Parathyroid gland	Parathyroid hormone (PTH)	Bone, kidneys, intestine	Increases blood calcium levels; vitamin D activation
Thymus	Thymosin	Immune system	Stimulates T-lymphocytes
Pancreas	Insulin	Muscle, liver, adipose tissue	Lowers blood sugar
	Glucagon	Liver and adipose tissue	Raises blood sugar
Adrenal medulla	Epinephrine and norepinephrine	Sympathetic receptor sites in the body	Reinforce stress responses
Adrenal cortex	Cortisol: Glucocorticoid	Most cells, muscle and immune system	Increase blood glucose: emergency response by breaking down proteins and fats; constricts blood vessels and increases heartbeat
	Aldosterone: Mineralocorticoids	Kidney	Increases Na^{2+} reabsorption and K^{+} excretion
Pineal gland	Melatonin	Hypothalamus	Synchronizes body's biological clock and inhibits release of sex hormones and initiation of puberty
Ovaries	Oestrogen	Mammary glands, uterus, and female reproductive organs	Stimulate maturation of sex organs and development of secondary sexual characteristics in females
	Progesterone	Uterus	Prepare and maintain uterus for implantation for growth and development of foetus
Testes	Andogens: Testosterone	Multiple tissues and organs	Stimulate sperm formation and promotes development of secondary sexual characteristics in males

The speed of response in the endocrine system may also vary. Some are very rapid, and others are very slow. The variable speed of action of different hormones of the endocrine system can be demonstrated by the following examples.

Rapid action: Release of epinephrine and norepinephrine from adrenal medulla causes rapid change in behaviour in the body; quickly after its secretion, the heart rate increases, blood pressure rises, increased movement of blood to muscles, and increased breathing rate is observed. A sudden scare, or a collision, or a fall has helped all of us to experience these at some point in our lives.

Moderate action: Antidiuretic hormone (ADH) released from the posterior pituitary gland acts over a period of time more slowly and regulates the water balance of the body through the kidney. The effects of ADH secretion can be noticed within a few minutes to a few hours.

Slow action: Growth-stimulating hormone is secreted slowly by anterior pituitary over a period of years to achieve typical human growth. Levels drop after sexual maturity, and body growth stops. Similarly, sexual development by growth of specific tissues and organs by the sex hormones like testosterone in males and oestrogen in females happens over long periods, stimulating growth of sexual tissues, resulting in normal gender pattern of development.

Integration and Influence of Nervous and Endocrine System on Each Other

In certain instances, it has been observed that the nervous and endocrine systems interact with each other to modulate or enhance response. Few examples where such interaction is observed are shown below.

1. The pituitary glands are constantly receiving information from the hypothalamus (The posterior pituitary directly connected to brain, and the anterior pituitary receives continuous input of messenger molecules from brain.); sensory inputs to hypothalamus can affect and influence the endocrine system too. Menstrual cycle in women is mainly regulated by hormones of the endocrine system; however, it is well documented that stress, tension, or worry can interfere with the normal levels of hormones and thus delay or stop menstrual cycles.
2. In many animals, the changing length of day causes hormonal changes that induce reproductive outcomes. During spring, birds respond to lengthening of days and start producing reproductive hormones for the summer breeding season. The pineal gland acts as a light sensor and stimulates hormone secretion by pituitary, which, in turn, influences other reproductive hormones. This leads to changes in behaviour; for example, courtship, mating, and nesting activities are enhanced. Here, the endocrine system is influencing the nervous system, and hormone levels affect physical behaviour.
3. Use of anabolic steroids by humans to increase muscle mass causing mood swings and behavioural changes is another example of such an influence.

So, though the nervous and endocrine systems appear separate and distinct, they are also interconnected in some ways. Both the systems cooperate to bring about appropriate responses to environmental challenges. To understand in a nutshell, the nervous system is functioning as a specialized system to receive and send short-term messages and immediate response; the endocrine system, on the other hand, regulates activities that require long-term, growth-related changes.

Interesting Fact 12.3

Endocrine system—molecules, mimics, and disruptors

- Endorphins are the natural pain killers of the body produced by pituitary glands. They associate with receptors in the brain that perceive pain. Hence, the feeling of pain is lost, and euphoric feeling takes over. This action of endorphins is similar to opiate drugs and morphine.

- Athletes and long-distance runners experience increase in endorphin release, leading to 'runner's high'. Similar outcomes in terms of increase in endorphin levels can also be seen with acupuncture needles and mild electric stimulation.
- Synthetic mimics are used to boost function in individuals with specific hormone deficits. For example, hormonal decline due to ageing like mitigating oestrogen drop in menopausal women.
- Anabolic steroids are used to treat gradual decline of testosterone in men.
- Similarly, drugs can be synthesised to inhibit hormones. For example, Tamoxifen inhibits action of oestrogen by binding to oestrogen receptors in breast cells and acts as a therapeutic action.
- Athletes use anabolic steroids to build muscle mass.
- Endocrine disruptors are natural or man-made chemicals that may mimic or interfere with the body's hormones, known as the endocrine system.
- Endocrine disruptors are found in many everyday products, including some cosmetics, food and beverage packaging, toys, carpets, pesticides, and flame retardants. For example, bisphenol, dioxin, phthalates, polychlorinated biphenyls (PCBs), polybrominated diphenyl ethers (PBDEs), etc.

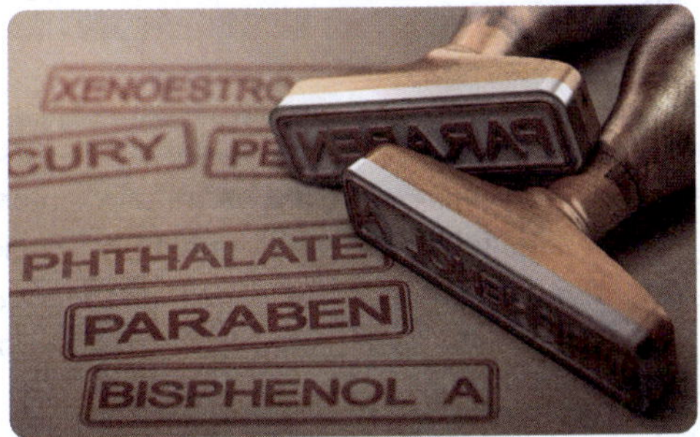

SUMMARY

- The basic unit of nervous system is neuron; its structural details and functional roles were studied.
- The classification and organization of the nervous system into different components and details of each system is understood.
- Mechanism of generation of nerve impulse and information transfer across neurons was elaborated.
- Structure of brain parts and their associated functions were discussed.
- Endocrine glands, their secretions, and their functions were explained and understood.
- Examples of integration and influence of nervous and endocrine systems to provide appropriate responses were highlighted.

EXERCISES

A. Objective-Type Questions

1. ______________ is the space between adjacent neurons.
   (a) Axon (b) Dendrite
   (c) Myelin sheath (d) Synapse
2. ______________ are ductless glands that secrete their products directly into the bloodstream.
   (a) Exocrine (b) Endocrine
   (c) Digestive (d) Salivary
3. ______________ is the gland that secretes insulin and glucagon.
   (a) Pituitary (b) Salivary
   (c) Pancreas (d) Adrenal glands
4. Which of the following is part of the brain stem or base of the brain?
   (a) Medulla oblongata (b) Cerebellum
   (c) Cerebrum (d) Pineal gland
5. The fluid surrounding the brain and spinal cord is ______________.
   (a) Sweat (b) Cerebrospinal fluid
   (c) Endorphin (d) Meninges
6. Which combination is incorrect?
   (a) Ovary–oestrogen (b) Thyroid–thyroxine
   (c) Pancreas–oxytocin (d) Adrenal medulla–epinephrine
7. ______________ of the neuron secretes neurotransmitters like acetylcholine inside the synaptic cleft.
   (a) Dendrite (b) Axon
   (c) Exon (d) Intron
8. Which of the following statements is incorrect?
   (a) Neurons have a resting membrane potential of –70 mV.
   (b) Change in membrane permeability causes depolarization.
   (c) Depolarization causes Na^{2+} ions to move from inside the neuron to outside.
   (d) Depolarization at one point causes the adjacent point to change its permeability and depolarize.

9. Which is the wrong combination?
   (a) Autonomic nervous system: sympathetic and parasympathetic
   (b) Sensory neurons: carry impulse from sense organs to brain
   (c) Peripheral nervous system: brain and spinal cord
   (d) Motor neurons: efferent nervous system
10. ____________ is not the endocrine gland of the human body.
   (a) Thyroid (b) Thymus
   (c) Pons (d) Pineal gland

B. Short-Answer Questions

1. What is a blood–brain barrier, and how does it function?
2. What will happen if acetylcholinesterase does not destroy acetylcholine at synapse?
3. Explain how the speed of response of endocrine hormones varies with suitable examples?
4. Why is hypothalamus called the control centre of the human endocrine system?
5. Identify the hormones of the pituitary and their target organs.

FURTHER READINGS

1. https://www.mayoclinic.org/diseases-conditions/epilepsy/in-depth/brain/art-20546821: How the brain works?
2. https://www.healthdirect.gov.au/central-nervous-system; Central nervous system.
3. https://www.endocrine.org/patient-engagement/endocrine-library/hormones-and-endocrine-function
4. Endocrine disrupting chemicals - https://www.endocrine.org/topics/edc

USEFUL WEBLINKS

1. https://www.youtube.com/watch?v=NKpkU4-Sf9A
2. https://www.youtube.com/watch?v=b6m35PlH7t8
3. https://www.youtube.com/watch?v=0GSRbmcNh3A

CHAPTER

14 Physiological Systems of Humans: Immune System

LEARNING OBJECTIVES

After reading this chapter, you will be able to:

- To understand the basic tenets and types of immunity in humans
- To understand the different cells and organs of the immune system
- To elaborate on the innate immune mechanisms of the body
- To assess the need and basic features of adaptive immunity
- To delineate the humoral immune response and its mechanisms
- To understand the components and functions of cell-mediated immunity
- To learn how the immune system fails—hypersensitivity, autoimmunity, and immunodeficiency

CHAPTER TAKEAWAY Immunity—immune system and immune response

This chapter deals with how the human body responds against a foreign onslaught and defends itself. The details of the whole immune process and the multiple players in the human defence mechanism called the immune system are explained.

PRELUDE

Whenever we think of the words 'defence' and 'protection' what comes to our mind is a multi-level security system that involves barriers and patrolling security men offering protection; what would also be inbuilt within such a system are alert mechanisms, specific units specializing in different aspects of combat, and quick and effective decision-making and control at the highest level. It is interesting to note that the animal body is also well-protected from invaders and intruders by a multi-level protection system called the immune system. On the surface, we may think that animals with heavy plates of armour on their body are well-protected, such as turtles; however, that is far from true. These heavy outer surfaces, however, offer limited protection against the major invaders of body—the bacteria and viruses; and to counter them a well-evolved immune system is needed that strategically and specifically protects the body from the onslaught of any foreign intrusion by both—unwavering prevention and effective elimination. The microbes in environment are diverse, omnipresent, omnipotent, and in numbers that far outweigh the cells of the human body; under this scenario, constant infections and diseases should have been the natural state to be for the human body; however, we rarely fall sick, and the reason for this is the highly effective immune system that tirelessly works behind the scenes to offer protection.

Immune System of Humans

Immune system is essential for fighting the harmful substances and germs that enter the body and cause disease. It protects the body from external agents as well as from internal changes.

Basically, the immune system helps to keep us healthy; we would be sick more often without an immune system. The immune system is a vast and interconnected system made of cells, organs, and proteins that work together to protect the body from illness. Effective defence mechanisms in the form of the immune system against constant onslaught of pathogens is a critical reason for our sustenance and survival.

Immunity in Humans

Immunity is the body's ability to defend against foreign agents. The protection offered by the immune system makes us resistant to diseases. This resistance to diseases is called 'immunity'. Often, we find that this immunity is not uniform, and some individuals have greater resistance to infection as compared to others. Imagine a scenario where three friends are walking by and one of the friends sneezes; out of the two other friends one may catch the infection from this friend quickly while the other may be immune and hence resistant to the invading pathogen. Thus, immunity differs among individuals and is due to the variation in level of protection offered by the immune system.

Immunity can be defined into various types and explained using the following terms. Some of the terms may overlap for the same type of immunity but described differently.

1. ***Innate immunity***: It is the natural immunity or general protection that each one of us are born with. The innate immune mechanisms are inherited and are active from the time an individual is born. This type of protection is due to the physical barriers and other mechanisms that all of us possess to prevent and fight infections. For example, the skin, the eye's cornea, and the mucous membrane that lines the respiratory, gastrointestinal, and genitourinary tracts, bodily fluids, and cells and tissues of the body offer innate immunity. These keep harmful agents away from entering the body and are usually the first line of defence against invaders.

   Natural immunity: The innate mechanisms that the humans are born with and are part of their natural being are called natural immunity. Innate immunity is also called natural immunity.

   Non-specific immunity: The innate immunity responds the same way to all germs and pathogens. Hence, innate immune response is also referred to as non-specific immune response.

2. ***Adaptive immunity***: It is a specialized system that takes over if the innate immune system fails to destroy the pathogens. The adaptive immune response is specific to the germ or intruder causing the infection. The effector cells of adaptive immunity are the lymphocytes. This system gets activated after the innate system; hence it is slower as compared to the innate system but is more specific and accurate. Adaptive immunity also has the ability to remember invaders and ability to provide a more robust, faster response when it encounters the same invaders again.

   Acquired immunity: As adaptive immunity is against a specific pathogen, it needs to recognise the pathogen and develop tools for its elimination. As the body learns to recognise and attack specific antigens by developing the range of effectors against specific pathogens over a period of time, it acquires the ability to tailor its attack to a specific pathogen; hence, adaptive immunity is also called acquired immunity.

   Specific immunity: Adaptive immunity is also called a specific immune response as it is targeted against a specific germ or pathogen that invades the body. Hence, such a specific response against one microbe makes adaptive immunity to be synonymized as specific immunity too. Adaptive immunity is of two types:

2 (a) ***Humoral immunity*:** It is also called antibody-mediated immunity and is a subtype of adaptive immunity where the effector molecules are the B-lymphocytes that secrete antibodies and protect against freely circulating antigens found in the blood and other body fluids outside the cells. The term 'humor' refers to body fluids. This type of immunity is called humoral immunity as antibodies are secreted in blood, protect through fluids, and remove pathogens from fluids.

Effector molecules of humoral response: The 'antibodies' against a pathogen can be obtained by the body in two ways: a) either by actively exposing the body to the specific microbe for generating antibodies or b) by passively procuring the readymade antibodies specific to a pathogen from outside, which helps in recognizing and targeting the pathogen.

Active immunity: When the pathogen directly activates the immune system to produce antibodies within the body, the immunity that results is called active immunity. For example, exposure of pathogens through infection or vaccination and triggering the body's own immune system to produce antibodies and memory cells for long-lasting protection.

Passive immunity: A type of immunity when preformed ready-made antibodies are received from another source, rather than producing them through their own immune system. For example, an infant receiving antibodies from mother through placenta is naturally passive; on the other hand, it becomes artificial passive immunity when antibodies from other sources are injected. Passive immunity is short-term immunity achieved through transfer of antibodies.

2 (b) ***Cell-mediated immunity*:** It is another subtype of adaptive immunity that deals with intracellular pathogens like viruses and infected or abnormal cells of the body. This type of immunity is mediated by T-lymphocytes, which work together with macrophages to eliminate the invader. Cell-mediated immunity is usually active as it involves cells and cell products of an individual's own immune system.

Overall, to summarize, the multifaceted and multilayered natural or acquired immunity mechanisms offered by the immune system of humans functions effectively in ensuring protection against foreign intruders (Figure 14.1).

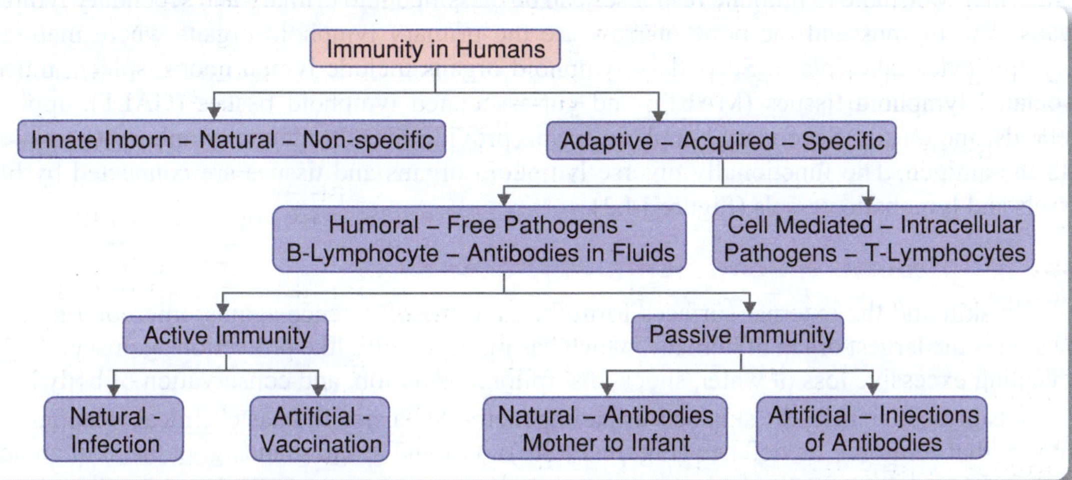

Fig. 14.1 Immunity and types

Cells and Organs of the Immune System

The components of the immune system form part of the multi-level protection mechanism operating inside the body and are analogous to any other advanced security system. The human body is defended by the immune system in the same manner as forts and cities are defended.

The multilevel security of a city would include physical barriers like walls and manned gates that prevent entry; if this is crossed or breached, then roaming patrols inside spot strangers and remove them; however, if there is doubt or if they are powerful individuals, they are directed to specific officers who ascertain specific identity and initiate appropriate action.

Similarly, the outermost layer of vertebrate body—the skin—functions as a physical barrier that prevents entry of microbes; mucous membranes in respiratory, digestive, and urogenital tracts act as manned gates that block intruders; the battery of cells and proteins in blood stream like phagocytic cells, macrophages, and neutrophils along with proteins like complement and interferons act as roaming patrols that mount a counter attack on foreign entrants; and, finally, the B- and T-lymphocytes of adaptive response act as specific officers that ascertain specific identity and initiate appropriate response to eliminate freely circulating foreign cells and viruses or intracellular pathogens, respectively.

Table 14.1 Three Lines of Defence in Human Immune System

Non-Specific Defences (Innate Immunity)		Specific Defences (Adaptive Immunity)
First Line of Defence	**Second Line of Defence**	**Third Line of Defence**
• Skin	• Phagocytic leukocytes	• Lymphocytes
• Mucous membranes	• Antimicrobial proteins	• Antibodies
• Secretions of skin and mucous membranes	• Inflammatory response • Fever	• Memory cells

Lymphoid organs

Organs that contribute to immune responses can be classified into primary and secondary lymphoid organs. The thymus and the bone marrow are the primary lymphoid organs where maturation of lymphocytes takes place. Secondary lymphoid organs include lymph nodes, spleen, mucosa-associated lymphoid tissues (MALT), and gut-associated lymphoid tissues (GALT), appendix, adenoids, and tonsils. Secondary lymphoid organs provide sites for mature lymphocytes to interact with the antigen. The functionally diverse lymphoid organs and tissues are connected by blood vessels and lymphatic vessels (Figure 14.2).

Innate Immune Mechanisms of the Body

The skin and the external surfaces form the first line of defence against microbial intrusion. The skin is the largest organ of humans, which has multiple roles like protection against pathogens, preventing excessive loss of water, shock absorption, insulation, and conservation of body heat.

In its protective role, the skin acts as an impenetrable barrier and has chemical weapons on its armoury that reinforce its defence (Figure 14.3(a)). Oil and sweat glands give the skin an acidic pH of 3–5 to inhibit growth of organisms; sweat also contains an enzyme lysozyme that destroys bacterial cell walls and offers protection. The other potential routes of entry of microbes—the digestive tract, respiratory tract, and urogenital tract, which are open to the outside, also need to

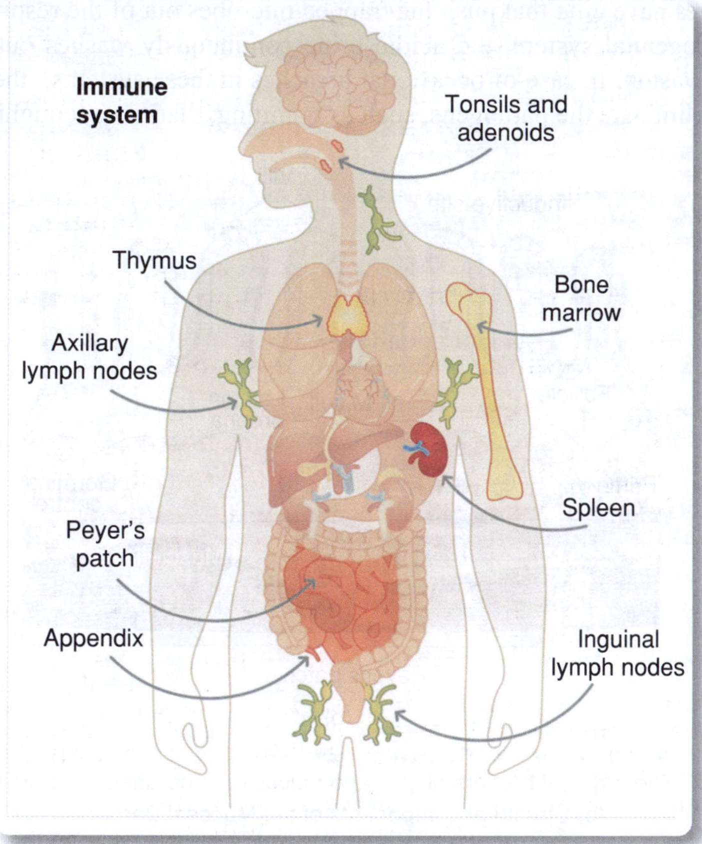

Fig. 14.2 Lymphoid organs of the immune system

be protected. Lysozymes in the saliva of the mouth, the acidic environment of the stomach, and digestive enzymes in the intestine guard against pathogen entry through the digestive tract. The microbes entering through inhaled air are trapped in the sticky mucus of the nose and the respiratory

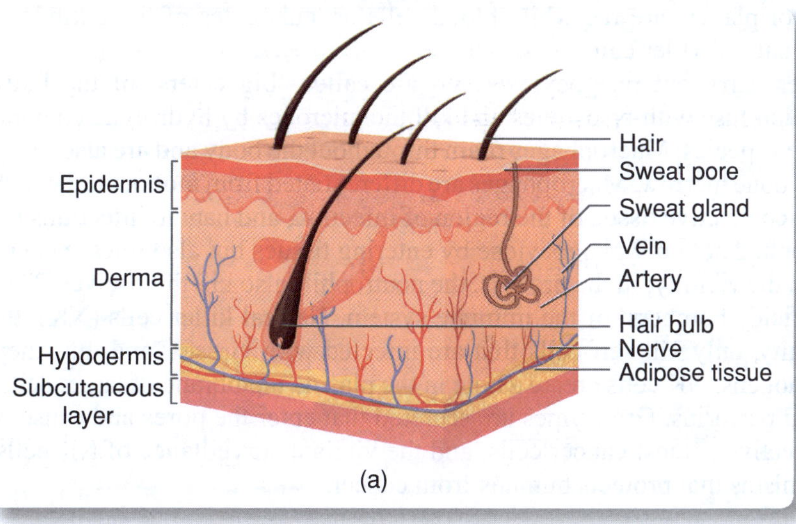

tract. These passages have cilia that push the trapped microbes out of the respiratory tract into the stomach. In the urogenital system, the acidic urine continuously washes out pathogens, which prevents foreign invasion. In case of occasional breaches in these systems, the body responds by other measures to eliminate the pathogens, such as vomiting, diarrhoea, coughing, sneezing, etc.

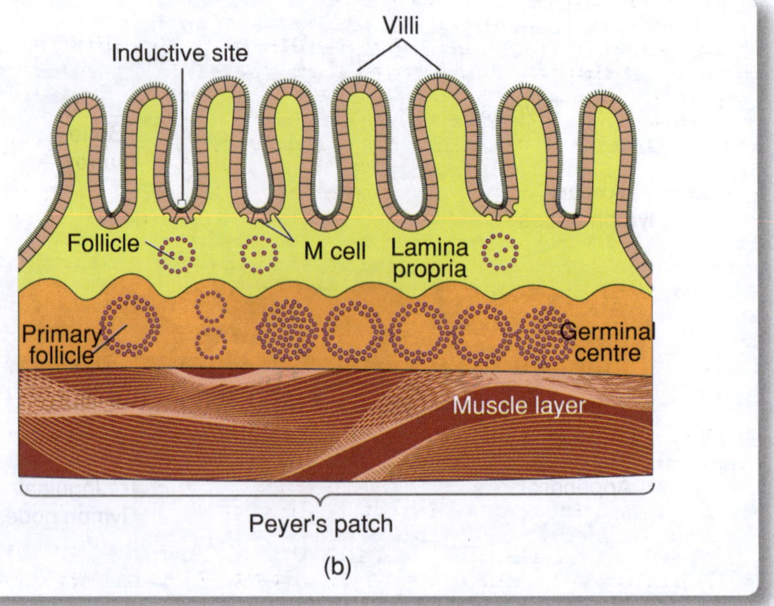

Fig. 14.3 The physical barriers of skin and mucous membranes: (a) Skin Anatomy; (b) Structural components of the Mucosal Barrier

Beyond these too, sometimes the pathogens can cross the surface defences and enter the body. At this point, the second line of defence consisting of the cellular and chemical devices are used by the body to defend itself. These components act on any microbial invasion without having the need to ascertain the identity and form the non-specific response of the body. These cells are present throughout the body but are concentrated in the lymphatic system and lymph nodes, where foreign invaders are efficiently identified and eliminated as body fluids pass through. Under this category of cells, the major players are the white blood cells or leukocytes of three kinds: macrophages, neutrophils, and natural killer cells.

Macrophages carry out phagocytosis and are called 'big eaters' of the body. They form phagosomes, which fuse with lysosomes and kill the microbes by hydrolytic enzymes and release of reactive oxygen species. Macrophages roam throughout the body and are also localized in lungs, liver, spleen, and bone marrow. Macrophages are differentiated from monocytes, which squeeze into capillaries, enter connective tissue of the region of infection, and handle infections in all regions of the body. Neutrophils not only phagocytose by entering tissues but also release chemicals that kill all the bacteria in the vicinity; in the process, the neutrophils also kill themselves. Thus, neutrophils function like 'suicide bombers' of the immune system. Natural killer cells (NK), though are part of innate immunity, only kill self-cells that are infected with viruses, and thus they take care of intracellular pathogens. NK cells create a hole in the plasma membrane of target cells with the help of proteins called perforins. Granzymes are secreted that enter the pores and cause cell death. NK cells are also effective against cancer cells, and the vigilant surveillance of NK cells is one of the strongest mechanisms that protects humans from cancer.

Apart from these cellular defences, there is also an effective chemical defence that enhances the chances of microbial elimination. Upon contact with bacterial or fungal cells a system of 20 different proteins in blood called the complement system aggregate themselves to form a membrane attack complex (MAC) that inserts itself into the foreign cell and makes the cell osmotically fragile due to pore formation, leading to its destruction. Another class of proteins with a key role in innate defence are the interferons, which exist in three types: alpha, beta, and gamma. Alpha and beta interferons, produced by almost all cells, protect their neighbouring cells from viral infection by preventing viral multiplication and viral assembly in these cells. Gamma interferons are produced only by select lymphocytes and NK cells and are of great help in cancer and infections.

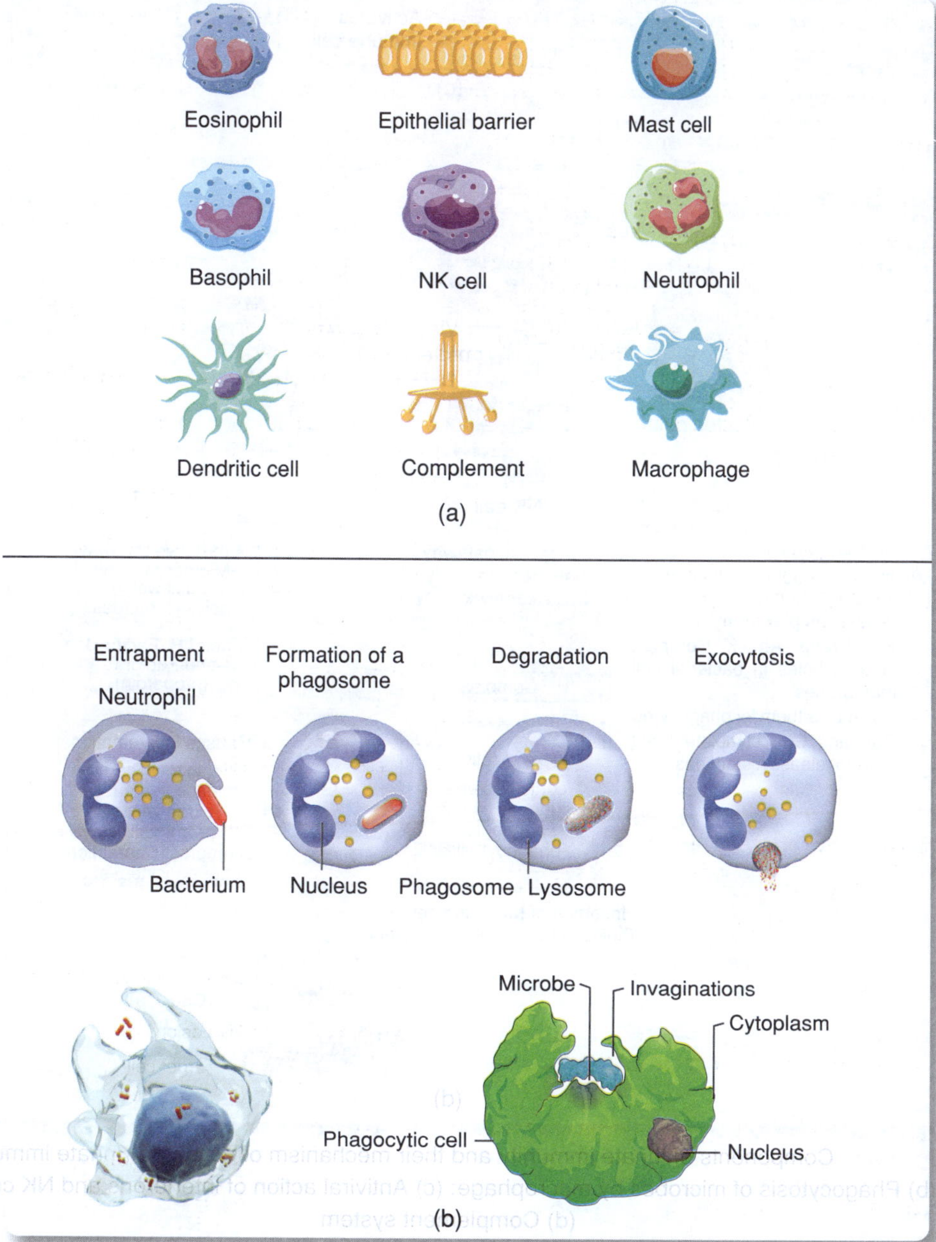

Eosinophil
Epithelial barrier
Mast cell
Basophil
NK cell
Neutrophil
Dendritic cell
Complement
Macrophage

(a)

Entrapment
Neutrophil
Formation of a phagosome
Degradation
Exocytosis
Bacterium
Nucleus
Phagosome
Lysosome
Microbe
Invaginations
Cytoplasm
Phagocytic cell
Nucleus

(b)

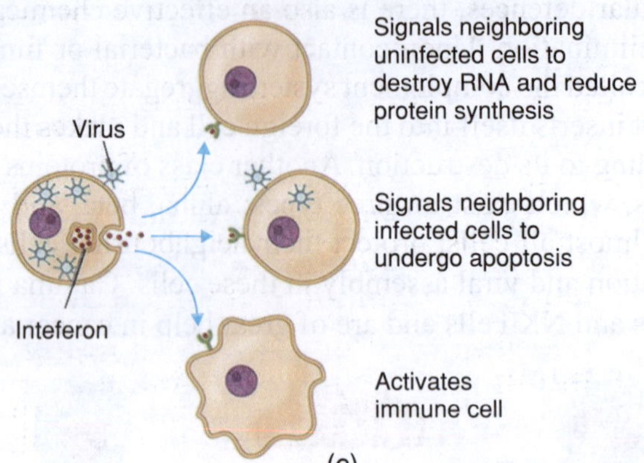

Signals neighboring uninfected cells to destroy RNA and reduce protein synthesis

Virus

Signals neighboring infected cells to undergo apoptosis

Inteferon

Activates immune cell

(c)

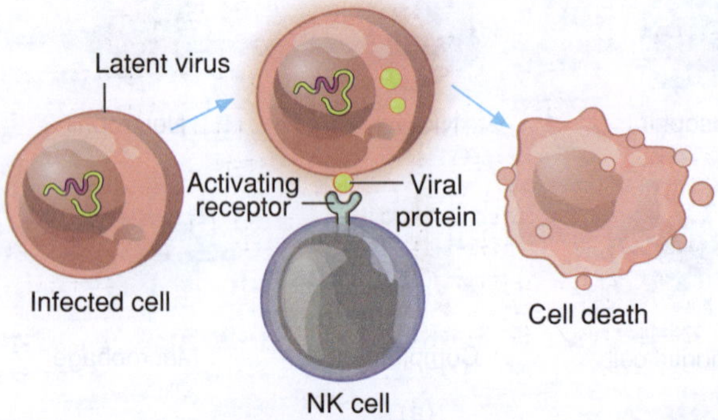

Latent virus

Activating receptor

Viral protein

Infected cell

Cell death

NK cell

- > 20 plasma proteins
- Activation triggers cascade of chemical reactions
- Molecular complex form:
  - Membrane Attack Complex creates holes in bacterial cell membranes
  - C3b marks them for phagocytes
  - C3a and C5a stimulate mast cells to release histamines

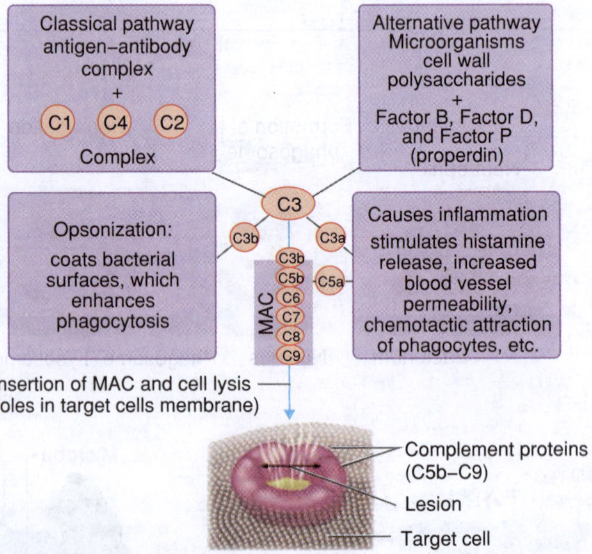

Classical pathway antigen–antibody complex + C1 C4 C2 Complex

Alternative pathway Microorganisms cell wall polysaccharides + Factor B, Factor D, and Factor P (properdin)

C3

C3b

C3a

Opsonization: coats bacterial surfaces, which enhances phagocytosis

Causes inflammation stimulates histamine release, increased blood vessel permeability, chemotactic attraction of phagocytes, etc.

MAC: C3b C5b C6 C7 C8 C9

C5a

Insertion of MAC and cell lysis (holes in target cells membrane)

Complement proteins (C5b–C9)

Lesion

Target cell

(d)

Fig. 14.4 Components of innate immunity and their mechanism of action: (a) Innate immunity; (b) Phagocytosis of microbes by macrophage; (c) Antiviral action of interferons and NK cells; (d) Complement system

Source: Adapted from https://www.lecturio.com/concepts/interferons/

Interesting Fact 14.1

Inflammation and Fever—What Do They Signify?

- All of us have smashed a finger or had a wound at some point in our lives. What do we see as reaction from the body when this happens? We see swelling, warmth, redness, pus, etc. Why do these happen? It is because of inflammation.

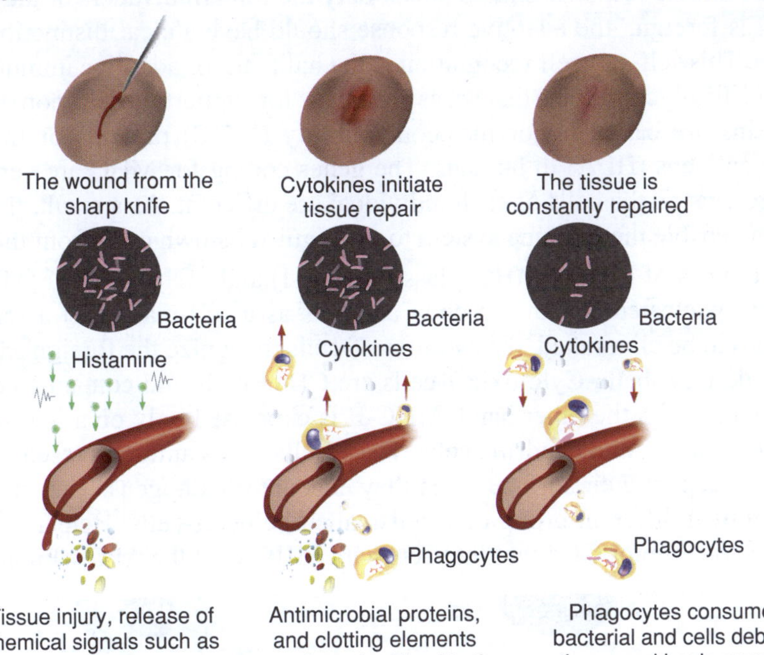

- The injured cells release alarm signals called histamines and prostaglandins. They produce dilation of blood vessels and blood flows to the wound, causing swelling, redness, and warmth. Permeabilities of capillaries (oedema) bring phagocytes (macrophages and neutrophils) into the extracellular fluid from blood, where they attack bacteria. Spillage of neutrophils kills bacteria, tissue, and themselves, and these dead cells and debris together form pus. Macrophages follow, and slowly all the dead cells and tissues are engulfed, clearing the place, and finally the wound heals. Therefore, inflammation, swelling, redness, and pus are all indicators of the innate cells in action.

Bacterial endotoxins as well as IL-1 released from macrophages direct hypothalamus to raise the body temperature, causing fever. Fever enhances phagocytosis and directs the liver and spleen to pull back iron, which is essential for bacterial growth. Thus, fever up to a certain level is a healthy indication of macrophages functioning and the bacterial infection being controlled.

Adaptive immune response

It is the third line of defence of the immune system in humans. Lymphocytes are the major effector molecules of adaptive immune response. They have specific receptor proteins with the help of which they identify the specific microbes and mount a specific response. The two types of lymphocytes that are critical to specific immune response are the B-lymphocyte for humoral

immune response and the T-lymphocyte for cell-mediated immune response. The surface molecules of pathogens (protein or polysaccharide) that provoke a specific immune response are called **antigens**. These proteins are foreign to the body and may contain several sub-parts called epitopes or antigenic determinants. These can act as independent antigens and stimulate different specific immune responses.

The adaptive immune response should attack only the foreign invaders or the abnormal cells. To identify 'what is foreign,' the adaptive response should have a clear distinction between self- and non-self-cells. This self–nonself recognition is the hallmark of adaptive immune response. All self-cells have specific glycoproteins that act as alert flags for supporting self–non-self-recognition. These glycoproteins are called major histocompatibility (MHC) proteins or more specifically human leukocyte antigens (HLA) in humans. The genes coding for MHC are very polymorphic, and thus MHC proteins on the cells of each individual are different. As a result, these proteins act as self-markers and enable the immune system to distinguish its own cells from the foreign cells.

There are two types of MHCs: MHC Class I (MHC-I) and MHC Class II (MHC-II). MHC-I is present on every nucleated cell of the body and acts as a self-marker. If a self-cell becomes abnormal and needs to be eliminated, the cytotoxic T-cells recognize the foreign antigen presented with MHC-1 and destroy them. Cytotoxic T-cells are CD8+ cells and contain a co-receptor CD8 that recognizes MHC-1. On the other hand, MHC-II is expressed only on a few specific types of cells: B-cells, macrophages, and dendritic cells. These cells act as antigen-presenting cells (APCs) and present the antigens on their surface when they encounter pathogens. The antigen from these cells is recognized by a different group of T-cells called helper T-cells. Helper T-cells are CD4+ cells and contain a co-receptor CD4 that interacts with MHC-II of the APCs (Figure 14.5).

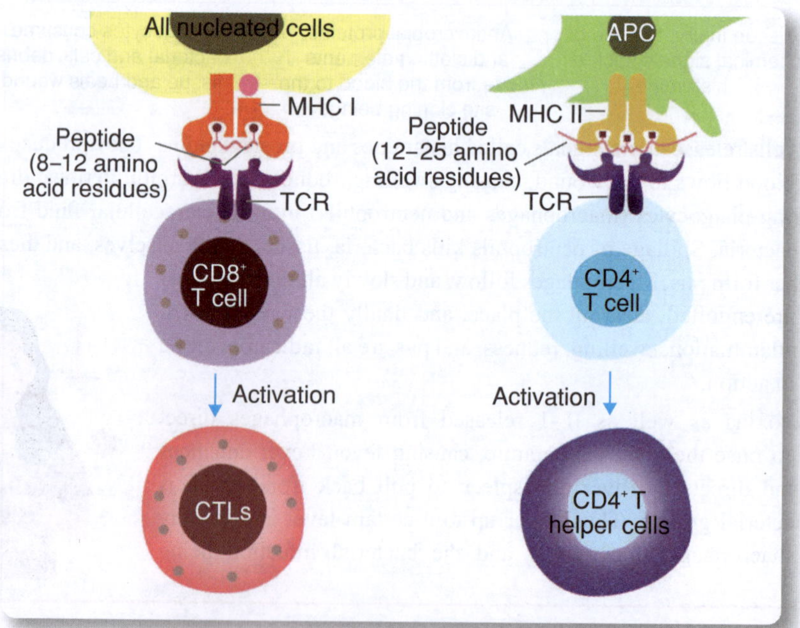

Fig. 14.5 MHC recognition by specific T-cells

Source: Adapted from https://www.researchgate.net/figure/Differences-in-recognition-and-downstream-processes-between-CD8-and-CD4-T-cells-CD8-T_fig2_350174562

In the context of the above and based on the functioning of immune response, let us understand the fate of a pathogen when it enters the human body. First, they are prevented from entering or being captured by mucus membranes; if they cross the mucus membranes then the pathogen comes in contact with macrophages in the bloodstream, which phagocytose them and present their antigens with MHC-II. The macrophages also secrete IL-1 (Interleukin-1), which acts as a chemical alarm signal. The T-helper cells recognize the antigen on the macrophage with the CD4 co-receptor, respond simultaneously to the released Interleukin-1, and gets activated. The T-helper cell activates two parallel lines of immune system defence: (a) Humoral response carried out by B-cells and (b) the cell-mediated response carried out by T-cells. Thus, the T-helper cell acts as a central molecule, connects the innate and adaptive immune system, and activates other immune cells in response to infection.

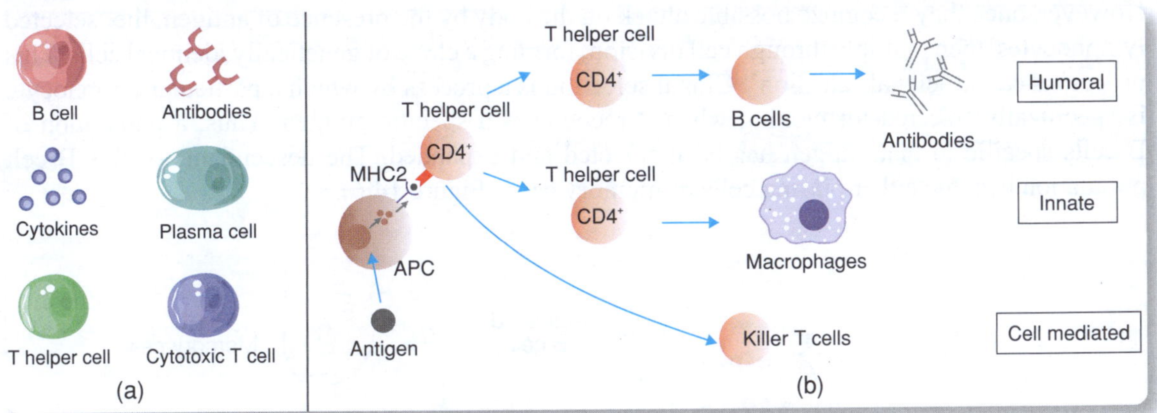

Fig. 14.6 Adaptive immunity: (a) Components of adaptive immune response; (b) Role of T-helper cells

Source: Adapted from https://www.wikidoc.org/index.php/T_helper_cell

The major functions of the T-helper cells are:

1. Detects infections by recognizing the foreign antigens presented on MHC-II by macrophages and other APCs.
2. They send signals to other immune cells, such as macrophages and dendritic cells, to fight infection through the release of cytokines and chemokines. Macrophage colony-stimulating factor and gamma interferon promote the activity of macrophages.
3. They activate B-cells to produce antibodies against the pathogen by secreting IL-4.
4. They activate the cytotoxic T-cells to kill infected cells with the help of IL-2.
5. Once the infection passes, suppressor immune cells terminate the response.

Overall, by coordinating all the arms of the immune system, the T-helper cells contribute towards tailoring the most appropriate and effective response against each pathogen.

Humoral immune response

The type of immune response mediated by the B-lymphocytes is called the humoral response or antibody-mediated immune response. Humoral response acts primarily against the freely circulating extracellular pathogens.

B lymphocytes mature in the bone marrow and contain receptors on their surface that recognize surface proteins or antigens on the pathogen. Each B cell has one type of receptor called antibody

displayed extracellularly in large numbers on the surface of the B cell, which is specific to antigen or antigenic determinant. There are innumerable B-cells, each with a unique antibody that allows the recognition of a diverse variety of pathogens. It is important to note that those B-cells that have specificity to the self-antigens are destroyed during maturation, thus preventing the immune system from attacking the host. Mature B-cells migrate to lymphatic organs such as lymph nodes and spleen, where they may encounter pathogens.

When a B cell encounters an antigen, the B-cell receptors or antibodies on the surface of the B cell recognize the antigen and bind to it. This complementary binding between the antigen and the antigen-specific receptor on the B-cell is specific similar to the 'lock and key' binding of enzyme and substrate. Other B-cells that do not have complementary receptors are not affected. Initially there are few B-cells in the body with complementary receptors to this specific antigen. However, once they recognize possible attack on the body by the presence of antigen, the 'selected lymphocytes' then multiply through cell division, forming a clone of genetically identical cells. This process is called 'clonal selection'. Clonal selection is a process by which a particular lymphocyte is specifically selected for cloning when it recognizes a specific antigen. Thus, a population of B-cells specific to that antigen has been selected and expanded. The descendants of this B cell population become either plasma cells or memory cells (Figure 14.6).

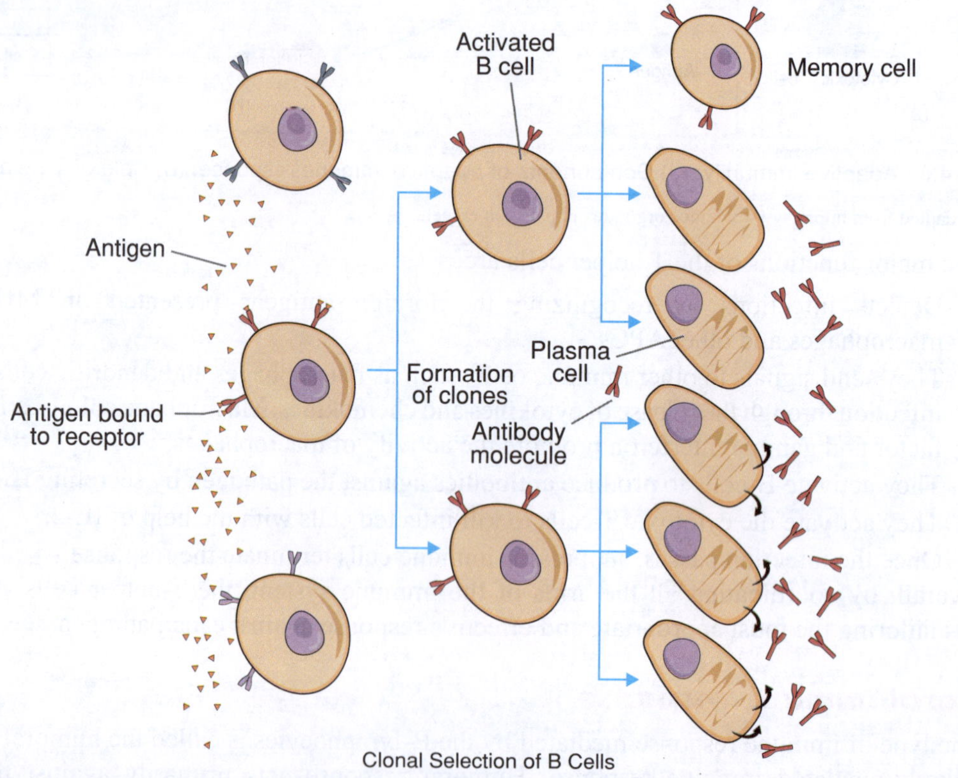

Clonal Selection of B Cells

Fig. 14.7 Activation of B-cells in humoral response to form plasma cells and memory cells

Plasma cells then act as short-lived effector B-cells that contribute to the immediate elimination of antigen by secreting antibodies into the blood and lymph. Antibodies are highly specific protein molecules produced by B-cells that recognize, combine with antigen, and help in clearing the pathogens from the body. Antibodies do not directly kill antigens or pathogens; instead they opsonize them and attract other blood molecules such as complement and phagocytes that eventually destroy the pathogen. Antibodies can also neutralize the toxins produced by pathogens by binding to them and preventing them from carrying out their damaging action (Figure 14.7).

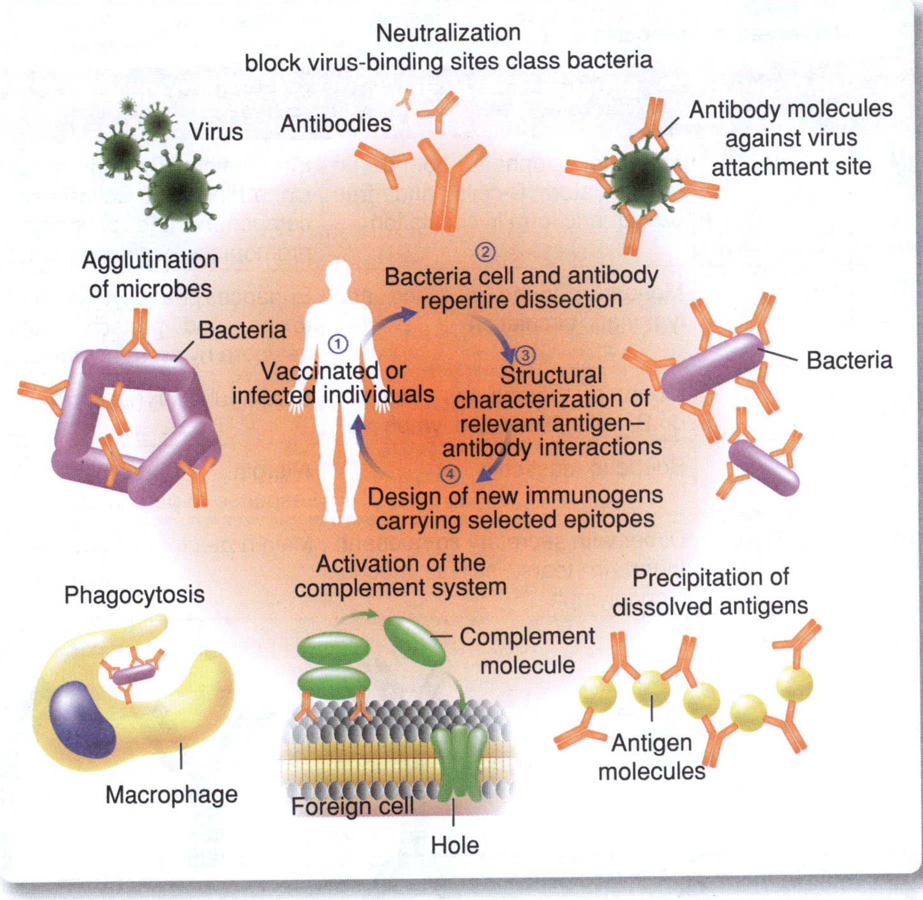

Fig. 14.8 Role of antibodies in elimination of pathogens

Antibodies are the core molecules in B-cell-mediated humoral response. They belong to a class of proteins called Immunoglobulins. Each antibody has a quaternary structure of four polypeptide chains with two heavy and two light chains. The chains are joined by disulphide bonds, and there is a constant region and a variable region within the chains. Variable regions help in diversity of antibodies and their ability to recognize different antigenic determinants. Based on the different types of constant regions, there are five classes of antibodies: IgM, IgG, IgA, IgE, and IgD.

Interesting Fact 14.2

All about Antibody Structure and Types

1. Antibody is also known as immunoglobulins. It is a Y-shaped structure with two polypeptides: two heavy chains and two light chains. Fab fragment is the antigen-bonding region consisting of one constant and one variable region of heavy and light chain. Fc region—tail of antibody—helps activate other arms of immune system containing Fc receptors and complement.
2. There are five classes of antibodies. Distinguishing features are given below (Table 14.1).

Table 14.2 Five Classes of Antibodies

Class	Ig-Constant Region	Structure and Presence	Functions
IgM	μ	Monomer suphate bound on unstimulated B-cells and free pentameric form in circulation	First antibody in primary response. Upon binding to antigen stimulates component and phagocytes, and promotes agglutination reactions.
IgG	γ	Monomer; found in blood and lymphatic circulation	Enhances phagocytosis, neutralizes toxins, and protects foetus and newborn by crossing placenta
IgD	δ	Monomer; membrane receptor on B- cells, blood and lymph	Serum function not known
IgE	ε	Bound to mast cells	Allergic reactions; immune response to parasitic worms
IgA	α	Dimer with secretary component; found in tears, saliva, mucus, blood, lymph	Main protector in mucosal surfaces

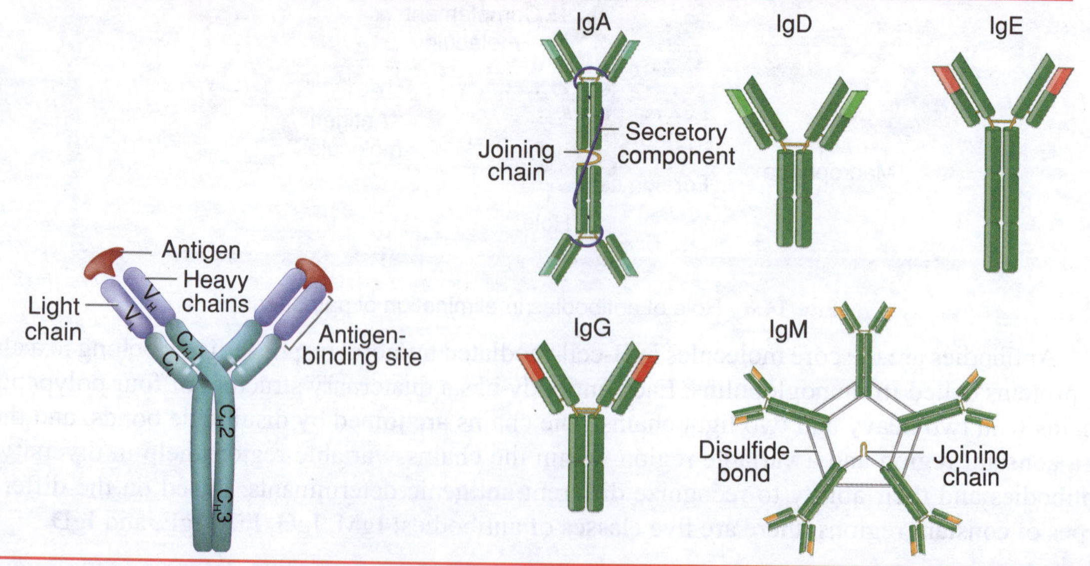

Till the infection is taken care of, there will be a higher number of plasma cells secreting antibodies and fighting the pathogen. Once the trouble is brought under control, the number of plasma cells drops. However, the clonal selection also produces the long-lived memory B-cells.

Memory cells are like reserve soldiers who have encountered the antigen and are familiar with them. They are present in the lymph nodes and can be triggered to attack if the same antigen were to infect again. Upon subsequent exposure, memory cells rapidly give rise to new effector plasma cells and help in generating a heightened rapid response. B-cells also act as antigen-presenting cells that express the antigen on their surface and activate T-helper cells, which, in turn, release cytokines to activate more B-cell response. So, this collective effort across the different arms of immune response eventually helps in complete elimination of the pathogen.

Production of plasma cells and memory cells by B-cell activation during the first exposure to antigen takes several days for clonal selection to produce effector cells. This is the primary immune response, which is slow and takes about 2–3 weeks to peak. The second exposure is faster and greater due to the presence of memory cells. Less antigen is needed to elicit a secondary response than the primary response. IgM is the predominant antibody in primary response, while IgG is the main antibody in secondary response (Figure 14.9) (Table 14.2).

Table 14.3 Difference Between Primary and Secondary Responses

S. No.	Primary Immune Response	Secondary Immune Response
1	Occurs during first exposure to antigen	Occurs during second and subsequent exposures
2	Naive B- and T-cells respond and become effector cells and memory cells	Pre-formed memory cells respond to activate effector cells
3	Peaks within 7–10 days due to longer lag phase. Immunity takes longer time.	Peaks within 3–5 days. Immunity takes shorter time.
4	Major antibody IgM, small amounts of IgG.	Major antibody IgG, along with IgD, IgE, and others.
5	Overall, slower and smaller response	Overall rapid heightened response.

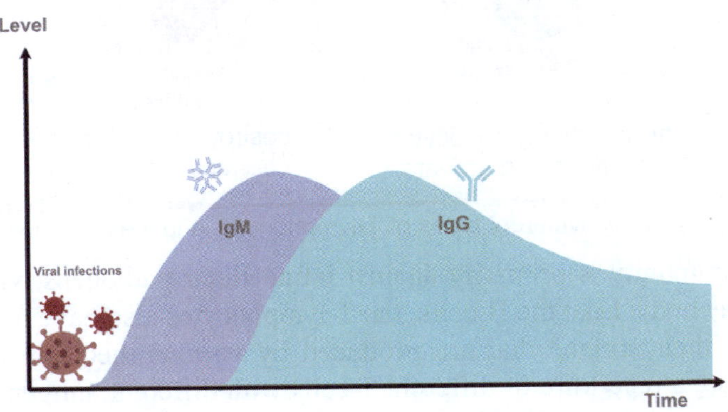

Fig. 14.9 Level of response and type of antibody in primary and secondary response

Cell-mediated immune response

The effectors of cell-mediated immune response are the T-lymphocytes. T-cells are formed in bone marrow but mature in thymus and are found in blood, lymph, and lymphatic tissues. The naive T-cells mature to different types of T-lymphocytes (Figure 14.10).

There are four major types of T-cells.

1. Helper T-cells: Other arms of the immune system like macrophages, B-cells, and dendritic cells act as antigen-presenting cells, and they activate helper T-cells to the effector and memory cells for performing its function of stimulating other immune cells by secreting cytokines. Helper T-cells are CD4+ cells and recognize antigens with MHC-II in APCs.
2. Cytotoxic T-cells: They are CD8+ cells, and their primary function is to target and kill infected/toxic/abnormal self-cells of the body. It initiates apoptosis and causes destruction of abnormal cells that present antigen with MHC-I.
3. Regulatory T-cells: Their main job is to downsize the immune response once the threat has been eliminated. They also prevent autoimmune diseases, inflammation, and suppress excessive and unrequisite immune response.
4. Memory T-cells: They are long-lived T-lymphocytes that remain after the pathogen is eliminated. Similar to the secondary response of B-cells, they have the ability to expand into large numbers of effector T-cells upon exposure to familiar antigens. It helps in building immunological tolerance.

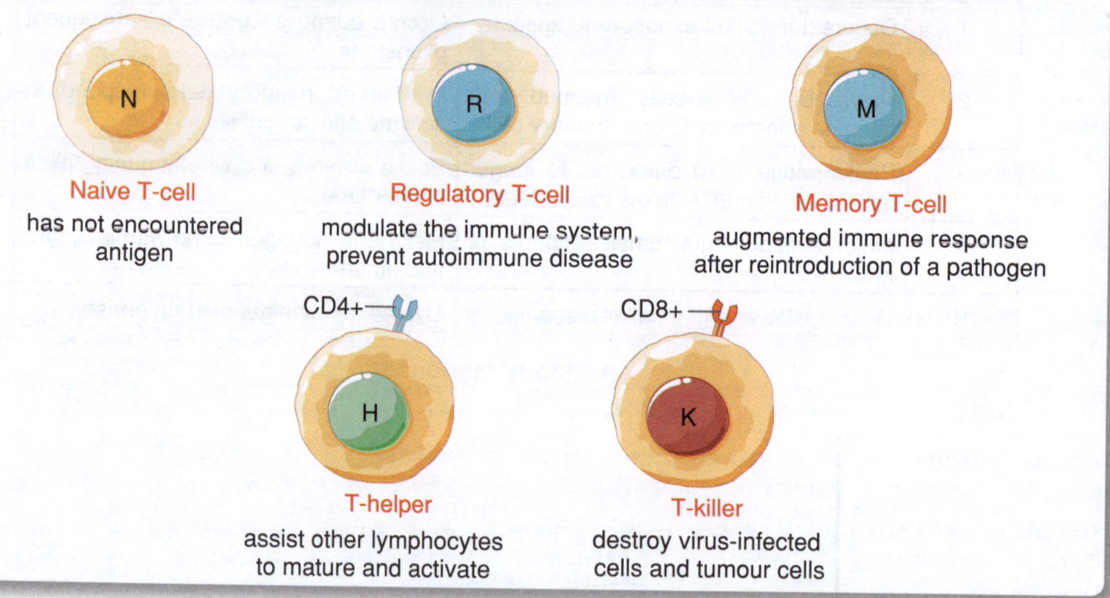

Fig. 14.10 Different types of T-lymphocytes and their functions

Cell-mediated immunity is primarily against intracellular pathogens, viruses, and abnormal cancerous cells of the body. Like the B-cells, the T-lymphocytes also have unique T-cell receptors against antigens on their surface that are produced by various mechanisms to ensure a huge diversity of TCRs. So, a repertoire of different T-cells with different antigenic receptors or TCRs is present in the body.

Unlike B-cells that directly bind antigens, these TCRs are capable of reacting with antigen peptides only when presented with MHC molecules. When they bind to antigens either on APC or target cell of the body, they clonally propagate by clonal selection and rapidly proliferate to produce large clones of thousands of identical T4- and T8-lymphocytes, similar to B-cells. The effector T-lymphocytes eventually successfully eliminate the intracellular pathogens or abnormal/

toxic cells. Cytokines produced by effector T4-helper lymphocytes further enable stimulation and proliferation of cytotoxic T-cells by clonal expansion. In this way, even though only a few T-lymphocytes in the body may have TCR molecules able to fit a particular epitope, eventually many thousands of cells are produced with the right specificity. These lead to an effective response against the intrusion and protects the body (Figure 14.11).

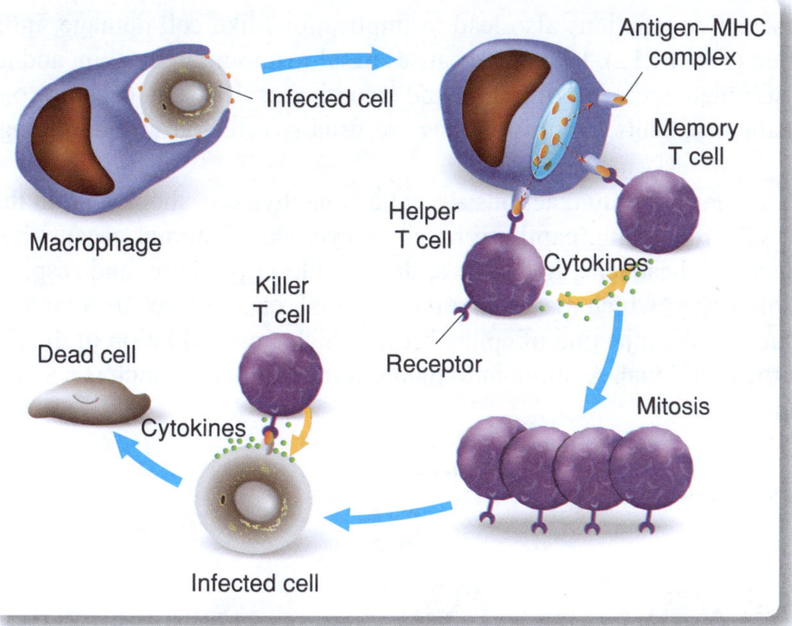

Fig. 14.11 Cell-mediated immunity

Immune Disorders

The immune system is a critical component for maintaining health in humans. However, sometimes it malfunctions. If the complex interplay between the innate and adaptive immune response goes awry, problems can arise in the body, ranging from mild irritation to deadly outcomes. There are three different types of immune disorders that can result, depending on whether the immune system is compromised, or overreacting, or improperly reacting.

Allergies or Hypersensitivity

An allergy is a heightened and abnormal immune reaction to an antigen. Antigens that cause allergies are called allergens. Pollen, dust, certain food items, and molds are common and well-known allergens. Allergic symptoms may be immediate or can be delayed for hours, depending on the nature and type of allergen. Allergic reactions can be mild or lead to life-threatening reactions called anaphylaxis. Allergens can be inhaled, ingested, or can enter through the skin.

Type I hypersensitivity: Common reasons for allergic reactions against pollen, food etc., are linked to IgE class of antibody. IgEs will have specific receptors for the allergen. When a susceptible individual is exposed to the allergen, the body produces large amounts of IgE antibodies. This is similar to normal response, except that plasma cells produce IgE instead of IgG. The Fc region of IgE antibody has a receptor on the surface of tissue mast cells and blood basophils. Therefore,

the IgE antibody binds to FcRI of mast cells and basophils. This binding of IgE to mast cell and basophil is known as Sensitization, and the mast cells and basophils are said to be sensitized. This is the sensitization phase of allergic reaction. When the allergen enters the second time, allergen crosslinks the IgE molecules, inducing them to secrete histamine that causes severe inflammation due to degranulation. Subsequent exposures will produce greater allergy as the allergen binds to antibodies on mast cells, which release histamine and trigger allergy symptoms faster and quicker. These hypersensitive reactions also lead to implications like cell damage, inflammatory damage and destruction (Fig. 14.12). Histamine causes nasal irritation, itchy skin, and tears; and others, in some cases, stimulate secretion of mucus and smooth muscle spasm causing constriction of airway as seen in asthma patients. Antihistamines are usually effective in countering the symptoms of histamine.

Extreme hypersensitivity reactions are called anaphylactic shock and, in this situation, within minutes, the skin reddening, capillary break in eye, skin, internal organs, burning sensation in mouth and genitals, headache, hot flashes, drop in blood pressure, and respiratory failure due to constriction of airway, which may sometimes be fatal, can also occur. Anaphylactic shock can be reversed by immediate injection of epinephrine, which causes dilation of the air passages in lungs and protects the individual. Allergic individuals carry 'Epipen', which is a single-shot syringe that contains epinephrine.

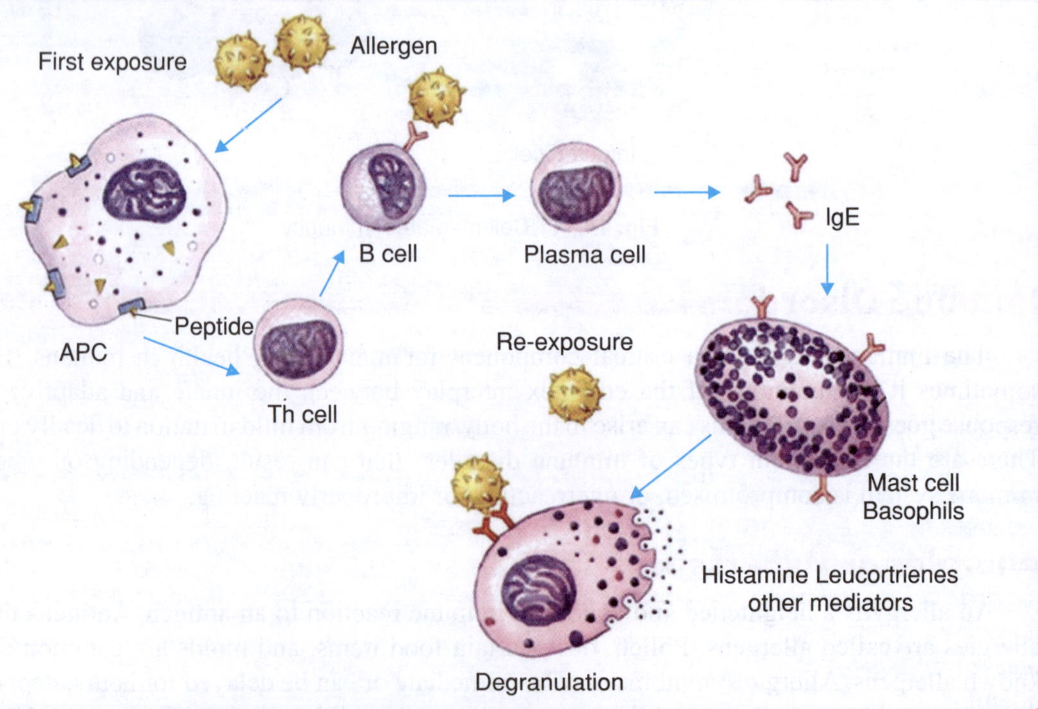

Fig. 14.12 Types of hypersensitivity and effector molecules

Other types of hypersensitivity include: Type-II hypersensitivity mediated by IgG, type-III hypersensitivity mediated by immune complexes, and Type-IV hypersensitivity, also called delayed hypersensitivity, mediated by cells (Figure 14.12). These hypersensitive reactions also lead to implications like cell damage, inflammatory, and nerve damage and destruction (Figure 14.13 a, b, c, and d)

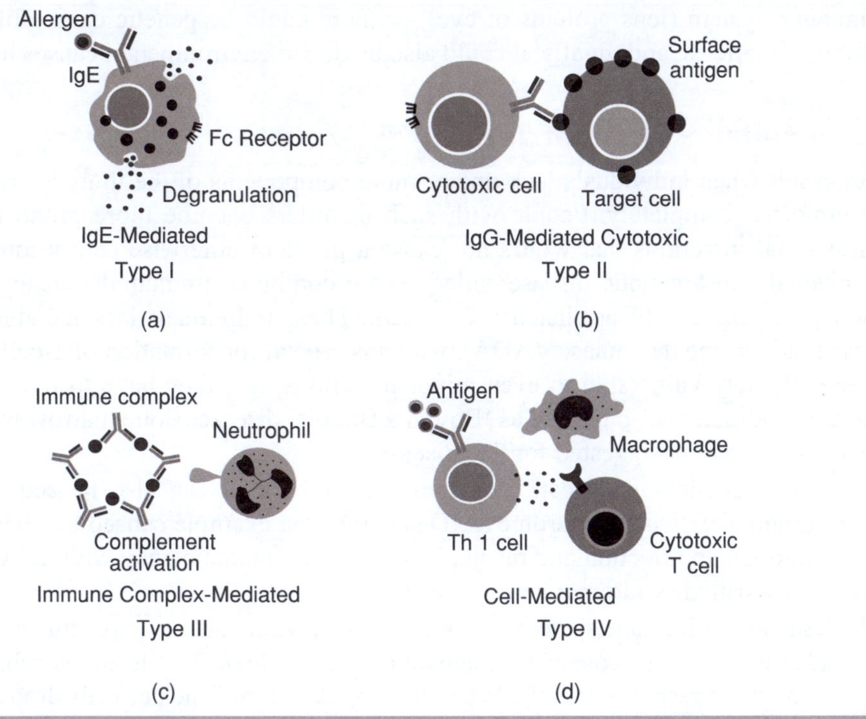

Fig. 14.13 Types of hypersensitivity and effector molecules

Source: Adapted from https://www.onlinebiologynotes.com/type-i-hypersensitivity-reaction-mechanism-and-clinical-manifestation/

Autoimmune Disorders

One of the critical factors in the success of the immune system is its ability to distinguish the foreign cells from self-cells and then mount a response against the invaders. The MHCs act as the unique self-protein markers that help in the recognition of self-cells and as a guiding force to prevent any attack on them. However, sometimes the immune system may falter in this distinction. In autoimmunity or autoimmune disorders, the immune system reacts to self-antigens and turns against the body's own molecules. In such cases, the immune system sees healthy cells as harmful and attacks them. In many cases, the effects of such anti-self-antibodies are not serious enough to cause any recognizable disease. Type-I diabetes, systemic lupus, rheumatoid arthritis, etc., are some examples of autoimmune disorders.

In Type-I diabetes, the insulin-producing cells of the pancreas are targets of cytotoxic T-cells; in systemic lupus B-cells make antibodies against many molecules, including proteins and DNA released by normal breakdown of body cells. Rheumatoid arthritis results in damage and painful inflammation of cartilage and bones of joints due to antibodies produced against them by the B-cells that destroy them. Autoimmunity in neurons where the T-cells wrongly attack neuronal proteins is the cause of multiple sclerosis (MS).

One of the key reasons is that immunological tolerance—a complex inbuilt mechanism in immune protection—fails, leading to an autoimmune reaction. Though it is very difficult to pinpoint an exact cause to what triggers an autoimmune reaction, there could be a few reasons like injuries that expose the body cells or proteins to immune system that were normally inaccessible

to the immune system (lens proteins of eye), or there could be genetic causes like mutations and combinations of alleles, and, finally, it could also be due to environmental causes like chemicals and drugs.

Immunodeficiency Disorders

They result when individuals lack one or more components of the immune system and, hence, are not protected completely. People with such disorders become more prone to infections and cancers. Normal infections that would not cause a problem otherwise can become challenging in such individuals. In a genetic disease called severe combined immunodeficiency disease (SCID), both the B-cells and T-cells are inactive or absent. These individuals lack the ability to synthesize the enzyme 'adenosine deaminase' (ADA), which is crucial for formation of T-cells. As people with SCID are extremely vulnerable to even minor infections, and they have to live behind protective barriers, the condition is also termed as 'Boy in a Bubble' disease. Bone marrow transplantation is a therapeutic intervention suggested for this disease.

Though immunodeficiency can be genetic and inborn, it can also be acquired later in life. Acquired Immunodeficiency syndrome (AIDS) is the best example caused by HIV infection. AIDS is acquired through an infection and results in severe immunodeficiency with individuals showing a set of symptoms called syndrome; hence, it is named so.

HIV destroys the immune system, leaving the body vulnerable to most infections. It attacks the Helper T-cells, which are responsible for activating the cytotoxic T-cells and B-cells, thus hampering the adaptive immune response greatly. When the population of T-helper cells drops significantly, the body fails to defend against microbial infections and loses the ability to destroy abnormal tumour-causing cancerous cells. Thus, in most cases, AIDS patients die due to secondary infections or cancers that cannot be defended against without a functional immune system (Figure 14.14).

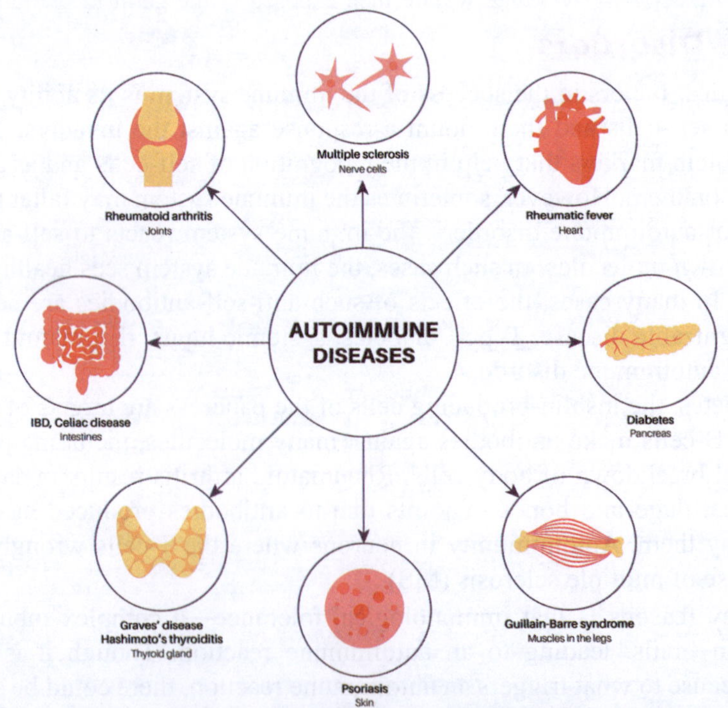

Fig 14.14a Autoimmune diseases

WHAT IS HIV?

Human Immunodeficiency Virus (HIV) is a virus that attacks cells that help the body fight infection.

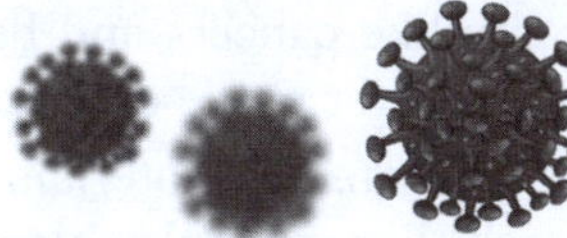

There's no cure, but it is **treatable** with medicine.

Fig 14.14b Types of autoimmunity and immunodeficiencies

Sources: https://www.hiv.gov/hiv-basics/overview/about-hiv-and-aids/what-are-hiv-and-aids
Sources: https://microbenotes.com/autoimmune-disease/

Experimental Insight 14.1 Vaccines: Importance and Working

1. Vaccines protect us from getting a disease. Vaccines work by imitating an infection to engage the body's natural defences.
2. When we encounter microbes, the body responds and produces memory cells that gives a more rapid and heightened response during subsequent exposures and eliminate the microbe. Instead for waiting for a natural infection to trigger our immune response, vaccines are simple, safe, and effective way to use your body's natural defences to build resistance against harmful diseases, before you come in contact with them.
3. When you get a vaccine, the immune system recognises the microbe and raises antibodies. It remembers the pathogen, and when we are then exposed to the germ in the future, out immune system will quickly destroy it before we become unwell. It is a form of acquired active immunity.
4. The active ingredient in all vaccines is an antigen, which could be either a weakened or killed bacteria or viruses; bits of their exterior surface or genetic material or bacterial toxin are treated to make it non-toxic.
5. Vaccines can be given in the form of injections (shots), liquids, pills, or nasal sprays that you take to teach your body's immune system to recognise and defend against harmful germs.

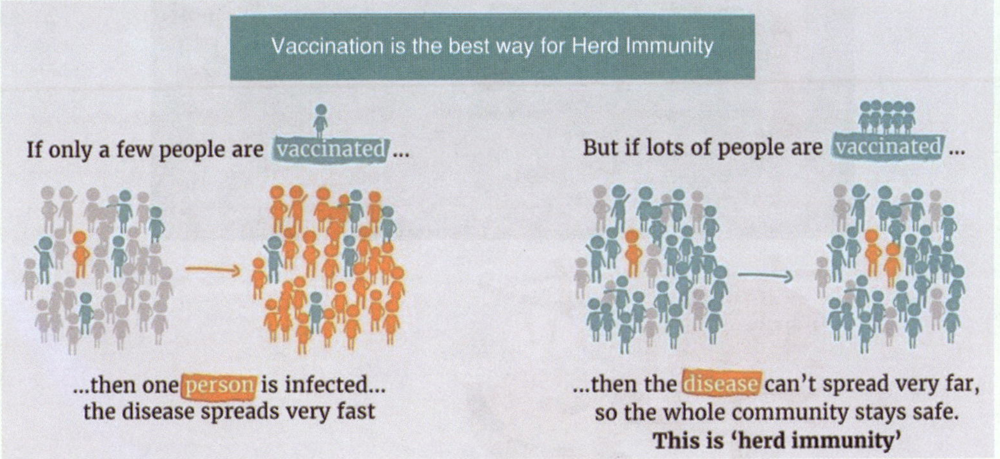

Source: Adapted from https://www.vpslakeshorehospital.com/blog/what-is-herd-immunity-and-can-we-really-achieve-it-with-coronavirus

Immune Engineering

It is a new and emerging field of biology that combines the principles and tools of engineering to investigate and modify the immune system. Power of the immune system is harnessed to treat complex diseases like cancers and help in healing. Research in immune engineering will create innovative solutions for tissue engineering, drug delivery, and medical devices.

Designing biomimetic materials to emulate and interact with immune system is one major focus of research. For example, biomaterials that signal the immune system to stop the attacks on 'self' and to allow healing to take place in autoimmune disorders is one solution. Immunoengineering approaches that disguise the surface as self, either by coating immune cell membranes on particle surfaces or conjugating biomimetic peptides to particle surfaces, enables synthetic particles to become camouflaged from the immune system and offer possible solutions for sustained drug delivery.

Immunoengineered vaccines offer another hope for diseases like cancers. Cellular engineering of immune cells ex vivo for adoptive T cell immunotherapy like chimeric antigen receptor (CAR) T cells are also promising approaches that are being experimented with.

The emerging field of regenerative immunology leverages immunoengineering approaches to probe immune cell interactions and design-improved biomaterials for healing and immunomodulatory outcomes (Figure 14.15). Based on these emerging possibilities, foundations have been laid since 2012 to identify immunoengineering and establish itself as a new discipline. Several major universities in the United States have now added tracks or focus areas associated with immunoengineering. In future, this multidisciplinary stream will attract more interest among students, researchers, and funding agencies to foster and catalyse a translational impact of this area.

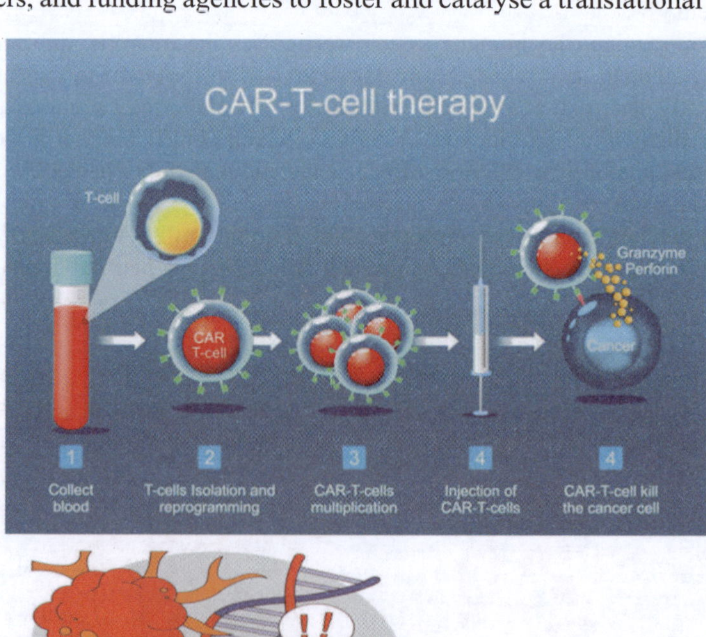

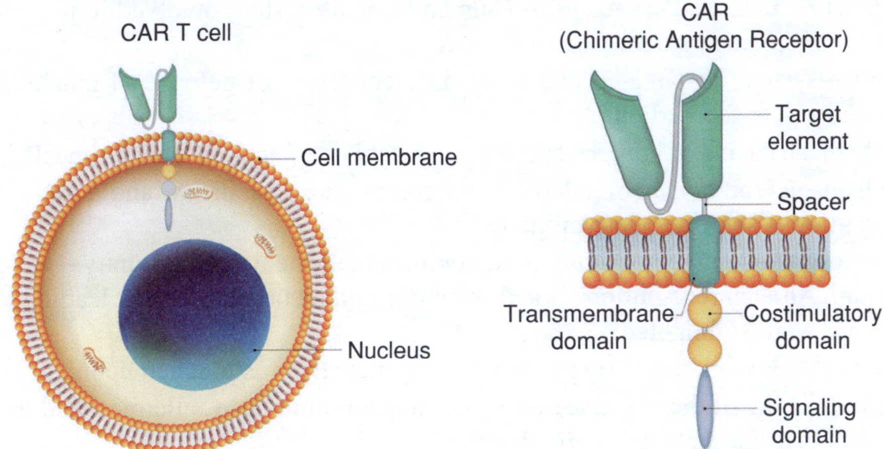

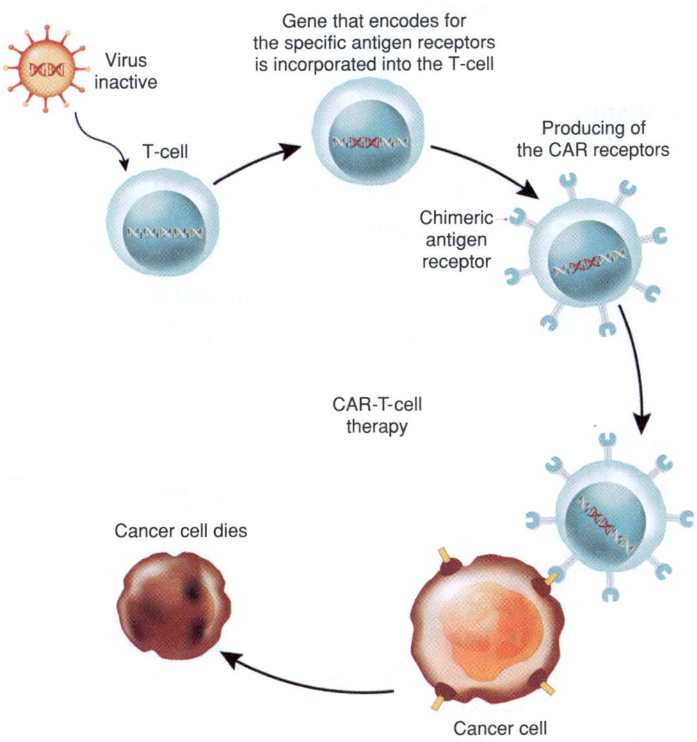

Fig. 14.15 Immunoengineering option: CAR-T-cell therapy

SUMMARY

- Concept of immunity and its ten different types were studied.
- The various cells and organs of the human immune system involved in protection against invaders were understood.
- Immune system demarcated into three different levels of defence of innate and adaptive immunity were focused on.
- Innate mechanisms of humans and their mechanism of actions were delineated.
- Mechanisms for tenets of adaptive immunity like specificity and how self–non-self-recognition is achieved were discussed.
- Two types of adaptive immunity—humoral and cell-mediated immunity—were elaborated in detail. Antigen recognition, clonal selection, and recurrent memory leading to secondary response were highlighted.
- How vaccines work based on immunological memory was touched upon.
- Malfunctioning of the immune system causing autoimmunity, allergies, and hypersensitivities and their mechanisms were identified.
- Immunodeficiency disorders and their implications were explained with specific examples of SCID and AIDS.
- Immune Engineering was explored.

EXERCISES

A. Objective-Type Questions

1. Numerous antigenic sites present on a foreign molecule are called as _______________.
   (a) Epitopes (b) Immunoglobulin (c) MHC (d) Antibody
2. Which of these cells do not contribute to cell-mediated immunity?
   (a) Macrophages (b) T-helper cells
   (c) Cytotoxic T-cells (d) Antibodies
3. Which of the following cell has no role in antibody production?
   (a) B cell (b) Plasma cell
   (c) T-helper cell (d) Natural killer cell
4. Which of the following statements is wrong for humoral immunity?
   (a) Humoral immunity is mainly for protection against extracellular bacteria and toxins.
   (b) Mature B-cells after activation by antigens secrete antibodies
   (c) Antibodies directly kill the pathogen by phagocytosis
   (d) Memory cells are produced in humoral immunity.
5. IL-1, IL-2 and IL-4 are examples of _______________.
   (a) Complement proteins (b) Interferons (c) Cytokines (d) Immunogens
6. Which of these are mechanisms by which antibodies work?
   (i) Activate plasma proteins called complement;
   (ii) Enhance phagocytosis by acting as opsonins;
   (iii) Bind to bacterial toxin and neutralize its harmful effects
   (a) only (i) (b) only (i) and (ii)
   (c) only (i) and (iii) (d) (i), (ii), and (iii).

7. Which of this is not a class of Ig?
   (a) IgM (b) IgG (c) IgA (d) IgJ
8. When the immune system is activated by a particular antigen to produce antibodies inside the humans, it is called as
   (a) Passive immunity. (b) Active immunity.
   (c) Non-specific immunity. (d) Innate immunity.
9. AIDS is an example of ______________ disorder.
   (a) Hypersensitivity (b) Allergic
   (c) Autoimmune (d) Immunodeficiency
10. IgE has a role in ______________.
   (a) Type I hypersensitivity. (b) Rheumatoid arthritis.
   (c) SCID. (d) None of the above.
11. In which cells are the MHC-II molecules found?
   (a) B-cells (b) Macrophages
   (c) Dendritic antigen-presenting cells (d) All of the above
12. T-cells mature in ______________.
   (a) Thyroid gland (b) Thymus (c) Spleen (d) Bone Marrow
13. Plasma cells are ______________.
   (a) Long-lived memory cells (b) Derived from monocytes
   (c) Mature antibody-secreting cells (d) A type of T-regulatory cells

B. Short-Answer Questions

1. What is passive immunity, and provide two ways by which it can be attained?
2. What do you understand by the term 'immunological tolerance'?
3. The characteristic of immunological memory is why vaccines work effectively. Justify this statement.
4. How many Immunoglobulin classes are recognized, and what is the basis and differences among them.
5. Explain the role of two proteins or chemicals involved in innate immunity.
6. What are the three ways by which our immune system can fail?

FURTHER READINGS

1. An introduction to immunology and immunopathology
2. https://aacijournal.biomedcentral.com/articles/10.1186/s13223-018-0278-1
3. Innate Vs. adaptive response https://www.technologynetworks.com/immunology/articles/innate-vs-adaptive-immunity-335116
4. How vaccines work? https://immunizebc.ca/immunization-basics/how-vaccines-work
5. Humoral Vs. cell mediated immunity https://www.technologynetworks.com/immunology/articles/humoral-vs-cell-mediated-immunity-344829

USEFUL WEBLINKS

1. Introduction to the immune system https://www.youtube.com/watch?v=gR84OJ_EGxk
2. 5 ways immune system protects our body https://www.youtube.com/watch?v=npHZSOBqOcY
3. Immunodeficiency - T-cell and B-cell disorders https://www.youtube.com/watch?v=CF6_4brCXU0
4. Hypersensitivity - overview of 4 types https://www.youtube.com/watch?v=jXTW4F-8jd4
5. Immune tolerance and autoimmunity https://www.youtube.com/watch?v=vDwNpDT-8L0

PART III OTHER IMPACTS AND OUTCOMES ON HUMANS AND LIVING BEINGS

PART [illegible]: OTHER IMPACTS AND OUTCOMES ON HUMANS AND LIVING BEINGS

CHAPTER

16 Discovering the Unseen World of Microbes

LEARNING OBJECTIVES

After reading this chapter, you will be able to:

- Become better aware of the world of microbes
- Study the tools and techniques to grow, identify, and manipulate microbes
- Pay attention to human–microbe interactions
- Enlist the harmful effects of microbes on humans and the environment
- Appreciate the beneficial contributions of microbes in nature
- Present mechanisms for microbial control
- Understand how microbes can be engineered to solve human problems

CHAPTER TAKEAWAY The microbial world and its implications on life

This chapter deals with the different aspects of the microbial world and the implications on human health and environment.

PRELUDE

An unseen world of invisible tiny creatures that impinge on every aspect of human existence is an interesting and critical domain to explore. The microbes are omnipresent, omnipotent, and omniscient beings that have, both, beneficial and harmful effects. Their influence on the planet is huge, to the point where all life on Earth depends on microbial activities. The excitement in studying microbes stems from the modern-day utility of microbes in contributing to solutions for various human problems and providing valuable industrial and pharmaceutical products. Understanding microbes has huge implications for health and disease too. Within the human body itself, there are more microbial cells than human cells; thus, even our own body is not entirely our own; we co-habit with microbes. Or, is it the microbial cells among which our cells are present? As the very first life on Earth, they are also presumed to be the ancestors of all other life forms.

Their power is immense and huge; hence, they just cannot be ignored and, thus, are a major component of human existence. *Microbiology* is the study of microscopic organisms (microbes), which comes from the Greek words: *mikros* = small; *bios* = life, and *logos* = study.

The Microbial World

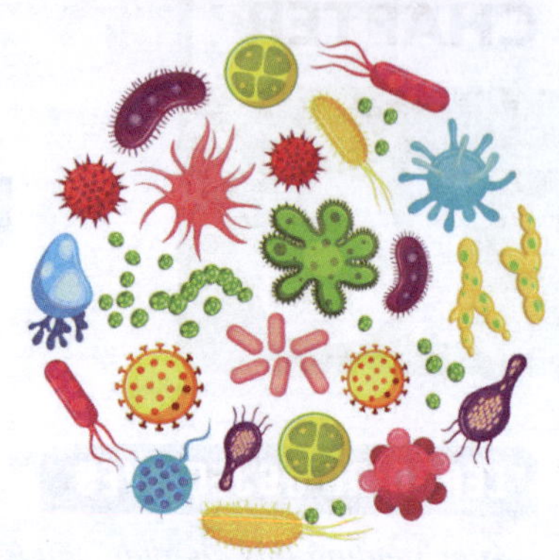

Microbes or micro-organisms are minute living things that are too small to be seen with an unaided eye. Normally, when we think of microbes, we assume them to be germs that cause disease. However, only a minority of microbes are pathogenic. Therefore, a detailed knowledge of the existence and practical ways of handling microbes is necessary for our survival and productive co-existence with them. Although microbes have been in existence for millions of years, they were unknown to us till the invention of microscope. Once they were observed and identified, the field of microbiology exploded and opened up immense opportunities for their control as well as utilization.

Microbial Diversity

The world of microbes is diverse and includes acellular entities such as viruses, prokaryotic cells like bacteria and archaea, and eukaryotic members like fungi, yeast, molds, protozoans, and microscopic algae. In terms of numbers, microbes far exceed every other group of organisms on this planet. It is estimated that microbes contain 50% of biological carbon and 90% of nitrogen on Earth. Microbes are found everywhere and in every habitat of the planet—ocean depths, geothermal vents, arctic ice, and deserts to the human body, skin, internal organs, etc. The microbes are the major contributors for recycling elements in nature and are at the base of ecological food chains and webs. Hence, they are indispensable to the functioning of the biosphere.

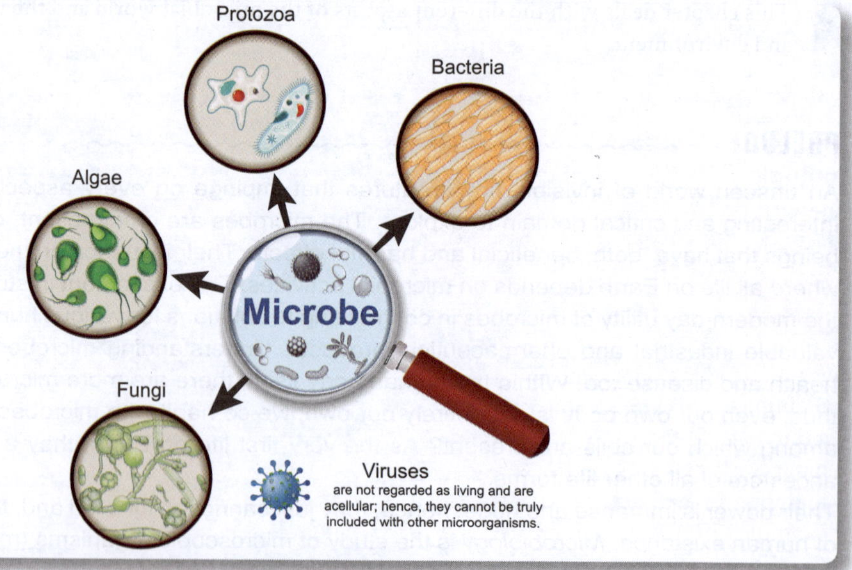

Fig. 16.1 Types of microbes

The discovery of micro-organisms was fueled by their suspected role in diseases and were later identified in the 17th Century by Leeuwenhoek and Robert Hooke with the invention of microscope. Once the detailed structures of microbial cells were studied, they were classified into eubacteria, fungi, and protists. The nomenclature of microbes follows the same scientific convention of genus

and species—binomial nomenclature. Genera name is first, starts with a capital, and the second name is species name. The scientific names are usually italicized or underlined. For example, *Escherichia coli, Saccharomyces cerevisiae, Helicobacter pylori, Entamoeba histolytica*

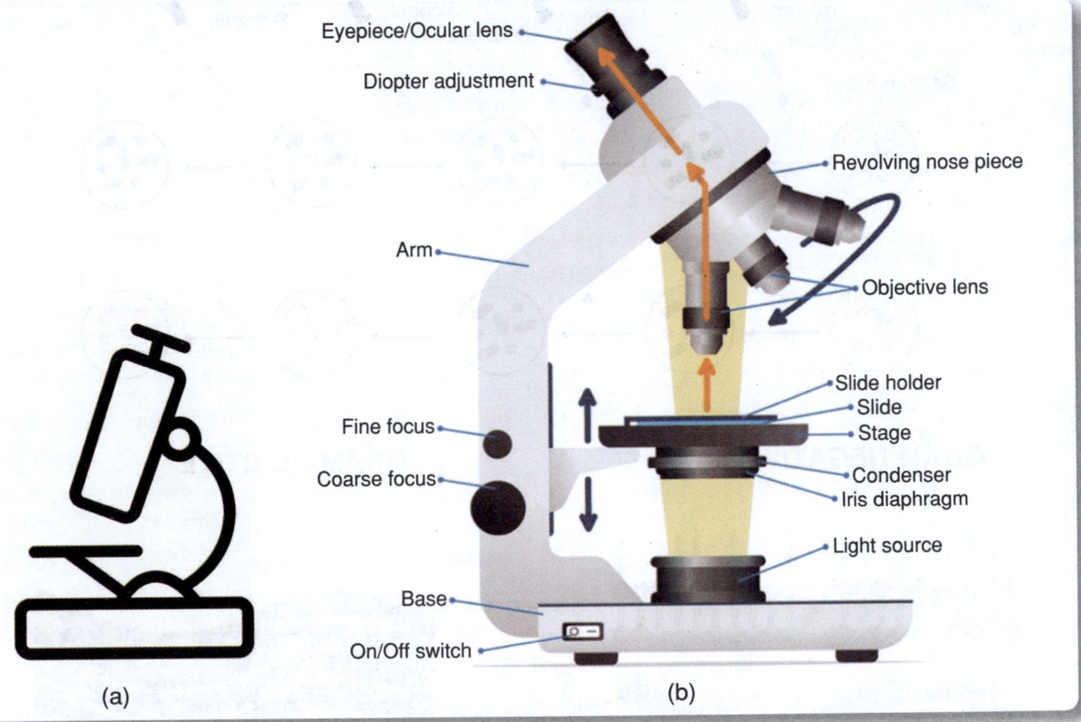

Fig. 16.2 Compound light microscope: (a) Simplistic view; (b) Parts detailed

Preparation of specimens by *staining* helps in better visualization. Staining can also be done in different ways such as: (a) simple staining with a single dye versus differential staining with multiple dyes, (b) staining the cells directly or staining the background surrounding the cells (negative staining) to make the cell visible, increase contrast, etc.

Interesting Fact 16.1

Significance of Staining

Staining means coloring an organism with a dye. Microbes are fixed to microscopic slide as a smear or a thin film. Simple stain is a single dye used to color the microbe. Mordants can be used occasionally to intensify the stain.

Staining methods can also be used to diffentiate microbes and are called differential staining. Two examples are gram staining and acid fast staining. **Gram staining** classifies bacteria into two groups – Gram +ve and Gram –ve. The difference is due to composition of the cell envelope of these bacteria and the differential manner in which the stains react with them. At the end of the staining process, gram +ve bacteria stain purple while gram –ve bacteria stain pink. Organelles like endospores and flagella can also be visualized by staining.

Contd.

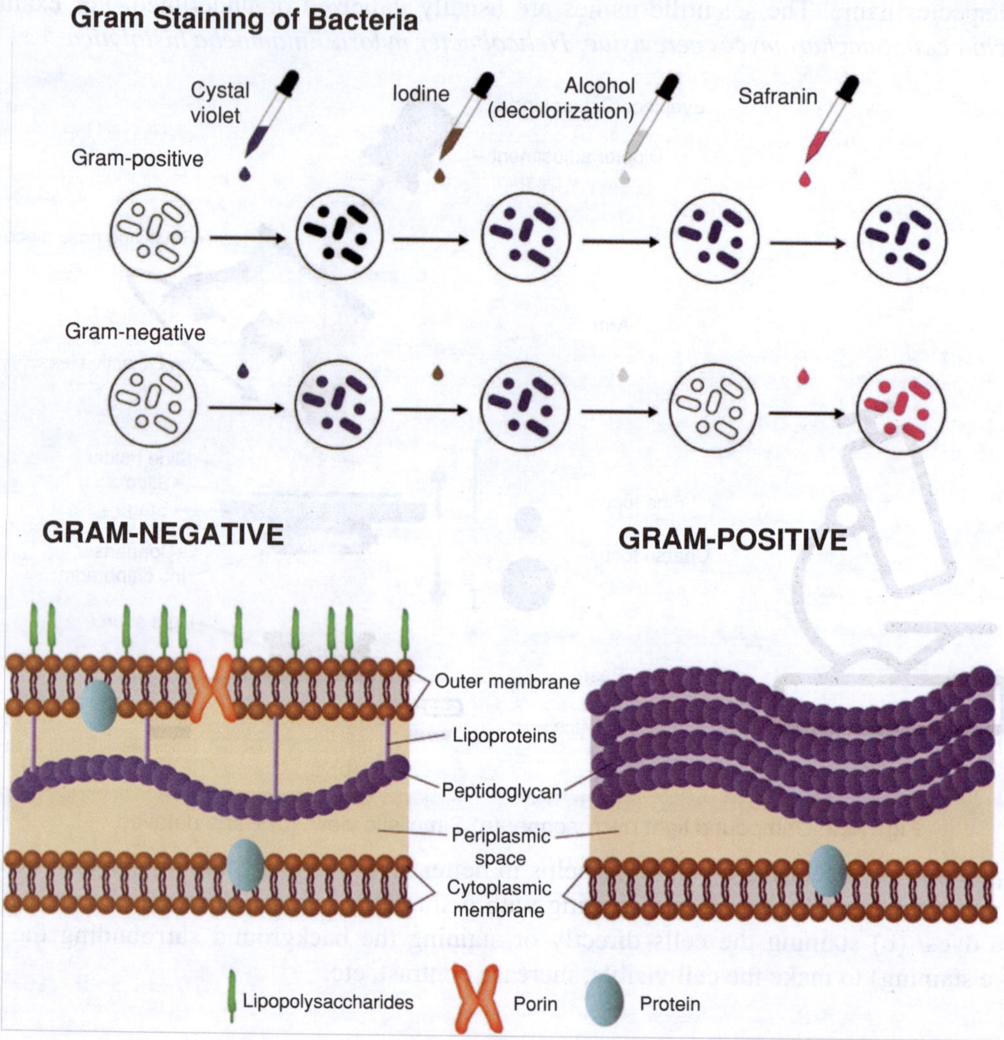

Gram Staining of Bacteria

Cystal violet

Iodine

Alcohol (decolorization)

Safranin

Gram-positive

Gram-negative

GRAM-NEGATIVE

GRAM-POSITIVE

Outer membrane

Lipoproteins

Peptidoglycan

Periplasmic space

Cytoplasmic membrane

Lipopolysaccharides

Porin

Protein

Tools and Techniques to Study Microbes

Microbes, being tiny, require specific tools and techniques to be studied. Based on the specific objective, different methods and procedures are adopted to work with microbes.

Visualizing Microbes

Microbes range in size from 10 nm to 100 µm. Viruses are the smallest, which are in the range of 10–100 nm, whereas eukaryotic microbes are bigger, between 10 and 100 µm. For such tiny organisms, to visualize them, you need microscopes, which help to magnify the image and show the details. Modern microscopes provide great magnification and resolution and help in detailed visualization of microbial structures. For bacteria and larger microbes, various types of light microscopes serve the purpose. There are six different types of microscopes; all using light as the source of illumination with increasing magnification and resolution, starting from the common compound bright-field microscope, dark-field microscope, phase-contrast microscope, differential interference contrast (DIC), fluorescence microscope, to the high-end confocal scanning laser microscope (CSLM). In light microscopy, visualization of cells is typically enhanced by the contrast between the specimen and the surrounding environment.

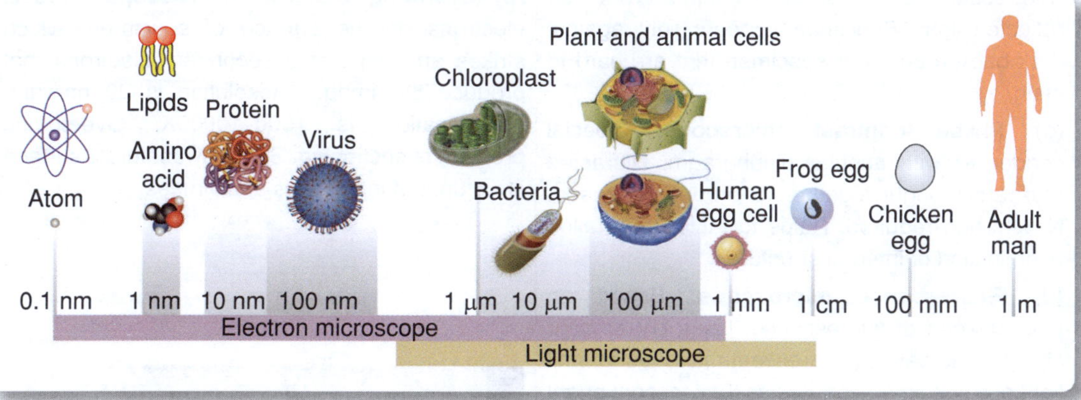

Fig. 16.3 Microscope types used to see different cells based on relative sizes

Preparation of specimens by *staining* helps in better visualization. Staining can also be done in different ways such as:

(a) Simple staining with a single dye versus differential staining with multiple dyes,

(b) Staining the cells directly or staining the background surrounding the cells (negative staining) to make the cell visible, increase contrast, etc.

Electron microscopy uses beams of electrons instead of light and, both, the transmission electron microscope (TEM) and scanning electron microscope (SEM) provide means to visualize microbes smaller than 0.2 µm, such as viruses and internal structures of cells. The resolving power of electron microscopes is greater than other microscopes due to the shorter wavelength of electrons. Scanning probe microscopes examine the specimen at very close range and help in studying atomic and molecular shapes, as well as magnetic and chemical properties.

Experimental Insight 16.1 Evolution of Microscopes

The beginning of microscopy was with Anton van Leeuwenhoek, who ground lenses with such precision that with a single lens, he obtained a magnification of 300X. Robert Hooke, a contemporary of Leeuwenhoek, then built compound microscopes with multiple lenses. From there, the imaging of cells and the domain of microscopy took over and many new improvisations for new variations were produced. Each of these came with features that enhanced the visualization experience of the tiny cells and provided greater resolution and clarity.

Contd.

Light Microscope	Electron Microscope
(a) Bright-field compound microscope: Specimens till 0.2 μm can be resolved. Bright background and stained specimen. Inexpensive, easy to use in lab. **(b) Dark-field microscope:** Special condenser with opaque disc that blocks light. Works on reflected light. Specimen appears light against dark background. For specimen that are hard to stain. **(c) Phase contrast micrscope:** Special condenser with annular diaphgragm. Diffracted rays are brought together to produce a image. No staining required. Helps to facilitate detailed examination of internal structures. **(d) Fluorescence microscope:** Based on phenomenon of fluorescence. Uses UV or near Uv light and causes fluorescent microbes to emit light and specimen appears as luminescent bright objects. Fluorochromes can be used to detect specimen. **(e) Confocal microscope:** Uses laser light to illuminate on plane of specimen at a time. Helps to obtain 2D or 3D image of cells. **(f) Scanning acoustic microscope:** Used sound wave of specific frequency sent through the specimen; resolution is about 1 μm. 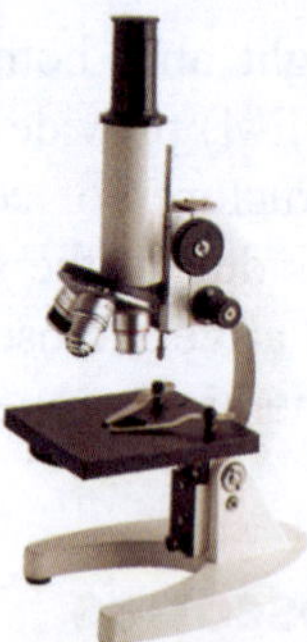	**(a) Transmission electron microscope:** Electron gun focuses beam of electron through ultrathin section of spacimen. Uses electromagnetic condenser to focus electrons. 10,000X–100,000X magnification with resolution of 2.5 nm **(b) Scanning electron microscope:** Directs electrons on the surface of specimen, which strikes and generates secondary electrons that produce the image. Resolution is 20 nm and magnification is 1000–10000X. Overcomes problem of sectioning. Specially useful for surface structures of intact cells and viruses. 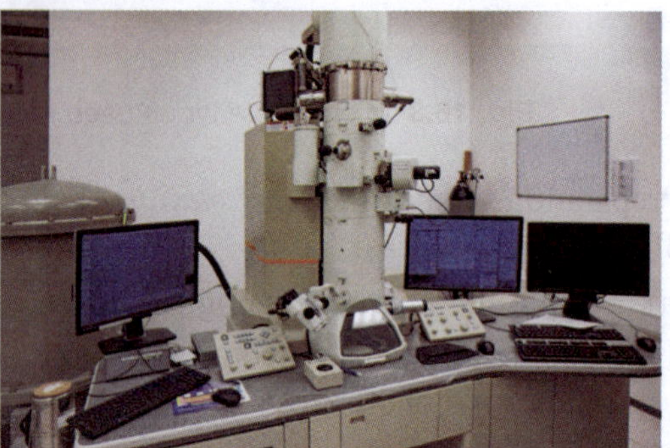

Growing Microbes

In order to grow microbes, there are certain physical and chemical requirements that need to be taken care of depending on the nature of microbes we intend to grow in the lab.

Physical aspects include the right pH, temperature, and osmotic pressure, whereas chemical requirements include nutrients that act as source of carbon, nitrogen, sulphur, phosphorus, trace elements, oxygen, and organic growth factors.

- Based on the physical requirements, such as temperature, the microbes can be either psychrophiles (need low temperatures), mesophiles (moderate temperatures), or thermophiles (heat-loving).
- Similarly, as per the pH needs, they can be classified as acidophiles, neutral, or alkaliphiles and salt-requiring bacteria called halophiles.
- Oxygen-requiring microbes are aerobic, and those that do not require are anaerobic; but those that can grow in, both, presence and absence of oxygen are said to be *facultative*.

To facilitate growth, it is important to know their growth requirements.

Nutrient requirements are met by using different types of culture media. Some microbes are not very fastidious and can grow in any media, but some require specific media. Addition of microbes into the culture media is called *inoculum,* and the growth of microbes in culture media is called microbial culture.

We all know that microbes are present everywhere, so how can we grow only one type of microbe as per our requirement. We need to eliminate all other microbes by sterilization, and then, once the media is sterile, which means it has no other microbes, we can inoculate ours and protect it from any further contamination, and incubate it to obtain a pure culture of our organism of choice.

Media can be of different types as follows:

- **Complex media**

Made up of nutrients such as extracts of yeast, beef, meat, plants, or digests of proteins, etc. These contain all the nutrients, but their individual concentrations are not well-defined. The exact composition will vary from batch to batch. For example, nutrient agar, Luria broth, etc.

- **Chemically defined media**

Exact chemical composition is known. The growth medium is prepared to include all the growth requirements in the right quantity and type. These media are usually reproducible in terms of ingredients and quantities, e.g., minimal media. It is specific to different specific microbes.

- **Selective media**

Ingredients are added in this media to selectively allow the growth of specific microbes while preventing the others, e.g., Bismuth sulphite agar for *Salmonella*.

- **Differential media**

This medium helps to distinguish colonies of the desired microbe from others. For example, blood agar for haemolytic bacteria such as *Streptococcus pyogenes*.

Microbes can be grown in these media in different forms such as solid agar stubs or slants, liquid solutions, or solid agar plates. To prepare solid media, solidifying agents such as agar are added to the liquid broth. Microbial growth in liquid media will show up as turbidity and as colonies in solid media. Pure cultures (only one type of microbe) can be obtained by streak plate and spread plate methods.

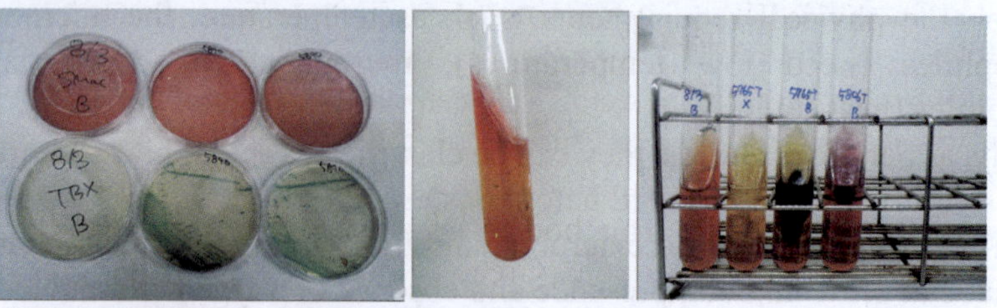

Fig. 16.4 Microbial growth in different media

Different features of colonies are studied as colony characteristics.

Interesting Fact 16.2

Bacterial growth curve—Phases of growth

When a bacteria is inoculated into a media, and if the population is counted at periodic intervals, we can plot a bacterial growth curve that tells us about the changes and dynamics of growth within the medium.

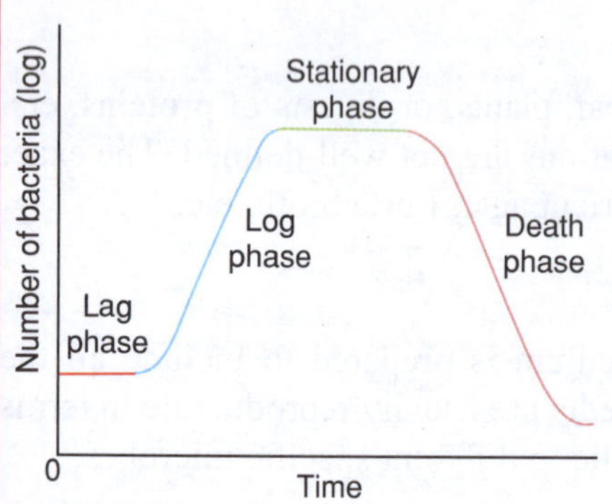

1. **Lag Phase:** Intially after inocultion, the number of cells does not change as the microbes do not immediately reproduce; instead the population is undergoing intense metabolic activity and preparing for division by acclimatizing and synthesizing eyzmes and other needs.
2. **Log Phase:** Active cell division, cell numbers rise exponentially, and generation time reaches a constant minimum. Hence a straight line for log numbers vs time. Metabolically very active too.
3. **Statinary Phase:** Cell growth continues, but number of microbial death balances the growth. Population stabilizes and reaches a plateau. Exhaustion of nutrients, accumulation of wastes, and changes in pH push the population from exponential to stationary phase.
4. **Death phase:** Death rate exceeds new cells formation, and populaton enters decline phase. Population is diminished eventually or completely dies out. Involution or morphological changes can happen in this phase in an attempt to survive.

Quantification or Counting of Microbes

Direct measurement of microbial populations can be done by various methods. Some methods will count cell numbers, whereas others will use the population mass. The source in which the microbial numbers are to be counted is usually serially diluted, and the numbers obtained per ml are then extrapolated to the counts in the whole sample. Direct methods are as follows:

- **Pour plate method**

Gives the count of viable cells in a sample based on their growth in agar after serial dilution and inoculated with a small volume (0.1 ml or 1 ml). Plate counts are then reported as colony-forming units (CFU) and extrapolated to the larger volume. Number of cells counted × Dilution factor/Total volume will provide the CFU/volume that can be then extrapolated.

- ***Spread plate method*** is also a variant that helps to count cells and extrapolate in the same manner.

- ***Filtration*** to capture microbes in a membrane filter followed by the growth in media also helps in detection and counting of microbes that are in very low numbers in the sample.
- ***Direct microscopic count*** using Petroff–Hauser counting chamber can also be done. Cell counters can be used to directly count the number of colonies in the plate and, assuming one colony is produced from one microbe, the numbers can be extrapolated.
- ***Most probable number (MPN)*** method is a statistical method on random dispersion of microbes per volume in a given sample. Greater the number of bacteria, more is the dilution required to have no growth in the indicator liquid medium. It helps to find the bacterial population within a certain range. The number is estimated by using reference statistical tables.

Indirect methods of microbial enumeration include the following:

- ***Turbidity measurements*** due to bacterial growth in liquid media using a spectrophotometer is an indirect method but more practical to measure growth. The amount of light that passes

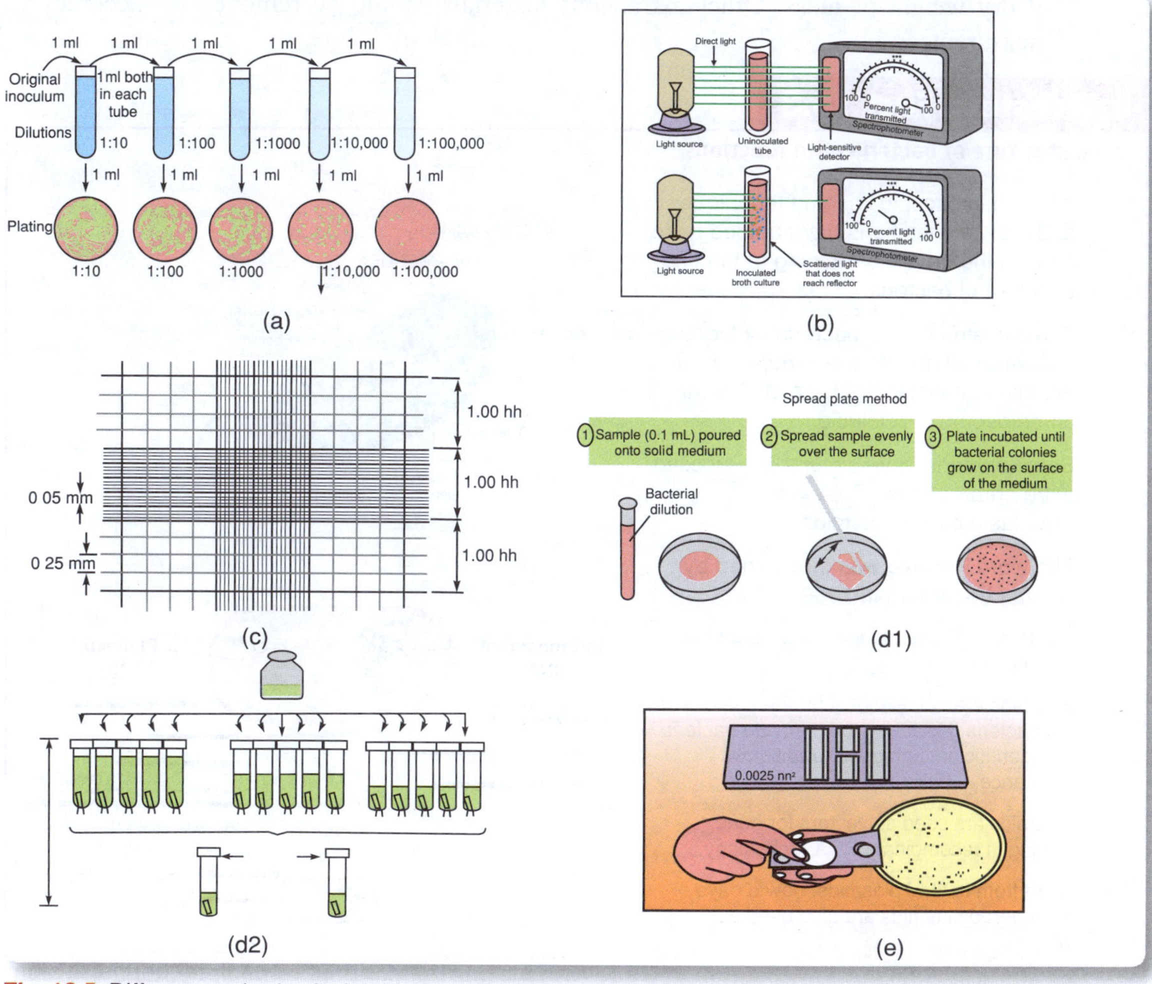

Fig. 16.5 Different methods of microbial enumeration (metabolic activity and dry weight are other methods not included here: (a) Pour plate method; (b) Turbidity measurement; (c) and (e) are Direct Microscopic count; (d1) Spread plate method; (d2) Most Probable number method)

Source: Adapted from https://www.slideshare.net/slideshow/mpn-and-indirect-methods-of-measurement-of-microbial-growth/120168601
https://conductscience.com/bacteria-enumeration/

through a bacterial suspension is measured—detection of light scatter in absorbance at 600 nm (OD_{600}). The optical density (OD) measurements of absorbance versus concentration is used to plot microbial growth with standard curves and detect the unknown. Turbidity measurements can be inaccurate because they count both live and dead cells, and culture density should be high enough to register turbidity. In some instruments, the direct linear relationship between density and light absorbance can deviate at higher OD.

- ***Metabolic activity*** of the microbial population is another indirect way to measure. The amount of metabolic products such as CO_2 or acid is in direct proportion to the number of microbes.
- ***Dry weight measurements*** are used in case of filamentous microbes, as regular methods do not serve the purpose. In this, the microbial mass is filtered, removed, and dried in a desiccator and then weighed. This also gives the increase in growth as a measure of dry weight or mass. Other extraneous materials should be removed for accuracy in measurement.

Interesting Fact 16.3

Fine structure of bacteria with functions

1. All the microscopic imaging techniques and growth conditions have helped us to deduce the details pertaining to the fine structure of bacteria.
2. The fine structure of bacteria includes a cell wall, cell membrane, cytoplasm, and genetic material in the form of circular chromosome or nucleoid.
3. Some bacteria have circular supercoiled extra-chromosomal DNA in the cytoplasm called plasmids.
4. The word Plasmid was first coined by Joshua Lederberg in 1952.
5. Plasmids replicate independently from the chromosomal DNA.
6. Plasmids are not essential for the survival of the bacteria. However, they provide bacteria with genetic advantages like antibiotic resistance.
7. Plasmids are used as vectors for gene cloning in recombinant DNA technology.
8. Apart from natural plasmids, now a days engineered plasmids are also used as vectors.
9. Plasmid numbers can vary in bacteria from one to hundreds.
10. Resistance plasmids, virulence plasmids, col plasmids, degradative plasmids, and fertility plasmids are different types of plasmids.

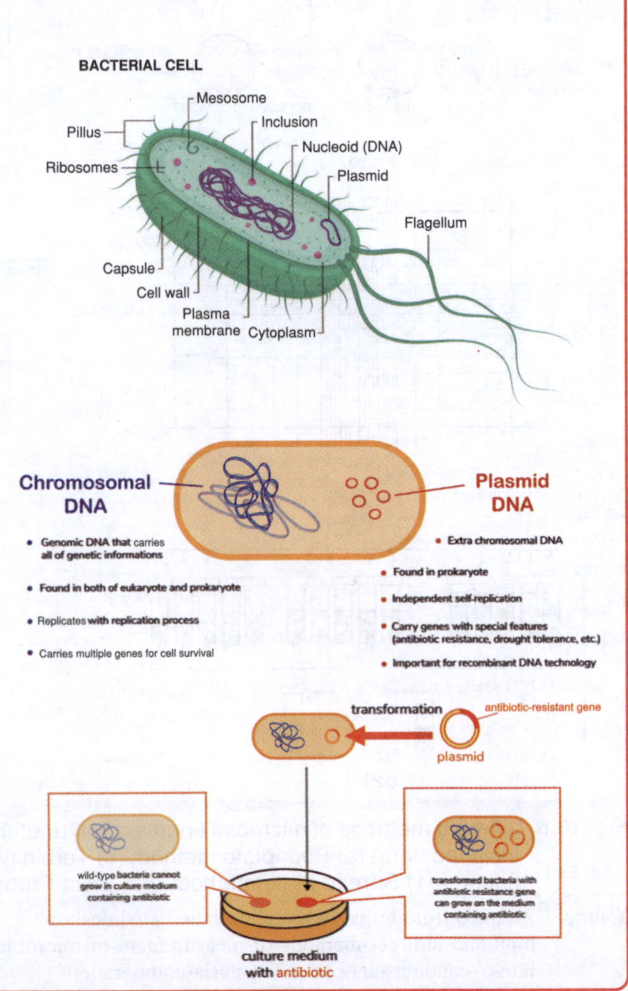

Microbial Interactions

Microbes can associate physically with many organisms and have interactions that influence their existence as well as of their partners. Symbiotic relationships are close associations between two or more biological species. Microbes are a part of different interactions such as the following:

(a) **Mutualism** In this, there is reciprocal benefit for both partners—Give-and-take interaction, where both are mutually benefited. This is an obligatory set-up in which absence of one can impact the survival of another. If it is non-obligatory, the association can be called cooperation.

(b) **Commensalism** An association where one partner benefits and the other is neither harmed nor benefitted.

(c) **Syntrophism** In this association, growth of one organism either depends or is improved by the nutrients and substrates provided by the other organism. The other organism might benefit or have a neutral outcome.

(d) **Predation** In this, there is a predator species that attacks and kills the prey.

(e) **Parasitism** In this relationship, one partner is benefitted while the other is harmed. Parasites co-exist with their hosts and harm them.

(f) **Amensalism** Here, there is an adverse effect of the presence of one organism or its products over the other, either by the release of compounds that negatively impact the other.

(g) **Competition** This relationship is defined when within a population or community different organisms try to acquire the same resources, either space or nutrients.

Association of Microbes with Humans

Understanding all these different types of associations that microbes can indulge in is important to study their interactions with humans and the nature of outcomes. There are a lot of microbes associated with the human body that form the normal flora or microbiota of humans. Environmental niche within the human body facilitates the growth of certain groups of microbes that form a long-term association inside the host. The human system also has a mechanism by which harmful organisms and foreign material are prevented from entering the human body. Microbes that cause disease in humans are called pathogens, and pathogenicity is the ability to produce disease. Infection is invasion or colonization of the body by pathogenic micro-organisms. In healthy humans, internal tissues such as the brain, blood, muscles, heart, etc., are free of microbes. Only surface tissues such as skin and mucous membranes are exposed to environmental microbes.

Microbial species that establish a more or less permanent residence inside the human body and do not produce disease under normal conditions are called *normal biota*. The distribution and composition of normal biota in the human body is dictated by various factors such as nutrients,

defense mechanism of the host, and other physical, chemical, and mechanical factors. Normal biota actually benefits humans by preventing the overgrowth of pathogenic microbes. By antagonism or competitive exclusion, they compete and prevent the establishment and propagation of the harmful microbes. They also maintain the conditions such that opportunistic pathogens are not able to survive. For example, the vaginal pH is maintained to 4 by normal biota, which prevents the growth of *Candida albicans*. Similarly in the intestine, the normal flora produces bacteriocins that prevent the pathogenic *Salmonella* and *Shigella*. Most of these normal biota either have a mutualistic relationship or are commensals. *E. coli* in the intestine produces vitamins such as K and B for the host, and, in turn, the large intestine provides nutrients for the survival of the bacteria. However, if these normal biota move from their normal niches to other locations, they can turn opportunistic pathogens. For example, *E. coli* can cause infections if it gains access to other locations such as urinary bladder, lungs, etc. It is also interesting to note that among the microbes residing within the human body, mutualism, competition, antagonism, cooperation, etc., can be seen between themselves, which impact their existence and numbers.

Harmful Effects of Microbes on Humans and Environment

Microbes are numerous, diverse, and are capable of causing multiple outcomes. While we have not yet identified all the microbes that inhabit this universe and have studied only a fraction of the total microbial population, some undesirable outcomes due to microbes are obvious and challenging, as follows.

1. Some of them cause diseases and are pathogenic to animals, plants, and human beings. There are numerous bacterial, fungal, and viral diseases that attack living beings and make them sick. Animal pathogens affect livestock and cause economic damage, while plant pathogens, especially those impacting crops and vegetables, cause great damage to food and feed reserves.
2. Many human diseases, such as typhoid, cholera, tuberculosis, meningitis, and pneumonia, are caused by bacteria, whereas measles, polio, hepatitis, and COVID are viral diseases. There are a few fungal diseases too, such as athletes' foot and ringworm, that can be quite challenging to curtail.
3. Bacteria and viruses can cause food spoilage and lead to food poisoning effects.
4. At the environmental level, microbes in soil and water produce greenhouse gases such as CO_2, methane (CH_4), and nitrous oxide (N_2O) when they decompose organic matter. These gases trap heat in the atmosphere, contributing to global warming.
5. In water bodies, too much of algal growth can lead to algal blooms that use up the oxygen and negatively impact the aquatic life.
6. Microbes in the environment can cause various types of allergies and skin conditions.
7. Microbes can spoil textiles and leather and cause financial loss.

Interesting Fact 16.4

Germ theory of disease

Ignaz Semmelweis

1. Many kinds of diseases are related to micro-organisms.
2. The idea that micro-organisms are the cause of various diseases is known as the 'germ theory of disease'. In early days, this was inconceivable as microbes were invisible and diseases were believed to be curse or punishment for individual crimes.
3. Ignaz Semmelweis demonstrated that if hands were not disinfected, it routinely transmitted infections. Pasteur also showed the first connection of microbes to animal diseases. Using these clues, Joseph Lister in 1860s used phenol as a disinfectant and reduced the incidence of infections and deaths.

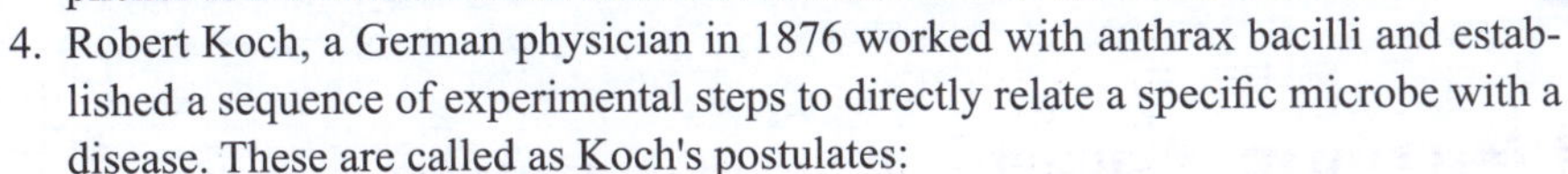

Robert Koch

4. Robert Koch, a German physician in 1876 worked with anthrax bacilli and established a sequence of experimental steps to directly relate a specific microbe with a disease. These are called as Koch's postulates:
   - Microorganism must be found in diseased individuals but not in normal individuals.
   - The microbe from the diseased individual must be isolated and cultured.
   - Healthy individuals if inoculated with the isolated microbe must develop the disease.
   - The microorganism must be re-isolated from the newly infected individual and must match with the original microbe.
5. This established the credibility of microbes as pathogens, and these four criteria proved the causal relationship between microbe and disease.

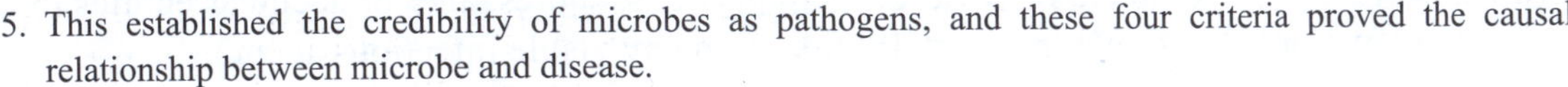

Beneficial Effects of Microbes

Microbes have a huge positive impact on us and our environment. In fact, the survival of the universe is dependent on the activities of microbes. Without microbes, the planet would be filled with corpses.

a. Microbes are the decomposers of nature. They break down (or decompose) dead organisms, animal waste, and plant litter to obtain nutrients.

b. They recycle nature's waste and release carbon, nitrogen, phosphorus, sulphur, etc., thus making them available for other life on Earth.

c. Microbes feed the world by being food for many animals. They are at the base of the food chain and help to sustain the aquatic and terrestrial food webs.

d. Micro-organisms are key components of soil and help in maintaining and replenishing the nutritional status and fertility of soil.

e. Normal biota in our bodies helps in digestion, releases valuable products such as vitamins, keeps pathogens away, and, overall, helps to keep us healthy.

f. They are responsible for various fermentation products that are of use to humans, e.g., alcohol, acids etc.

g. Because of their decomposing abilities, micro-organisms help us to manage our domestic and industrial wastes.

h. Photosynthetic bacteria (called cyanobacteria) living in lakes, ponds, and shallow oceans produce 50% of the oxygen we breathe.

i. Some micro-organisms produce antibiotics, which act as a tool for us against pathogens and infections.

j. Nitrogen fixation by some bacteria helps to convert gaseous nitrogen into ammonium and nitrate salts for other organisms to use.

k. Microbes are metabolically very talented. They can metabolize metals, acids, salt, methane, or even radioactive wastes. There seems to be a microbe for every pollutant.

Microbial presence is critical for the universe, and cohabitating with them, understanding them, and utilizing them is a useful endeavour. Considering that what we know is just a tip of the iceberg in terms of microbial diversity, identifying new microbes with the potential to solve many problems is an ongoing effort. However, regulating the harmful effects of microbes by effective control is also a crucial need for sustenance and successful survival for us. This is discussed in detail as follows.

Mechanisms of Microbial Control

Microbial control is the ability to control microbial populations and is of critical value in many settings. In some cases, it is necessary to completely eliminate all micro-organisms from an object, whereas partial elimination or destruction may be needed in some other situations. According to the depth of outcome, several terminologies are used to explain the nature of microbial control, as follows:

(a) *Sterilization* Complete removal of all living cells, spores, and/ or acellular entities from an object or habitat, making it totally free from any viable infectious agent.

(b) *Disinfection* Killing, inhibition, or elimination of micro-organisms that cause disease. This is aimed at particularly destroying potential pathogens, and overall microbial population is reduced.

(c) *Sanitization* It is closely related to disinfection and implies that the microbial load is reduced to levels that are considered safe by public health standards.

(d) *Antisepsis* It is the prevention of infection on living tissues using antiseptic agents.

(e) *Chemotherapy* This is the use of chemical agents to kill or inhibit the microbes in host tissue.

(f) *Germicide/bactericide/fungicide* These are agents effective in killing a specific group and accordingly the term is used.

(g) *Bacteriostatic/fungistatic* This also is effective in preventing the growth of a specific group such as bacteria, fungi, etc.

Micro-organisms can be controlled by physical and chemical agents. Physical agents include low temperatures, heat, and radiation. These are commonly used in food preservation, e.g., pasteurization of milk, freezers for food storage to prevent spoilage, and ionizing radiations are used to sterilize packed foods. Heat is commonly used in laboratories as dry heat in ovens or as moist heat in autoclaves for sterilization purposes.

Filtration can be used for heat-sensitive materials, such as enzymes, vaccines, and antibiotic solutions, that cannot be sterilized by heat. Membrane filters of 0.22 and 0.45 μm sizes are used for these purposes. Rooms and operation theatres filter air by the use of HEPA filters that trap micro-organisms larger than 0.3 μm. High pressure can also be used to kill microbes. Desiccation prevents microbes from growing or reproducing and cannot remain viable in the absence of water. Lyophilization and freeze-drying work on this concept. Osmotic pressure by the use of high sugar or salt concentration creates a hypertonic environment, which causes water to leave the cell and kills microbes. Jams, pickles, meat, and fruits are preserved by this mode.

Many different groups of chemicals have the potential to control microbes. Some of them are micro-bio-static, i.e., they inhibit growth, and some are microbicidal, i.e., they kill microbes. phenols, phenolics, alcohol, heavy metals, soaps, detergents, quaternary ammonium compounds, halogens, etc., are some well-known chemicals for microbial control. Ethylene oxide is a gaseous chemo-sterilizer, which is used to sterilize closed chambers.

Antimicrobial drugs must kill or inhibit the pathogen without any or minimal damage to the host. A drug that disrupts a microbial function not found in eukaryotes, like, humans, is a preferred choice. For example, penicillin inhibits bacterial cell wall peptidoglycan synthesis but has no effect on human cells as they lack cell wall. Hence, the therapeutic index of such drugs will be high. If the drug is effective only on a limited variety of pathogens, it is said to be a narrow-spectrum drug. Broad-spectrum drugs can attack different types of pathogens.

Broadly, antibacterial drugs fall under the groups of inhibitors of cell wall synthesis (penicillin, cephalosporin, vancomycin, etc.) or protein-synthesis inhibitors (tetracyclines, streptomycin,

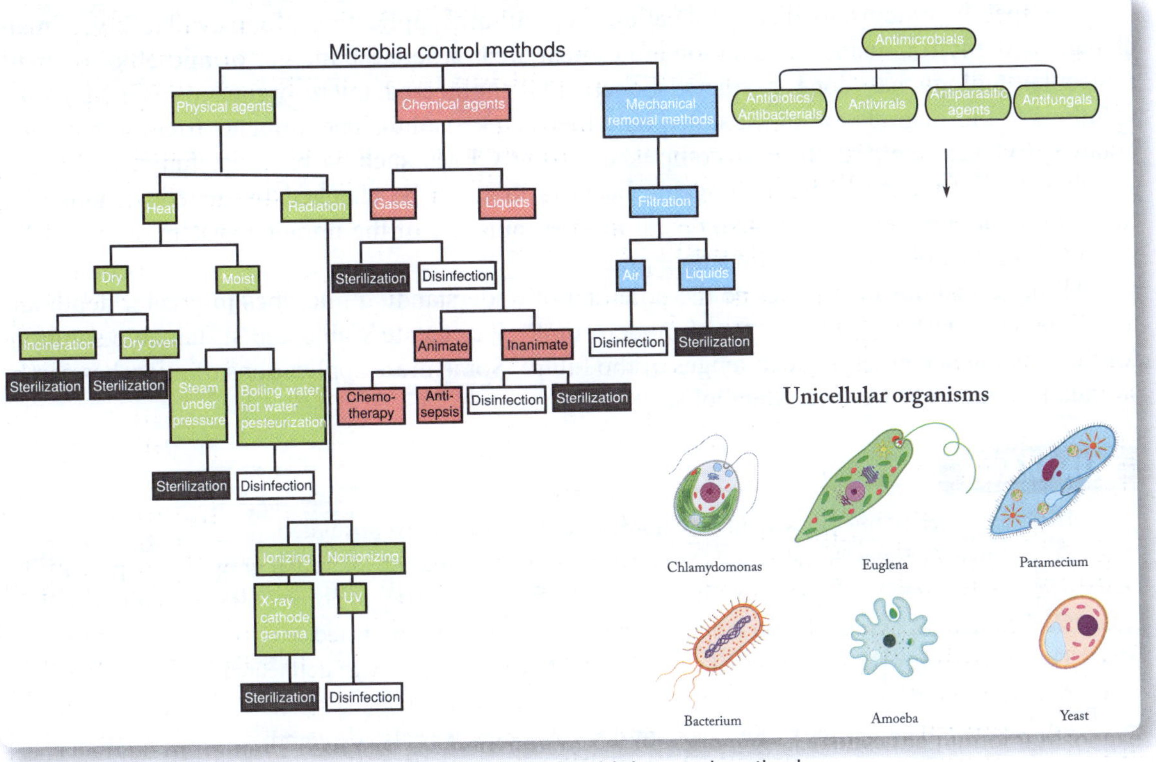

Microbial control methods
Physical agents
Chemical agents
Mechanical removal methods
Heat
Radiation
Gases
Liquids
Filtration
Dry
Moist
Sterilization
Disinfection
Air
Liquids
Incineration
Dry oven
Animate
Inanimate
Disinfection
Sterilization
Sterilization
Sterilization
Steam under pressure
Boiling water hot water pesteurization
Chemo-therapy
Anti-sepsis
Disinfection
Sterilization
Sterilization
Disinfection
Ionizing
Nonionizing
X-ray cathode gamma
UV
Sterilization
Disinfection
Antimicrobials
Antibiotics/ Antibacterials
Antivirals
Antiparasitic agents
Antifungals
Unicellular organisms
Chlamydomonas
Euglena
Paramecium
Bacterium
Amoeba
Yeast

Fig. 16.6 Summary of all microbial control methods

Source: Adapted from https://rbrlifescience.com/control-of-microorganisms/
https://courses.lumenlearning.com/suny-microbiology/chapter/mechanisms-of-antibacterial-drugs/
https://byjus.com/chemistry/antimicrobial-agents/

chloramphenicol, etc.). Some drugs can be metabolic antagonists, such as sulpha drugs, while others can be nucleic acid-synthesis inhibitors, such as quinolones. Many of these antibiotics are natural products of microbes that inhibit other microbes (e.g., penicillin from *Penicillium notatum*), and some can be synthetically prepared (e.g., quinolones). Although there are drugs effective against many pathogens and infectious diseases are in control, the spread of drug-resistance pathogens and the decreasing effectiveness of frontline drugs is a major public health threat in the 21st Century.

Applied Microbiology—Microbes to Solve Human Problems

Microbes, in the natural environment, perform processes that are useful to mankind. Microbes play a key role in wastewater treatment, bioremediation of pollutants, and generation of products useful to humans, such as antibiotics, hormones, solvents, organic acids, chemical feedstocks, amino acids, and enzymes of economic value. Initially, the traditional method was bioprospecting or 'mutations followed by selection for the strain' capable of providing a valuable solution. Microbes with the ability to decompose oil and their utility in oil spills is a case in point.

However, genetic engineering has replaced this classic approach, and recombinant molecules created by incorporating the gene of interest into a vector or phage and expressing the protein in a bacterial host have revolutionized the possibilities. Bacteria have contributed to the domain of genetic engineering by way of restriction enzymes, plasmid vectors, and *E. coli* model host systems, which facilitate modern-day solutions to important challenges.

Therapeutic proteins, medical applications, agricultural applications, food availability, climate change, and environmental remediation are critical areas where the merger of microbiology with molecular biology has done wonders. Genetically engineered micro-organisms (GEMs) offer solutions to bio-degradation of oil spills, halo benzoates, naphthalene, toluene, trichloroethylene, octanes, xylenes, etc., has been accomplished using GEMs such as bacteria, fungi, and algae. Another crucial reality is non-pathogenic bacteria that can be changed by genetic engineering to express antigens on their surface, resulting in stimulation of the immune system. Such GEMs provide long-lasting immunity and stable protection.

All these insights clearly define the potential of understanding microbes in greater depth and to utilize them with the other facets of modern biology to create viable and sustainable solutions for the existing and emerging challenges of the future. Some more applications of microbes can be studied in Chapter 8: Cellular Metabolism.

SUMMARY

- Different types of microbes and the features of their existence were studied.
- Tools and techniques that are used to visualize, grow, and measure microbial growth were presented.
- Microbial interactions of different kinds and with humans by way of normal biota were discussed.
- Harmful effects of microbes to humans and the environment were listed.
- The crucial role of microbes in nature and their beneficial impact was delineated.
- Agents of microbial control and their effective purpose was explained.
- Engineering microbes for various modern-day challenges was also explored.

EXERCISES

A. Objective-Type Questions

1. Which of these is not a micro-organism?
   (a) Bacteria (b) Fungi
   (c) Algae (d) Tapeworm
2. ______________ gave the germ theory of disease.
   (a) Louis Pasteur (b) Joseph Lister
   (c) Robert Koch (d) None of the above
3. Which of these is normally used for sterilization?
   (a) Antibiotics (b) Soaps and detergents
   (c) Autoclave (d) Desiccation
4. Streak plate method helps primarily in ____________ micro-organisms.
   (a) counting (b) killing
   (c) genetically engineering (d) obtaining a pure culture of
5. Mechanism of action of penicillin __________________.
   (a) inhibits protein synthesis (b) inhibits nucleic acid synthesis
   (c) inhibits cell wall peptidoglycan synthesis (d) acts as a metabolic mimic
6. Which statement about gram staining is true?
   (a) It is a simple stain and uses one stain.
   (b) It stains endospores.
   (c) It classifies bacteria into gram +ve and gram –ve.
   (d) It is a type of acid-fast stain.
7. Blood agar is an example of ______________.
   (a) Chemically defined media (b) Differential media
   (c) Minimal media (d) None of the above
8. Microbes are involved in recycling of elements such as C, N, P, and S. This statement is ____________.
   (a) Incorrect (b) Correct
   (c) Incomplete (d) Irrelevant to microbes
9. Which of these is a beneficial effect of microbes?
   (a) Polio (b) Food spoilage
   (c) Nitrogen fixation (d) Depletion of soil fertility
10. Fungistatic and fungicide are different because
    (a) fungistatic inhibits growth and fungicide kills
    (b) fungicide inhibits growth and fungistatic kills
    (c) fungicide works against fungus but fungistatic does not
    (d) Both are produced by fungus

B. Short-Answer Questions

1. How is the normal microbiota of humans developed and how is it useful?
2. If you are given 1 g of soil and asked to isolate the microbes, identify the steps that you will use.
3. How does turbidity measurement by spectrophotometer help in enumeration of microbes?
4. Koch postulates are specific for only one type of diseases and microbes and not universally applicable. Is this statement true or false? Justify your choice.
5. Write the different types of microscopes with their specific utilities.

C. Self-Study Questions

1. Analyse and discuss 3 ways each of microbial control and microbial benefits in our everyday lives.

FURTHER READINGS

1. Stark, L.A. Beneficial Micro-organisms: Countering microphobia; CBE Life Sci Educ, 2010 Winter;9(4):387–389.

USEFUL WEBLINKS

1. https://www.heraldopenaccess.us/openaccess/a-review-on-chemical-and-physical-methods-of-controlling-microbial-growth
2. https://www.britannica.com/technology/microscope/History-of-optical-microscopes

VIDEO LINKS

1. https://microbiologysociety.org/why-microbiology-matters/what-is-microbiology.html
2. https://www.youtube.com/watch?v=prTsBspXotg
3. https://www.youtube.com/watch?v=EhyvSNRzDCI

CHAPTER

17

Ecology: Environment and Evolution

LEARNING OBJECTIVES

After reading this chapter, you will be able to:

- Study ecosystems and the environment in the human context.
- Understand the distribution of energy and matter in our world.
- Define biological species and their uniqueness.
- See how species evolve and the factors involved in evolution.
- Explain the theory of natural selection and its role in evolution.
- Study population genetics and ways to evolution.
- Understand present day environmental challenges and their mandated solutions

CHAPTER TAKEAWAY **Interconnectedness of population and environment, and evolution**

This chapter deals with the different aspects of the ecology, environment, and evolution of populations and their interconnectedness for sustainable universe.

PRELUDE

The 21st century is an era of great technological developments and innovations, with new emerging and groundbreaking technologies being a frequent occurrence and unquestioned reality. However, this is also the time where we constantly hear about terms such as 'environmental destruction', 'ecological concerns', 'climate change', 'global warming', 'greenhouse gases', 'carbon footprint', and so on in many of our conversations across all levels. Threat to our environment and ecosystem seems more real and closer, necessitating proactive action at the earliest. Awareness among the millennials, Gen Z, and future generations about the basics of the environment and the associated challenges is paramount for them to do their bit for ensuring an environmentally healthy and robust future ahead. As this is one critical aspect impacting living systems, this book will not be complete without understanding the various facets of ecosystems and environment. How all organisms are interconnected in this universe and how human actions act as a stimulus to drive evolution will be the focus of this chapter. How the universe operates and matter and energy get distributed, how species evolve, and how natural selection operates in action, are the key facets that students and global citizens need to be exposed to. Though environmental science is a vast field, keeping in mind the basic purpose of the book, the concepts pertaining to ecology and environment are detailed in a simplified manner.

Ecology and the Environment

The branch of science that deals with the relationship between the organisms and their environment is called 'Ecology'. Environment can be broadly defined as anything that affects and influences the organism during its life. Just as the environment has an impact on organisms, they also, in turn, affect the environment. Influence of organisms, including humans, and their actions on the environment is a crucial factor, and the social interactions, behaviour, and their economic and political implications are studied under the domain of environmental sciences.

Environmental influences on organisms can be of two types: Biotic and Abiotic factors. Biotic factors are the living things that affect an organism, including all the other living species in the environment that can positively or negatively affect you. Abiotic factors include the non-living factors such as climate, breeze or wind, rain, atmosphere, minerals, soil, temperature, sunlight, altitude above sea level, etc. These factors are all interrelated and form a conducive environment termed as an ecosystem. For example, a tree will be affected by rain, the soil fertility, bees for pollination, humans who chop them, or birds that nest on them. This way, it is an interconnected world where multiple contributing factors help in creating a niche or habitat for every organism in the ecosystem.

Study of these complex interactions can be done at multiple levels. Some ecologists will focus on individual organisms and the impacts during their lifetime. Others focus on one type of organism in groups called populations and indulge in population studies, while some others focus on the community interactions within a specific geographic location. To categorize and study the entire ecosystem, ecologists have devised trophic levels and food chains to understand the pattern by which energy moves through the ecosystem.

Sun is the ultimate source of energy, and generally it moves from sun to plants and then to animals. This series of organisms where one organism feeds on another and obtains energy is called the food chain and involves various trophic levels. The different levels in an ecosystem are as follows:

Producers They are organisms that trap energy from sunlight and convert inorganic ions to organic compounds. They form the first trophic level. Energy flows from the sun to the living matter of these producers. For example, producers include plants, cyanobacteria, ferns, mosses, algae, trees, etc. Only producers can trap energy from sunlight, whereas all other organisms depend directly or indirectly on the producers and are called Consumers.

Consumers Consumers are categorised as herbivores, carnivores, and omnivores, depending on the way they obtain their food and energy. Herbivores obtain their energy by eating plants; hence, they are the primary consumers and are classified as the second trophic level. Carnivores are animals that feed on other animals. If the carnivore feeds on the herbivores, they are the primary consumers and fall at the third trophic level. Carnivores that eat primary consumers are secondary consumers and fall at the fourth trophic level.

Omnivores are animals that are sometimes herbivores and are carnivores at other times. They are classified based on the pattern that they present at a particular moment and are taken as primary or secondary consumers. Humans are omnivores and fall into this category. They are classified into the separate trophic levels, accordingly.

Decomposers They are a special category of organisms that feed on the organic matter from the wastes of living organisms or dead individuals. They are common to all trophic levels and break down matter from all producers and consumers. Most decomposers are micro-organisms such as bacteria and fungi. Decomposers convert back the organic matter to inorganic ions and are hence very critical components of the ecosystem.

Thus, with the sun as the main source of energy, the food chains, the trophic levels of organisms, and ecosystems help in continuous recycling of elements in nature. One such combination in the terrestrial ecosystem is given in the Figure 17.1 below. Similar food chains and trophic levels can be drawn for any ecosystem such as aquatic (freshwater, marine) and terrestrial (forest, grassland, desert, tundra).

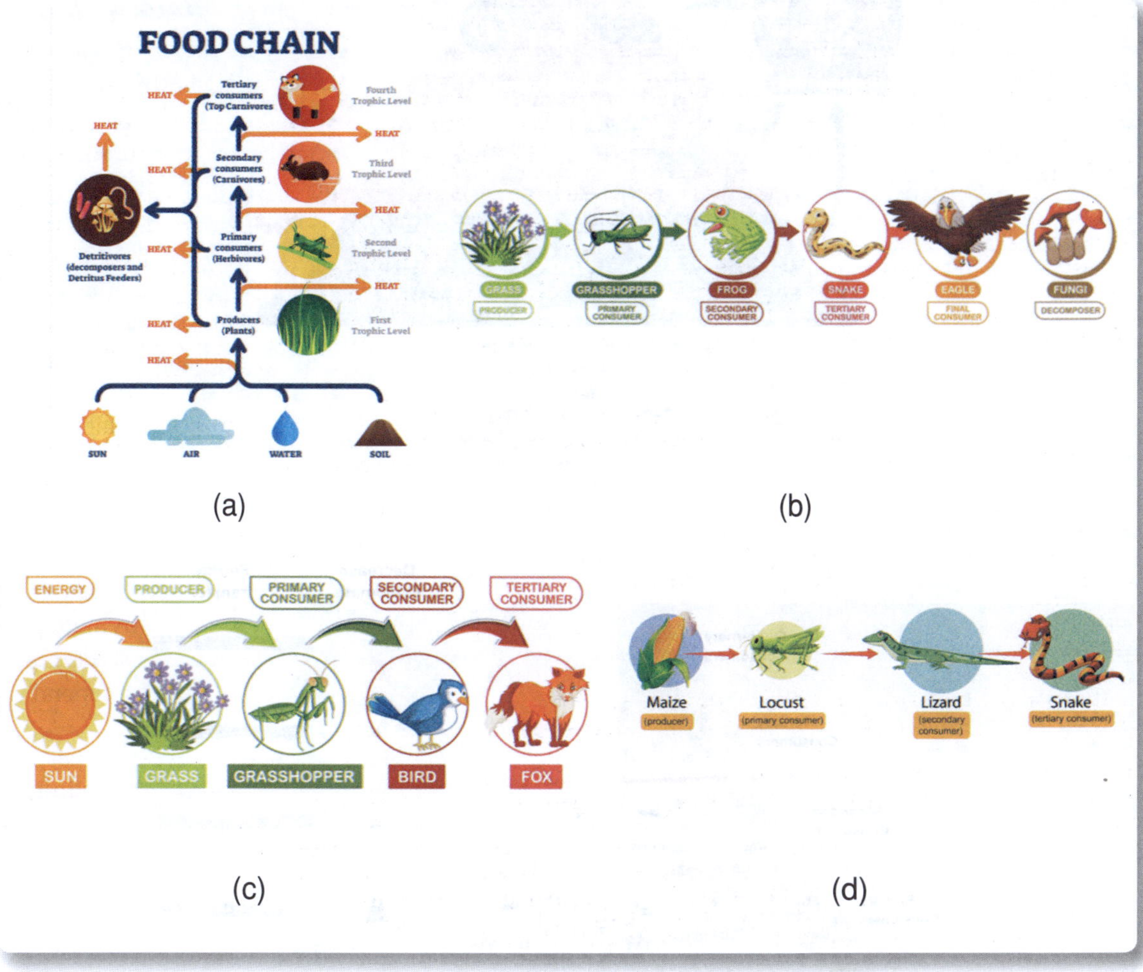

(a) (b)

(c) (d)

Fig. 17.1 Different combinations of food chain: producers–consumers–decomposers in a terrestrial ecosystem

Distribution of Energy and Matter

A pyramidal structure exists for relationships between various trophic levels in an ecosystem, i.e., large quantities of producers, then herbivores, and then decreasing level of carnivores, and subsequently at trophic level.

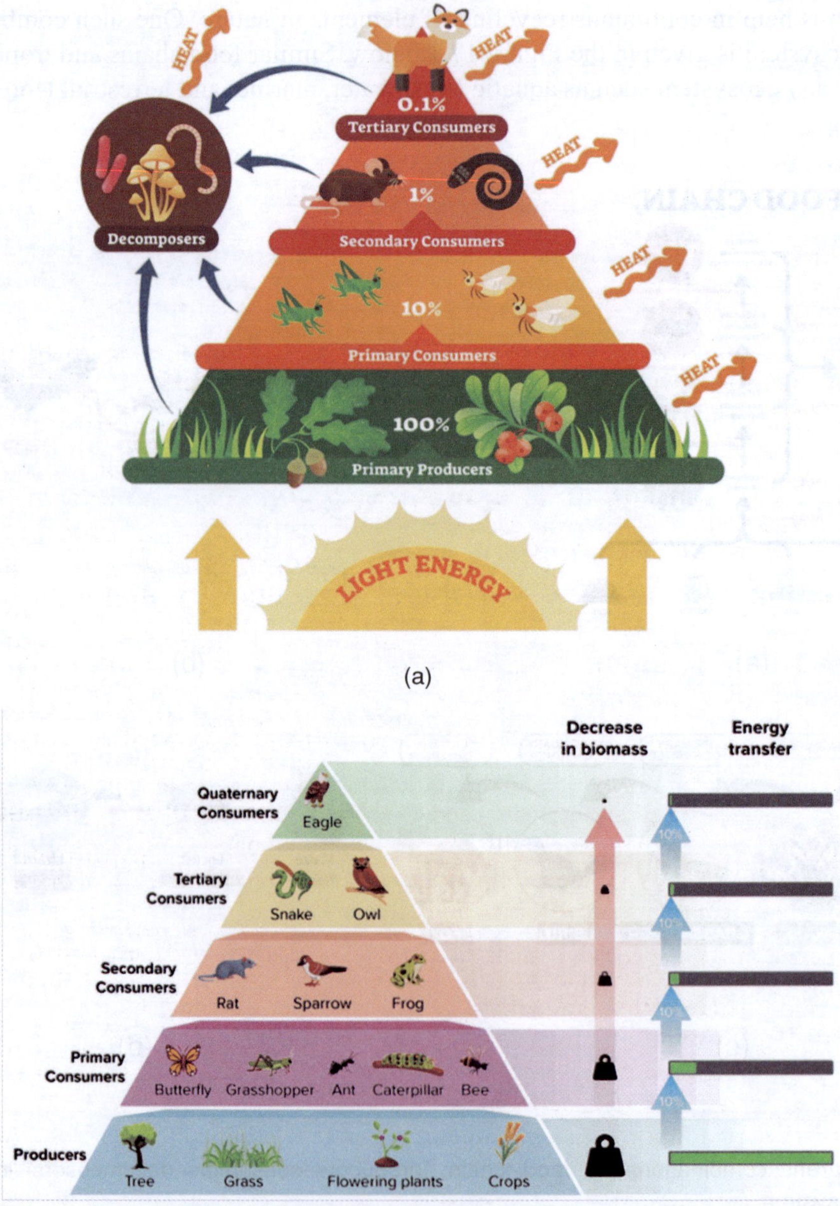

(a)

(b)

Fig. 17.2 Pyramid of energy and biomass

Source: Adapted from https://www.edrawmax.com/article/what-is-energy-pyramid-diagram.html
https://www.ck12.org/flexi/biology/trophic-level/define-ecological -pyramids/

The thermodynamic laws state that energy can neither be created nor destroyed; it is converted from one form to another. As energy travels through various trophic levels, it is passed on from one source to another; however, there is also some dissipation or loss of energy as heat. When the energy availability at each trophic level was measured, it was concluded that nearly 90% of energy is lost and only 10% is passed on to the next trophic level. Along with the heat energy lost, this could also be attributed to energy expended at trophic levels to reach their prey or chase their prey. As a rule of thumb to conclude, only 10% of all this energy gets retained at the next trophic level and 90% is lost. These numbers are also corroborated by a pyramid of biomass, which also shows a typical 90% reduction or loss at each trophic level (Figure 17.2). Pyramid of numbers at each trophic level was also measured; but it was not a good indicator as the size of organisms at each level varied. Therefore, it was not a conclusive indicator.

Evolution and the Biological Species Concept

After having understood the dynamics of environment and ecosystem at the broader level in terms of numbers and interrelationships, now let us focus on evolutionary aspects. Evolution is a change that happens at the population level to produce or evolve new variants of the species. When there is a change in frequency of genetically determined characteristics of a population over time, it is said to be evolving. When the genetic differences in allele frequencies are minor between populations of the same species, it leads to formation of subspecies and is called microevolution. Major differences over long periods of time that lead to generation of new kinds of species is called macroevolution. In both instances, changes in genetic information cause changes in characteristics over time across many generations. Evolution is a constant continuous phenomenon, which is happening even today.

By definition, a species is a population of organisms that interbreed naturally to produce fertile offspring. Individuals are not species but members of a species. In essence, the genetic similarities between individuals and populations along with successful gene flow from one generation to the next are factors that help to clearly define species. For example, though horses and donkeys have similarities, mating between the two results in mules, which are sterile; thus, they are not from the same species as the gene flow is disrupted, and they do not produce fertile offspring. Similarly, though lions and tigers are mated artificially in zoos, again the natural gene flow is disrupted.

Speciation is the process of generating new species. Geographic isolation, reproductive isolation, ecological isolation, polyploidy, and genetic divergence are some ways by which speciation is initiated. Study of fossils help evolutionary biologists to determine the pace, nature, and reason for evolution, thus explaining speciation. This whole process can be a gradual change over a long period of time called gradualism or large number of changes in short spurts of rapid change called punctuated equilibrium. Evolution can also be the result of a combination of both.

Chasing the evolutionary history of a particular species is a challenging endeavour as (a) fossil records can be incomplete and limited; (b) some intermediate changes may not be fossilized, and (c) fossils give only structural information but not much about behaviour and physiology or the natural historical progression. However, scientists and anthropologist are constantly studying, identifying differences, dating them, and redefining terms and names to arrive at evolutionary history.

Interesting Fact 17.1

Human evolution

1. Humans are recent additions and seem to have evolved 2.2–2.5 million years ago.
2. The scientific name is *Homo sapiens*. Those belonging to genus Homo have evolved large brains and moved from herbivores to omnivores.
3. The Out of Africa theory hypothesizes the emergence of this genus in Africa from where they moved to Europe and Asia.
4. Neanderthal was a separate species–sub-species found in Europe and Asia that was extinct due to inability to compete with *H*. sapiens.
5. Other species like Homo ergaster, *Homo habilis*, *Homo heidelbergensis*, small-size hobbits *Homo florensis*, *Homo erectus*—all evolved progressively, leading to Homo sapiens, but all are extinct, and *H*. sapiens is the only surviving species in this genus
6. Evolutionarily, the opposable thumbs, big toes, hug shoulders, and grasping hands of early primate (chimpanzees, apes, and monkeys) helped them to survive in forest and trees.
7. Walking upright was an adaptation, along with large brain and ability to use tools that propelled the hominid category

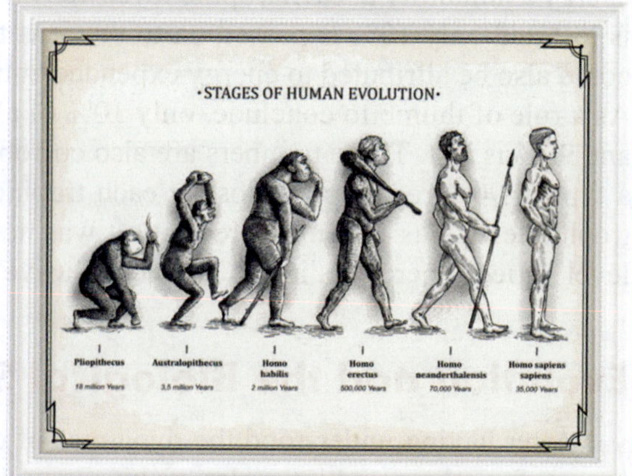

Domain: Eukaryota
Kingdom: Animalia
Phylum: Chordata
Subphylum: Vertebrata
Class: Mammalia
Order: Primates
Family: Hominidae
Genus: Homo
Species: *Homo sapiens*

Theory of Natural Selection and Evolution

The process of evolution is very slow and cannot be recognized during one's lifetime. So, there was a general lack of awareness about the change happening. Once the thought that animals change and evolve came by, naturalist Buffon and his student started their work on conceptualizing evolution. Lamarck proposed that acquired characteristics were inherited and explained the thought with the example of Giraffe's neck. Giraffes when stretched their necks to obtain food, the shorter necks became long and were inherited in the next generation. With every generation the neck became longer to reach the present-day version of the giraffe. However, the debate raged with disagreements, and, finally, Wallace and Darwin, in 1858, explained the Theory of Evolution Through Natural Selection.

The theory of natural selection states that individuals with genetic combinations that favour life in their existing habitat are likely to survive and reproduce, and hence they pass on their genes to the next generation as compared to others. This theory was explained by Darwin in his book '*Origin of Species by Natural Selection*'. The giraffe example was clarified based on the following observations:

(a) All giraffes compete for the same food resources.
(b) Some giraffes with relatively longer necks will have greater access to food; hence, they survive better and produce more offspring.

(c) As a result, succeeding generations would show an increase in individuals with longer necks.

Thus, over a period of time, the beneficial characteristic that favoured survival and life was selected, and the shorter neck was slowly weeded out. Through natural selection theory, it also became clear that only genetic traits are inherited, and acquired characteristics are not inherited and have no role in natural selection (Figure 17.3).

Lamarck-Acquired character longer neck

Darwin and Wallace-Theory of natural selection

Fig. 17.3 Lamarck understanding of evolution

In the modern context, it can be explained that selecting agents lead to genetic variations and new traits in the gene pool, and the individuals with the best genotypic allelic combinations pass on their genes more favourably to the next generation; thus favourable alleles are selected. Other terms like 'survival of the fittest' and 'struggle for existence' are often associated with natural selection. While survival is critical for beneficial genes to be selected, it is important that the organism not only survives but also reproduces. Only then the organism can contribute genes to the next generation. Struggle for existence does not mean individual antagonism or fights. Instead, it means the competition to utilize the limited existing resources and still survive and reproduce. For example, plants compete for sunlight, birds compete for nesting spaces, etc. When the organism utilises best resources and survives and reproduces, it can successfully contribute beneficial alleles by natural selection.

Three critical pillars of natural selection are: (a) Environmental factors that act as selecting agents or stimulus, (b) genetically diverse gene pool, and (c) sexual reproduction among individuals of gene pool. Genetic diversity is crucial; only then there can be a choice for selection. If the gene pool is homogenous and same, then the one who reproduces and passes on genes will have no impact as all are the same. Mutations, migrations, genetic recombinations, and degree of gene expression—all contribute to genetic diversity in a gene pool. Differential survival, differential mate selection, high reproductive rate, and success—all contribute to effective evolution by natural selection.

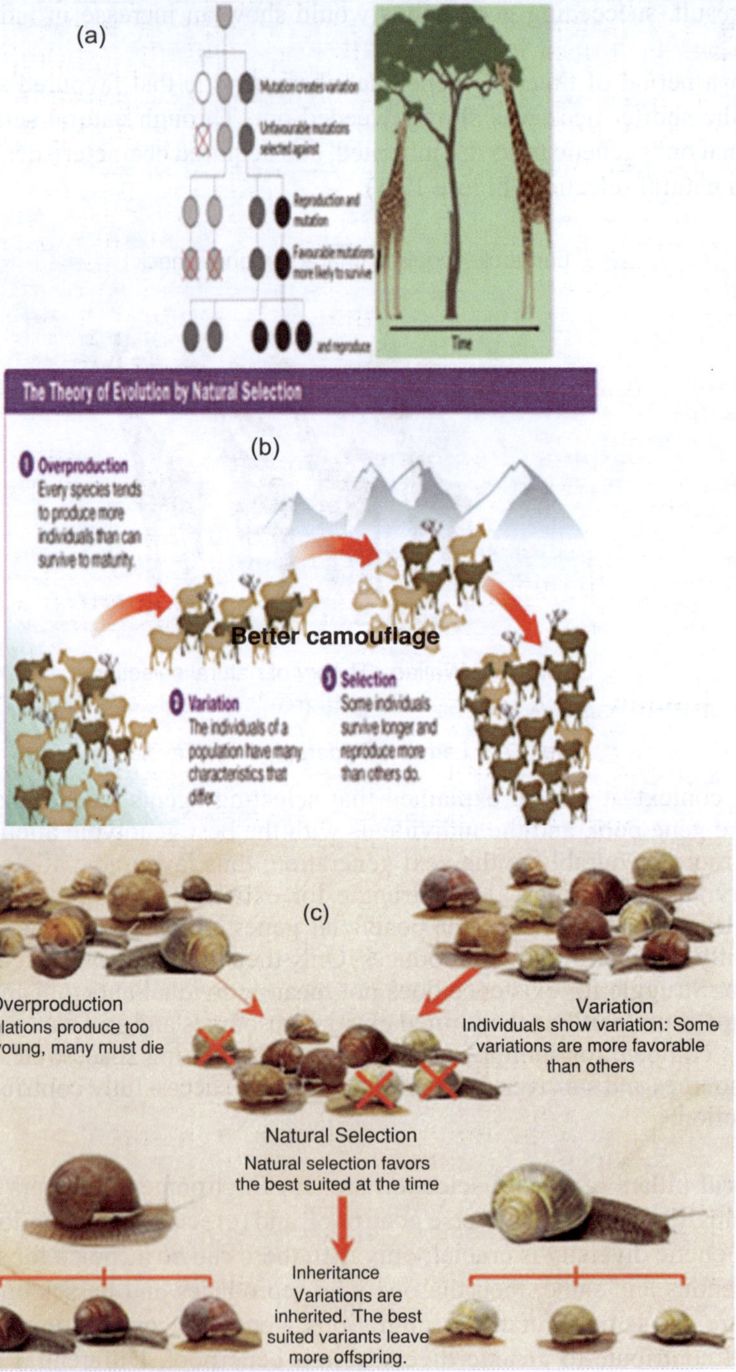

Fig. 17.4 Wallace and Darwin–Theory of natural selection and evolution

Source: Adapted from https://www.turito.com/learn/biology/process-of-evolution-selection
https://www.keystagewiki.com/index.php/Evolution_by_Natural_Selection;
https://blogs.ubc.ca/mrpletsch/2019/10/18/unit-2-3-mechanisms-of-evolution-and-natural-selection/
https://igcse-biology-2017.blogspot.com/2017/06/338-explain-darwins-theory-of-evolution.html

When one particular allele is selected for or against, it will manifest itself as a change in gene frequency in the population. Changing gene frequencies indicate that evolution is taking place.

Population Genetics and Way to Study Evolution

Population genetics is the study of genes, their numbers, and how these numbers change over time within a population. Genetic differences within a population are given by gene frequency or allele frequency that indicates how common a particular gene variant or allele is in a population. It is usually indicated as a percentage. The entire human population along with its diverse genetic combinations is a gene pool of the same species. But local populations due to environmental impact show distinguishing characteristics transmitted across generations. For example, let us take the phenotypes of skin colour, nose type, and hair structure and compare populations across Europe, Asia, and Africa. It is obvious that allele frequencies are different in these populations. Dark skin alleles are at a higher frequency in Africa; light skin alleles are of higher frequency in Europe; and Asia has a moderate frequency of both the alleles. Similar observations are also true for curly hair—the highest in Africa, the lowest in Europe, and moderate in Asia. It is important to note that this distribution has nothing to do with whether the allele is dominant or recessive. Straight hair and light skin are recessive but in high frequency and are common in the European population. Once mixing of alleles happens, such distinctions will also get distributed more evenly. So allele frequencies can be indicators of population status. Change in status, here, means evolution.

To study whether evolution is happening, Hardy and Weinberg gave a mathematical formula. They proposed that gene frequencies will remain constant if certain conditions are met, and the genotypes can be expressed by the formula:

$$p^2 + 2pq + q^2 = 1$$

Interesting Fact 17.2

Why genetically distinct populations exist?

1. Adaptation to local conditions: The genetic combination of individuals in a population is varied. Of these, some individuals may find the local conditions hostile and hence may leave or quit. They add their genes to the pool where they join and remove the genes from pool where they existed. Mutations for adaptations, migrations for non-adaptability—all these cause genetically diverse pools to exist.
2. Founder's effect: It is a form of genetic drift in which the gene pool of the population is made of few individuals who colonize and set up a new population—its founding members. This collection of alleles is likely to be different, limited, and smaller as compared to the large population from where they moved to establish a new population.
3. Genetic bottle neck: This is also a form of genetic drift in which there is a sharp reduction in population size due to a chance event like disease, predation, natural disaster, etc., which results in a reduction in genetic diversity in subsequent generations. When the size of a population is greatly reduced, some alleles will probably be lost from the population.

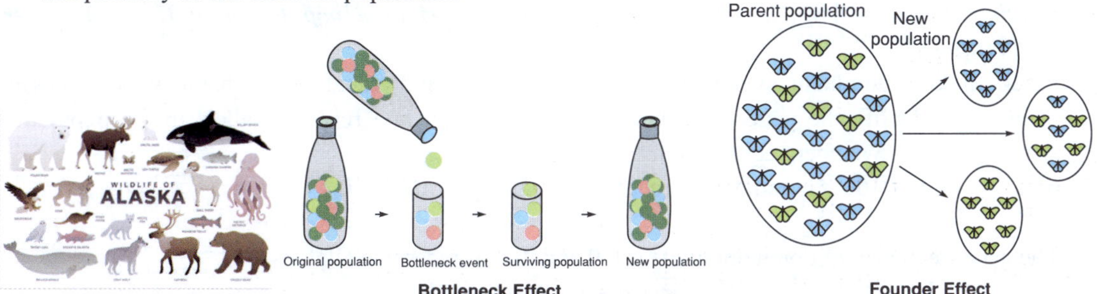

Bottleneck Effect

Founder Effect

In this, p = frequency of dominant allele and q = frequency of recessive allele, where p^2 = number of homozygous dominant individuals, $2pq$ will denote heterozygous individuals, and q^2 is the homozygous recessive individuals.

Constant gene frequencies as per this formula over several generations mean that the population is not evolving. This helps to compare two different populations across geographic locations in the same period or compare populations at different times to see whether populations are evolving.

We are aware that evolution is constantly happening. Hardy and Weinberg also gave five conditions under which the gene frequencies will remain constant and evolution will not happen.

(a) Populations should be very large. In smaller populations, the chances of change in gene frequency even due to one or two events is high. Genetic drift will happen with more ease.
(b) Mutations must not occur. The same version should continue to exist, and natural variations should not be seen.
(c) Mating must be completely random. There should be no choice in terms of preferred or bias towards any characters. Only then the frequency will be maintained.
(d) Migration in or out of the population must not happen. Or else it will lead to addition or deletion of certain alleles and can change frequencies, especially, in small populations.
(e) All genes must have an equal chance of getting passed on to the next generation, that is, no natural selection.

If these five conditions are met, then Hardy–Weinberg equilibrium will be applicable every time; and there will be no change in gene or genotype frequencies.

However, we are aware that Hardy–Weinberg conditions rarely exist. Random mating cannot happen due to geographical, social, cultural, and other barriers. Spontaneous mutations are a reality and happen often even during replication and due to other environmental triggers. Immigration and emigration are common in populations. Populations as expected are not infinitely large and natural selection is in action all the time, as some alleles have greater value for life than others.

Experimental Insight 17.1 Hardy–Weinberg Equilibrium

1. ***Finding genotype frequencies from gene frequencies in a population that is in Hardy–Weinberg equilibrium.***

   If the frequency of dominant allele $p = 0.6$ and recessive allele $q = 0.4$, find the genotype frequencies in a population of 1000 individuals.

   Solution: The three genotypes can be calculated by $p^2 = AA = 0.6 \times 0.6 = 0.36$; in a population of $1000 = 0.36 \times 1000 = 360$ individuals are homozygous dominant.

   $2pq = Aa = 2 \times 0.6 \times 0.4 = 0.48$; in a population of $1000 = 0.48 \times 1000 = 480$ individuals are heterozygous.

   $q^2 = aa = 0.4 \times 0.4 = 0.16$; in a population of $1000 = 0.16 \times 1000 = 160$.

   $360 + 480 + 160 = 1000$.

2. ***Finding the gene frequency from genotype frequencies in a population in Hardy–Weinberg equilibrium.***

   Black allele is dominant over white allele. If 1 in 100 individuals has a homozygous recessive phenotype, identify the gene frequencies of the dominant and recessive allele in the population. Assume Hardy– Weinberg equilibrium.

   Solution: $q^2 = 1/100$ = homozygous recessive = 0.01; $q = \sqrt{0.01} = 0.1$

   $$p = 1 - q = 0.9$$

   The gene frequency of dominant allele is 0.9 and that of recessive allele is 0.1;

   $$p + q = 1; 0.9 + 0.1 = 1.$$

Present day Environmental Challenges and their Mandated Solutions

Environmental studies gain great relevance in the present world, and it is critical to understand the challenges of today and find sustainable and mandated solutions for a robust future ahead.

(a) *Climate change is a global threat that humanity faces today.* Global warming due to CO_2 emissions is accelerating climate change and threatens our survival. Natural disasters are becoming more frequent and extreme; hence, measures to mitigate this remains a high priority.

(b) *Pollution problems and impact on health:* Air pollution in many metro cities of the world has risen to unprecedented limits, causing various health hazards that need immediate attention. Water treatment and elimination of recalcitrant wastes are imperative for a healthy and sustainable future.

(c) *Cleaning oceans of plastic:* They have become dumping grounds for plastic. Micro- and nano-plastics in water bodies and their health implications are real challenges today.

(d) *Capitalizing renewable energy sources instead of fossil fuels:* A cleaner, more accessible and efficient energy source will need considerable thought and investment for the future.

(e) *Providing food for all:* Tackling challenges of food security, shortage, and malnutrition will remain a top priority in the years to come.

(f) *Dearth of hydrologic sources and water scarcity looms large in future.* Conservation of water resources and provision of water for agriculture to meet food requirements are interlinked commitments that we need to focus on.

(g) *Overpopulation and waste management:* Waste reduction, recycling, and reuse for a circular economy will have a huge impact on health and environment.

So, knowledge, awareness, and investment of resources and manpower in this domain is important and imperative.

SUMMARY

- Broad tenets of ecosystem and environment were studied.
- Food chains, trophic levels, and distribution of energy and matter across these levels were explored.
- Natural selection as a means of evolution with all the associated factors was touched upon.
- Population genetics, speciation, and measurement of evolution by mathematical equation was explored.
- Details of Hardy–Weinberg equilibrium and the problems for gene and genotype frequencies were solved.
- Relevant environmental challenges of the future and need for solutions were emphasized.

EXERCISES

A. Objective-Type Questions

1. ______________ is the correct order of interdependence in food chain.

(a) Carnivore –Herbivore –Decomposer –Producer

(b) Consumer–Decomposer–Producer
(c) Producer–Herbivore–Carnivore–Decomposer
(d) Omnivore–Herbivore–Carnivore–Producer.

2. ______________ proposed that acquired characteristics can be inherited.
(a) Charles Darwin (b) Lamarck
(c) Buffon (d) Wallace

3. Few changes over a long period of time leading to evolution is called ______________.
(a) Punctuated equilibrium (b) Gradualism
(c) Struggle for existence (d) Biotic influence

4. Changes within a species leading to sub-species is called ______________.
(a) Speciation (b) Macroevolution
(c) Microevolution (d) None of the above

5. Which of these does not contribute to genetic diversity in a gene pool?
(a) Migration (b) Mutation
(c) Sexual reproduction (d) Abiotic factors

6. Which of these is not a condition to maintain the Hardy–Weinberg equilibrium?
(a) Population size should be small.
(b) Mutations must not occur.
(c) Mating must be random.
(d) Every gene must have an equal chance of passing to the next generation.

7. In Hardy–Weinberg equation, the heterozygous genotype is indicated by ______________.
(a) p^2 (b) $2pq$
(c) q^2 (d) 1

8. The scientific name of the existing human species is ______________.
(a) *Homo habilis* (b) *Homo erectus*
(c) *Homo sapiens* (d) *Homo heidelbergensis*

9. ______________ is an environmental problem of the future,
(a) Absence of natural selection
(b) Genetically homogeneous gene pool
(c) Micro- and nano-plastics in water bodies
(d) Humans converting from omnivores to vegans

10. Which of the following cannot contribute to speciation?
(a) Reproductive isolation
(b) Geographic isolation
(c) Reproductive fitness without genetic change
(d) Polyploidy

B. Short-Answer Questions

1. Explain why acquired characteristics are not relevant to evolutionary biologists.
2. What are the factors that influence natural selection? Explain with specific examples?
3. In Africa, both sickle cell anaemia and malaria are seen, and it has been found that individuals heterozygous for sickle cell anaemia show resistance to malaria. Sickle cell anaemia is a genetic disease. If 9% of an African population is born with a severe form of sickle cell anaemia (ss), what percentage of the population will be more resistant to malaria because they are heterozygous (Ss) for the sickle cell gene?

FURTHER READINGS

1. Most important environmental issues https://www.iberdrola.com/sustainability/most-important-environmental-issues
2. Understanding natural selection: Essential concepts and common misconceptions https://evolution-outreach.biomedcentral.com/articles/10.1007/s12052-009-0128-1
3. Environmental Ecology, Bio-diversity and Climate Change. https://iipa.org.in/upload/ecology.pdf

USEFUL WEBLINKS

1. Hardy–Weinberg equilibrium - Amoeba sisters https://www.youtube.com/watch?v=7S4WMwesMts
2. Examples of natural selection - Bozeman science https://www.youtube.com/watch?v=S7EhExhXOPQ
3. Food chains and food webs https://www.youtube.com/watch?v=yeSgimGfXP0
4. Speciation - Amoeba sisters https://www.youtube.com/watch?v=udZUaNKXbJA

CHAPTER 18

Biological Concepts to Engineer Modern Solutions

LEARNING OBJECTIVES

After reading this chapter, you will be able to:

- Understand molecular machines and molecular motors with examples.
- Study ATP synthase motors and cytoskeleton-associated motors.
- Focus on different genetic engineering techniques for gene manipulation.
- Identify medical applications using genetic engineering.
- Study how agricultural and environment needs are met using genetic engineering.
- Explain biosensors, biodetections, and bioremediation.

CHAPTER TAKEAWAY

The whole purpose of understanding biological concepts is to apply them for real-life challenges. This chapter will provide examples and help us appreciate how biological concepts can be used to engineer modern-day solutions.

PRELUDE

Most of our present-day challenges and their resolutions through sustainable solutions is a continuing endeavour for mankind. It is indeed obvious that our answers lie in being in tune with and true to nature. When our actions and activities are out of sync with nature, many challenges crop up as is evident today. Prominent diseases today are lifestyle diseases; many environmental impacts are due to human greed and unsustainable growth without balancing the ecosystem; and mental health challenges are due to disturbances in human behaviour, connections, and interactions. We need to reset and re-organize to find solutions to these within the realm of nature; in that context, understanding life processes and biological principles and application of these in modern-day solutions are plausible approaches that have paid good dividends. In this chapter, we will focus on modern developments that have their roots in biological concepts and how they have been leveraged to find answers to complex problems. The mechanistic aspects of life involving molecular machines will be the first part of the chapter, followed by applications like biosensors, genetic engineering applications for health, agriculture, and environmental impact such as disease management, food production, and bioremediation.

Structural Design Framework of Human Body

Study of the detailed anatomy of human systems in the previous section has helped us to gain an insight about how the structures of each component of various physiological systems are linked to and designed to meet the functional requirements. In totality, the human body is a collection of

multiple complex systems that not only function as effective individual units but are also integrated together to ensure proper functioning and efficient productive results for survival and existence of the whole organism. Each organ system and different individual components at their own levels can be compared to machines for their coordinated precision in function. In a sense, the human body can be considered as a machine with individual units as parts of the larger whole. To understand the mechanistic aspects of design, molecular machines and molecular motors within the human body are good examples to pursue and study.

Molecular Machines and Molecular Motors

Molecular machines are a group of molecular components that convert chemical energy into mechanical movement in response to a particular stimulus. Vital life processes, such as DNA replication, ATP synthesis, as well as hydrolysis, movement of flagella, and muscle contraction, etc., employ naturally occurring biological molecular machines. Molecular machines can take the form of multi-protein complexes inside living systems and include examples like myosin, dynein, kinesin, spliceosomes, and ribosomes. Molecular machines perform various complex functions within the biological systems. Myosin is responsible for muscle contraction; kinesin moves cargo along microtubules inside the cell as well as causes synchronous beating of cilia and flagella; ATP synthase harnesses the energy from proton gradient to form ATP; spliceosomes remove introns; polymerases synthesize nucleic acids; and ribosomes synthesize proteins. Attempts are made to study and understand these macroscopic events in order to adapt and produce synthetic smaller versions of the same. Such alternatives are needed as biological machines are unstable and cannot be directly employed. Their complex nanoscale dynamics of molecular machines can have potential applications in nanomedicine.

Studies on molecular machines to help in creating synthetic or hybrid machines combining the biological principles with synthetic components to produce robust stable alternatives that can be useful for various challenges is a pivotal tool being explored. Synthetic molecular machines mimic functions that happen at a macroscopic scale, and scientists are making attempts to miniaturize them.

Normally, in biological molecular machines, some inputs like chemical, light, or thermal energy is converted into kinetic energy to deliver the output. Similar considerations have to be taken into account during the design of synthetic machines. The direction of action, its speed, energy supply, nature of design, and size of machine with optimal efficiency are various factors that need to be optimized.

Biological Concepts to Engineer Modern Solutions

The first artificial molecular machine was designed in 1994 with a rotaxane having a ring and two binding sites. In 2016, Nobel prize in Chemistry was awarded for the design of molecular machines to Sir Frazer Stoddard, Jean-Pierre Sauvage, and Bernard Feringa.

Design of synthetic molecular machines falls within seven categories, and all have met with some success.

(a) *Molecular motors* These perform mechanical shift movements like rotation when energy input is provided. The energy can be in the form of chemicals or light. Normally, these motors facilitate unidirectional movement. For example, a prototype by Kelly and coworkers in 1999 has a three-bladed triptycene rotor and a helicene that was capable of performing a unidirectional 120° rotation.

(b) *Molecular propellers* These not only produce rotational movements but are also more fluid, act as propellers, and may consist of a higher structural design of seven blades around a shaft. For example, propellers designed by Prof. Petr Kral have blades made of planar aromatic molecules and a carbon nanotube shaft. They can be used as pumps in liquids.

(c) *Molecular switches* These are designed to exist in two alternative stable forms. These equally important outcomes can be varied by inputs such as electric current, pH, light, temperature. For example, crown ether switches, photochromic switches, etc.

(d) *Molecular shuttlers* These shuttle ions from one location to another with the help of a macrocycle and dumbbell-like molecule. For example, rotaxane with an axle in a dumbbell-shaped component where interlocked microcyclic parts can be displaced by external stimulation in a controlled manner.

(e) *Molecular tweezers* They are molecules that contain cavities created with the help of hydrogen bonds, metal coordination, hydrophobic, Van der Walls, or electrostatic interactions. For example, anthracene arms, photoswitchable molecular tweezers developed by Shinkai in 1981 for photocontrolled cation-binding.

(f) *Molecular sensors* These molecules sense particular analytes and create a signal. For example, molecular recognition-based glucose sensors, cation-selective chemosensors, pH sensors, etc.

(g) *Molecular logic gates* These need an input signal-like chemical and then help to produce output signal. For example, YES molecular logic gate: benzo–crown–ether connected to a cyano-substituted anthracene unit—output of fluorescence in presence of sodium.

Artificial molecular machines have captured our imagination with a clear potential to transform and enhance human life. Blueprints from Biology have led to various beginnings, but a multidisciplinary approach will fuel more advanced outcomes with diverse applications.

Let us focus specifically on understanding the functioning of Molecular Motors.

Molecular motors convert chemical energy from ATP into mechanical energy. They can be classified as linear and rotary. Linear motors move in a straight line, walking along a filament. Linear motors include: (a) kinesin: It moves along microtubules, transports vesicles within the cell, and is involved in synchronous beating of cilia and flagella; (b) Myosin: It moves along actin filaments and is responsible for muscle contraction; (c) Dynein: It moves along microtubule often in opposite direction as kinesin. Rotary motors spin along the central axis. Rotary motors include: (a) F1F0 ATP synthase: It generates ATP by rotating a protein complex within the mitochondrial membrane. (b) Bacterial flagellar motor: It enables bacteria to swim by rotating a flagellar filament. Details of naturally occurring molecular motors of each type are described below. They are of two types.

ATP synthase motors

They are enzyme–ion pumps–motor combinations that produce the primary energy storage molecule of the cell—ATP in all living organisms. It joins inorganic phosphate to ADP to form ATP. These rotary ATP'ases are central to biologic energy conversions (Figure 18.1). There are three types of ATP synthases—F-type and V-type in Eukaryotes and the bacterial ATPases (either F- or V-type or archeal A-type). F-type ATP synthases found in mitochondria and chloroplasts

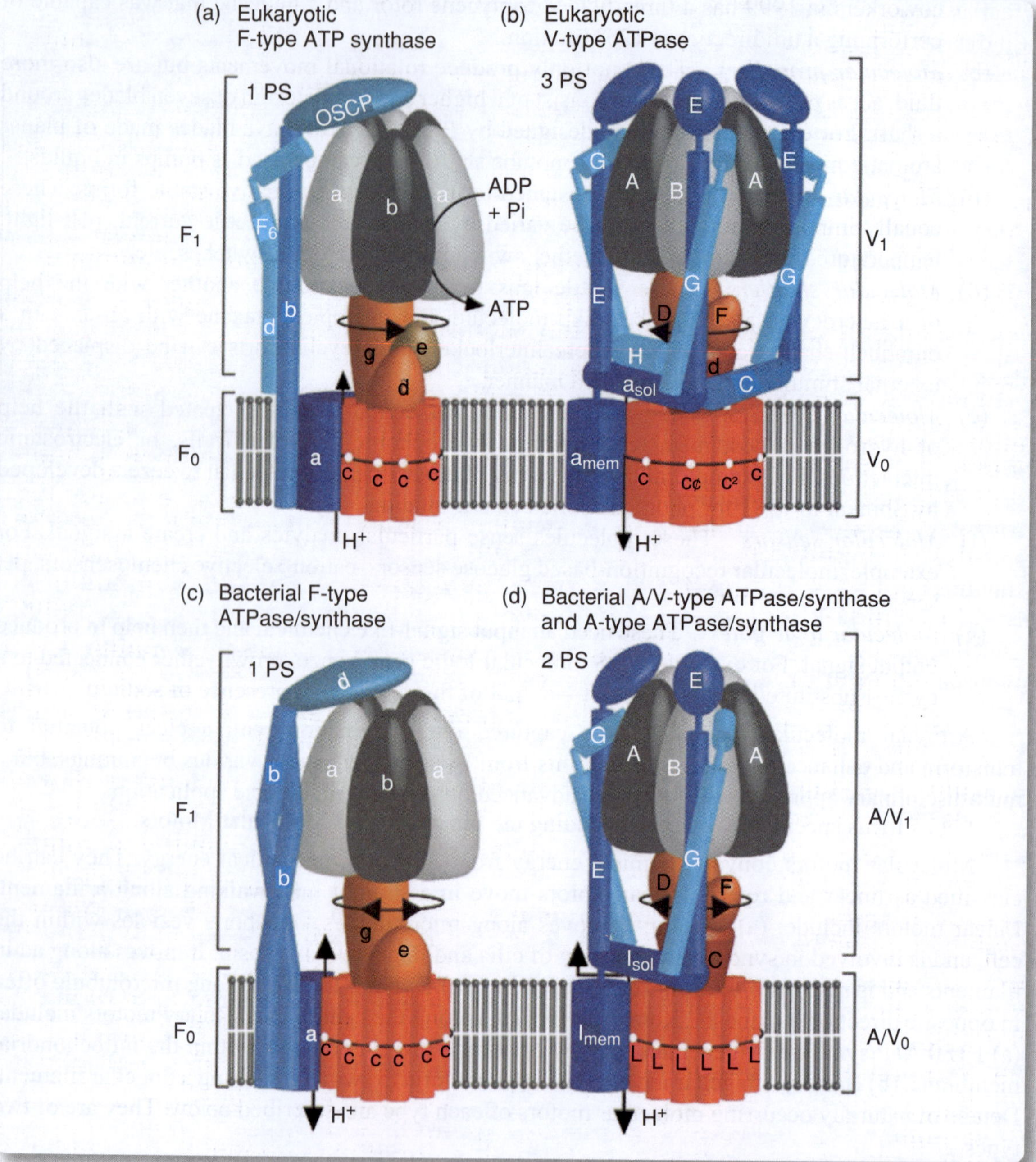

Fig. 18.1 Schematic diagram of rotary ATP sub-types and their parts

Source: Adapted from Alastair G Stewart, Elise M Laming, Meghna Sobti, Daniela Stock, "Rotary ATPases—dynamic molecular machines", *Current Opinion in Structural Biology*, Volume 25, 2014, Pages 40-48, ISSN 0959-440X, https://doi. org/10.1016/j.sbi.2013.11.013

act as biological power converters. These contain one peripheral stalk (PS) that helps in rotation movement. When protons flow from high to low concentration, due to proton motive force or electrochemical potential, they bind to a ring of ATP synthase and rotate it within the membrane. It now acts like a biological turbine, and the mechanical energy of torque pushes the central shaft to change the catalytic subunit into conformations that provide the correct chemical environment for ATP synthesis from ADP and Pi. This is analogous to water flow in hydroelectric generators that moves turbines to generate torque, which is converted to electricity. On similar lines, V-type ATP'ase (with three peripheral stalks) provides the chemical environment for ATP binding; then the power stroke causes hydrolysis to release the products ADP and Pi. Bacterial F-type synthase is bidirectional and contains one peripheral stalk, while the V- and A-type have similar architecture with two peripheral stalks. Studies with cryo-electron microscopy and cryotomography have shown that the concerted movement of subunits within the ATP synthase complex might provide regulation to fine tune their efficiency in cellular environments.

Cytoskeleton-associated motors

These are motors that transport cargo across microtubule filaments inside the cell. Three motor proteins involved in this are myosin, dynein, and kinesin. All these three cytoskeletal motors provide linear motion using ATP as a fuel. They use energy from ATP hydrolysis to move cargo unidirectionally.

Kinesin and dynein motors help in moving vesicles and organelles within the cell, with cilia and flagellar beating and action in mitotic and meiotic spindle for chromosome separation. Kinesin transports towards the positive end of microtubule, whereas dynein transports towards the negative end. The myosin motor superfamily is primarily responsible for cell-surface contractions, muscle contraction, and cytoplasmic streaming and acts on actin. Each of these three super families has conserved ATP-binding or force-producing domain along with unique polypeptide sequences to provide for their functional specificity (Figure 18.2).

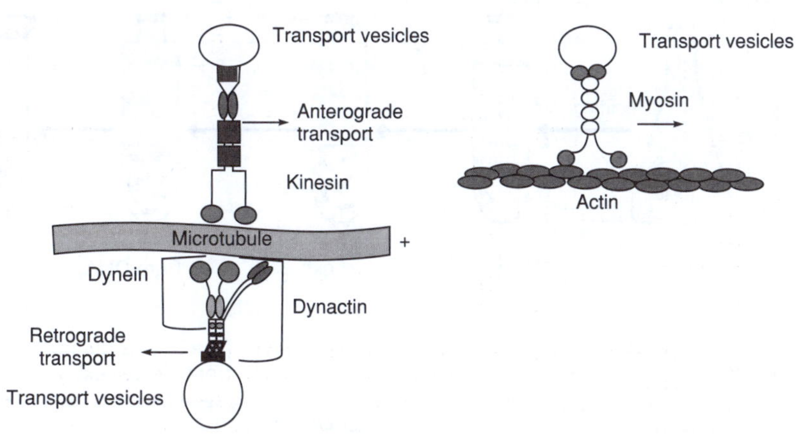

Fig. 18.2 Cytoskeletal-associated motors and their functions

Source: Adapted from https://www.researchgate.net/publication/279155721_Molecular_aspects_of_endosomal_cellular_transport.

Genetic Engineering Applications

The main workhorse in humans and living organisms are the proteins. As we are aware, proteins are coded by genes and are responsible for various phenotypes. In case proteins are not available or to use proteins and enzymes related to various actions, it is important to have the ability to produce proteins as per our need. This, in turn, will require the ability to manipulate DNA or genes.

Genetic Engineering is the field of biology that provides means to manipulate the information encoded by DNA and genes to engineer new products or processes aimed at solving various medical, agricultural, and environmental problems. Other synonymous terms that explain genetic engineering are gene cloning, gene manipulation, molecular biology, recombinant DNA technology, and modern genetics. The technology is now used in many areas of life, such as comparison of two DNA sequences, for production of useful proteins by cloning methods, for understanding gene structure and function of various genes, forensic analysis of scene-of-crime samples, paternity disputes, medical diagnosis, genome mapping and sequencing, for producing transgenic organisms, and for the understanding of biotechnology industry.

To achieve these, a few steps are mandatory. First, the ability to isolate DNA and specific genes, then to cut and join DNA to make new combinations, introduce it into any cell or host, express the protein, and finally purify proteins for their function.

Obtaining DNA for Genetic Manipulation

It is necessary to have enough DNA or nucleic acid material with us to carry out any kind of manipulation. The source of the genetic material can be any cell from where it is isolated. DNA isolation protocols involve three steps:

(a) opening up the cell from where DNA is to be isolated,

(b) second, separating the nucleic acids from other cellular components;

(c) and third, purifying the DNA for further use.

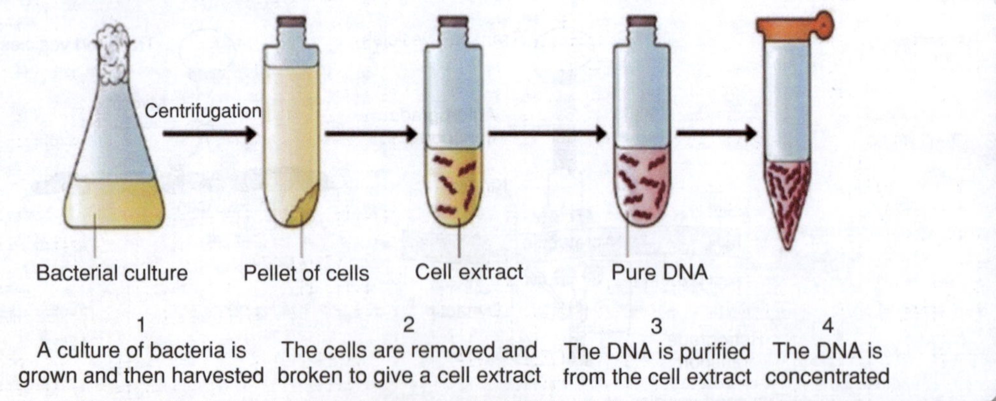

Fig. 18.3 DNA isolation

However, sometimes this method of isolation will only give a very small amount or low yield of DNA, and manipulation methods may require DNA in large quantities. This problem is solved by a technique called Polymerase Chain Reaction (PCR). PCR is a technique that helps to obtain large amounts of DNA from minute amount of template available. Small amount of the available source DNA is taken; then primers—short stretches of SSDNA complementary to the ends of DNA to be amplified—are added. These primers anneal to template, and they extend and create

more copies of the DNA in the region between primers when DNA polymerase is provided. This whole process is done in a thermal cycler machine and is basically an 'in-vitro replication' process. As DNA denaturation happens using high temperature, a special kind of polymerase called Taq polymerase from thermophilic bacteria called *Thermus aquaticus* is used. This polymerase can withstand high denaturation temperature. The basic steps of PCR include denaturation of double-stranded template followed by annealing of primers and extension by DNA Polymerase. Each cycle of PCR doubles the template, and, thus with multiple cycles, there is an exponential increase in the DNA amount. The principle behind high amount of DNA generated by amplification is shown below (Figure 18.4).

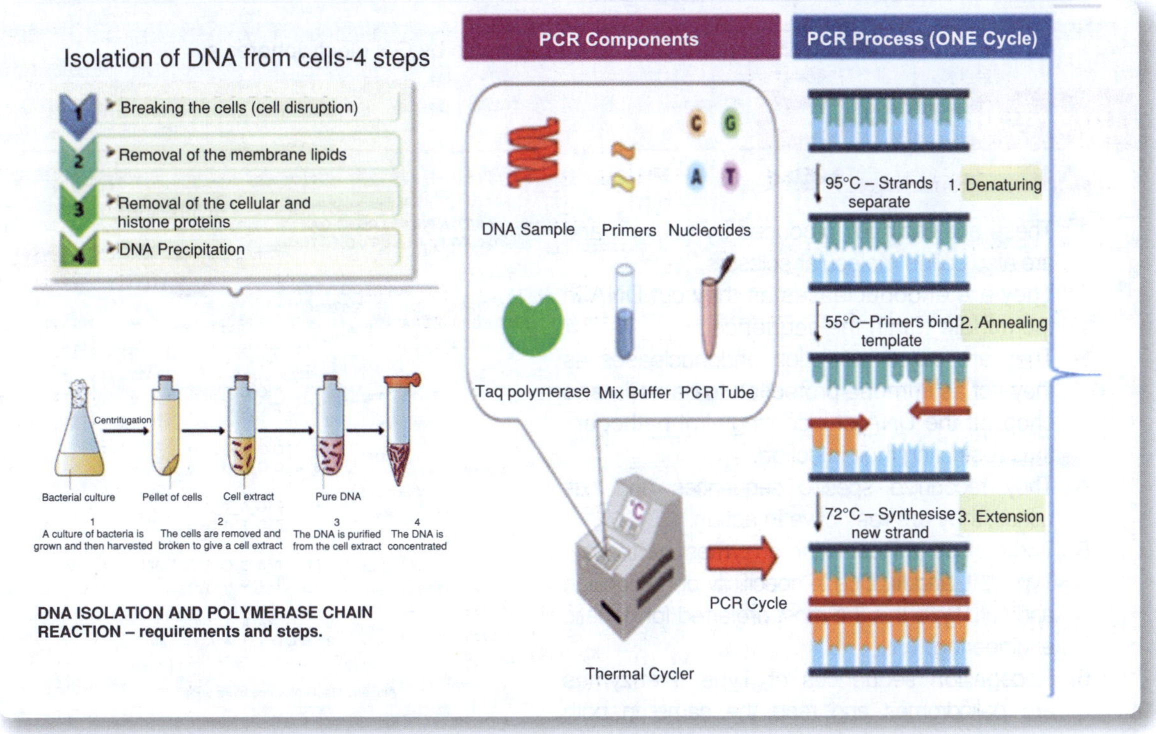

Fig. 18.4 Obtaining source DNA: DNA Isolation and Polymerase Chain Reaction

Source: Adapted from https://learnlifescience.com/an-overview-of-bacterial-genomic-dna-isolation/
https://old-ib.bioninja.com.au/standard-level/topic-3-genetics/35-genetic-modification-and/pcr.html

Once DNA is obtained, it can be visualized by a technique called Gel electrophoresis. Gel electrophoresis helps in separation of DNA fragments through a gel under the influence of electric current. As DNA is negatively charged, it moves towards the positive electrode and the fragments are separated according to their size and mass. The gel matrices commonly used are agarose and polyacrylamide, which retard the movement of molecules. A gel-running buffer (Tris-BorateEDTA) provides the ions for electric conductance and movement of DNA. The speed at which DNA moves through the gel depends on the length of the DNA molecule. Longer DNA molecules move more slowly through the gel matrix than do shorter DNA molecules. The fragments can be visualized easily as the migrating bands fluorescingin UV light when stained with ethidium bromide (Figure 18.5).

Thus, DNA is isolated, amplified, and visualized by gel electrophoresis and is also quantified to know the quality and quantity of the DNA available.

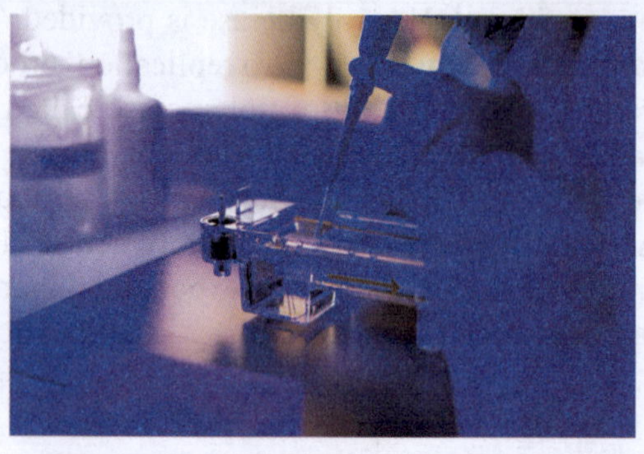

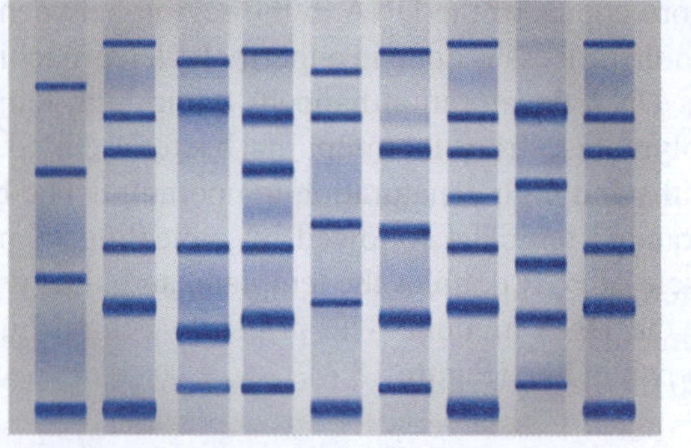

Fig. 18.5 Separation of DNA fragments by Gel electrophoresis

Interesting Fact 18.1

Restriction endonucleases 

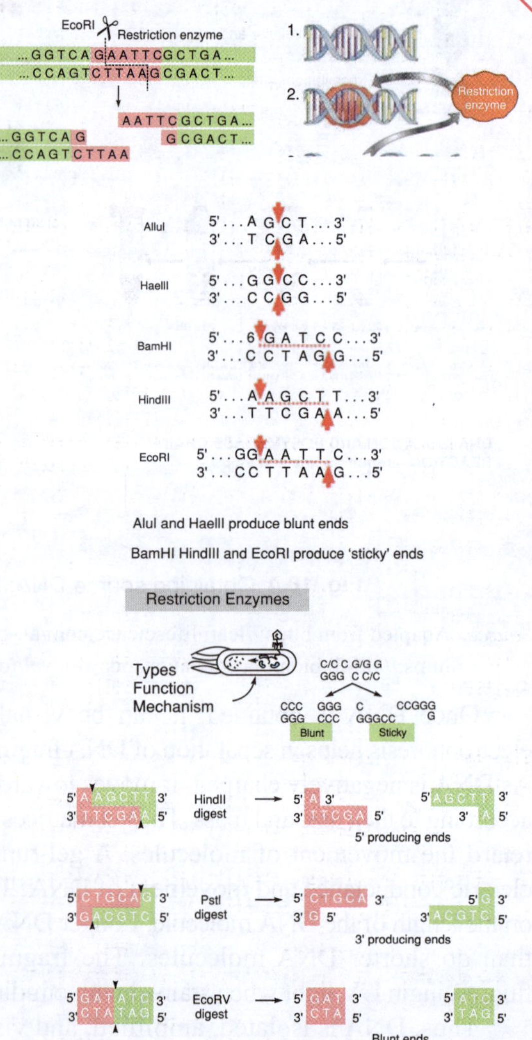

1. These are enzymes produced by bacteria and are also called molecular scissors.
2. They are endonucleases as they cut DNA in the middle within the sequence.
3. They are called restriction endonucleases as they act as immune protection to bacteria and chop off the DNA of incoming viral pathogen, thus restricting their infection.
4. They recognize specific sequences and cut; hence, they are restrictive in action.
5. Three class of Restriction enzymes are there – Type I, II, and III. Due to specificity of recognition and cut, Type II is the most preferred for genetic engineering.
6. Recognition sequences of Type II enzymes are palindromes and read the same in both orientations, thus cutting both strands of DNA at the same point.
7. Type II enzymes can produce blunt ends or sticky ends.
8. Complementarity of sticky ends helps in joining fragments specifically by ligase.
9. More than 1300 such enzymes are recognized in bacteria.
10. Enzyme produced by the bacteria will not cut its own DNA as it is methylated by the bacterial methylases. Methylation protects restriction digestion and are part of restriction system, hence called restriction-modification system.

Types of DNA Manipulations

Creation of Recombinant DNA and Gene Cloning

A recombinant DNA molecule is generated when DNA from different sources are combined in a lab setting. This process can create combinations that do not exist in nature. For example, human DNA that is joined to bacterial DNA and present inside the bacterial cell. This is done by splicing together the human gene or DNA fragment with a bacterial plasmid with a beneficial aim. The human DNA that you need to join is called the 'Gene of Interest'. To carry this 'Gene of Interest' into the bacterial cell, you need a vehicle or vector.

Plasmids are bacterial extrachromosomal DNAs that can replicate independently inside the bacteria. These are smaller than chromosomal DNA of bacteria and are easy to isolate and work with. These act as vectors or vehicles that carry human gene into the bacterial cell. Basically,the human gene is being cut and pasted in to the bacterial plasmid. To do this, we use enzymes called restriction endonucleases produced by bacteria. These cut DNA at specific nucleotide sequences along the length of the DNA. These nucleotide sequences are usually palindromes, and, hence, they simultaneously cut both strands of DNA. Once they cut the double-stranded DNA, they can leave sticky ends or blunt ends. These ends are unique and any two DNA, for example, plasmid DNA of bacteria and the human 'gene of interest', cut with same restriction enzyme and with the same sticky ends can be joined together to complete the 'cut–paste' process (Figure 18.6).

This generates a recombinant DNA molecule that can then be inserted into a bacterial cell by a process of chemical transformation. Once the plasmid with the gene of interest has gone inside, it will replicate, and the plasmid will be present in the new cell with every bacterial division. Thus, the gene has been cloned into many different bacterial cells. This creates millions of bacterial cells now carrying a human gene. This process is called DNA cloning or gene cloning.

The bacterial cell containing the gene of interest is screened out and selected using antibiotic resistance marker genes and visual phenotypes, such as, lacZ.

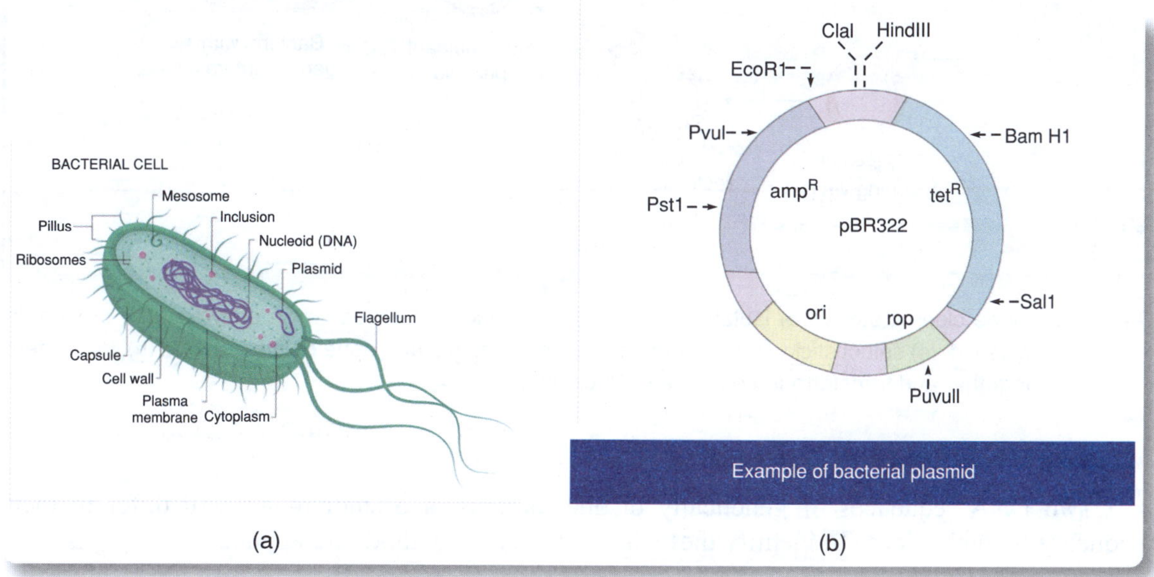

(a) (b)

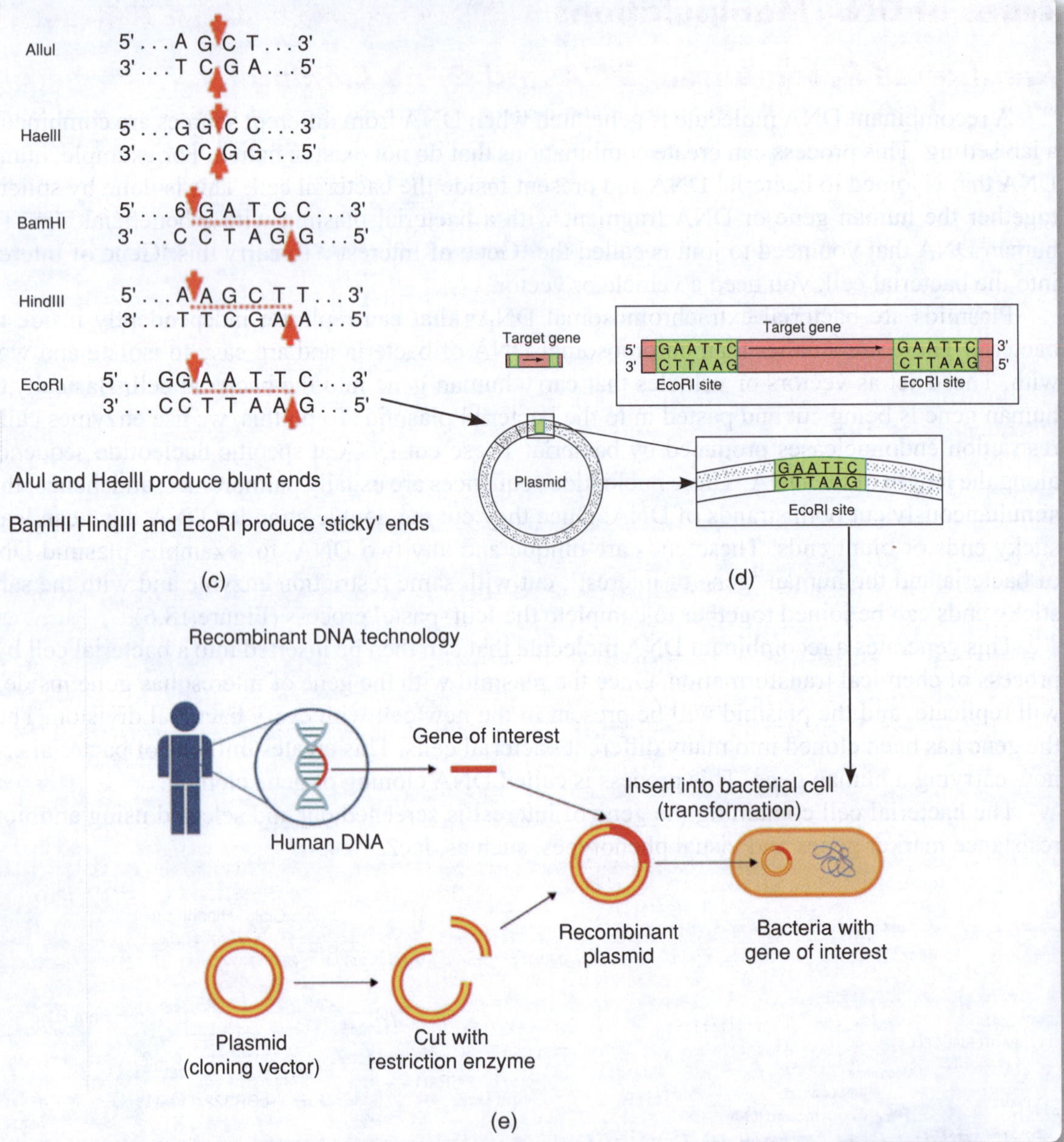

AluI 5' . . . A G C T . . . 3'
3' . . . T C G A . . . 5'

HaeIII 5' . . . G G C C . . . 3'
3' . . . C C G G . . . 5'

BamHI 5' . . . 6 G A T C C . . . 3'
3' . . . C C T A G G . . . 5'

HindIII 5' . . . A A G C T T . . . 3'
3' . . . T T C G A A . . . 5'

EcoRI 5' . . . G G A A T T C . . . 3'
3' . . . C C T T A A G . . . 5'

AluI and HaeIII produce blunt ends

BamHI HindIII and EcoRI produce 'sticky' ends

(c)

Target gene

Target gene
5' GAATTC GAATTC 3'
3' CTTAAG CTTAAG 5'
EcoRI site EcoRI site

Plasmid

GAATTC
CTTAAG
EcoRI site

(d)

Recombinant DNA technology

Human DNA

Gene of interest

Insert into bacterial cell (transformation)

Recombinant plasmid

Bacteria with gene of interest

Plasmid (cloning vector)

Cut with restriction enzyme

(e)

Fig. 18.6 Gene cloning steps: (a) isolation of plasmid, (b) selection of restriction enzyme sites present in plasmid, (c) select sticky end cutters, (d) cut both plasmid and gene of interest, and (e) ligate them together and transform to complete DNA cloning

Comparing DNA Sequences

Two DNA sequences of genetically distinct organisms or individuals will differ in their sequence of nucleotides. To identify these differences, two methods are applied.

DNA fingerprinting

It identifies different lengths of DNA fragments generated when cut with restriction enzymes. DNA regions containing repeat regions called variable number of tandem repeats (VNTR) are focused, and the pattern or fingerprint of DNA fragments and their banding patterns will be different depending on the differences in number of repeats. No two individuals will have the same nucleotide sequence; hence, the repeat variations will provide the clue to the differences. It is useful in paternity disputes and forensic identification of suspects.

DNA sequencing

In this method, each nucleotide of the DNA fragments is identified, and the entire nucleotide sequence is compared. Various methods of DNA sequencing like Sanger's dideoxy method, Maxam and Gilbert method, pyrosequencing, and nanopore sequencing are developed to see the nucleotide sequence of each DNA fragment. This has opened up a very significant field, and, in recent times, whole genomes of various organisms, including humans, have been sequenced (Figure 18.7).

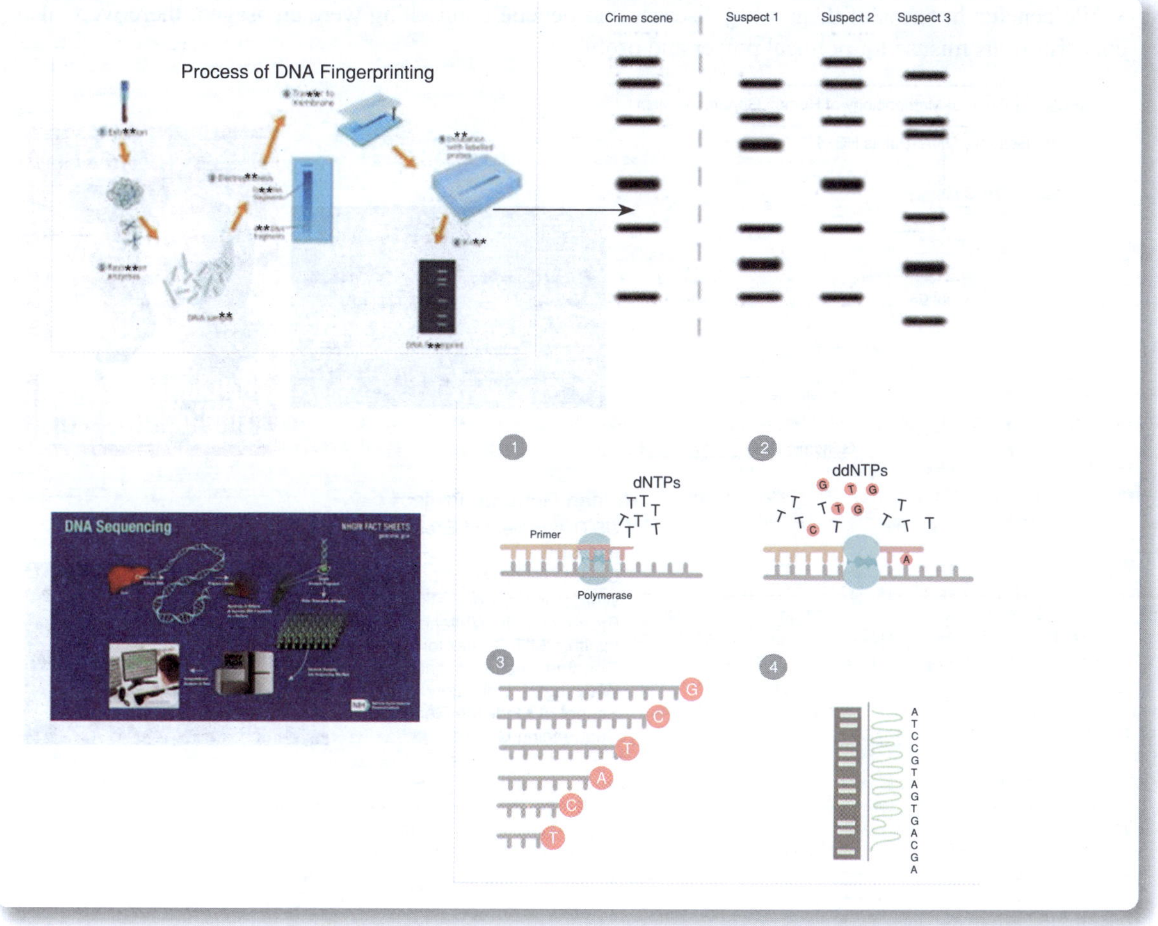

Fig. 18.7 Methods of comparing DNA sequences: DNA fingerprinting and DNA sequencing

Source: Adapted from https://microbenotes.com/dna-fingerprinting-principle-methods-applications/
https://courses.lumenlearning.com/wm-biology1/chapter/reading-dna-sequencing/

Experimental Insight 18.1

Human Genome Project: It is an effort of 13 years to determine the nucleotide sequence of healthy human genome. The Human Genome Project is one of the most ambitious projects ever undertaken in the biological sciences. The entire human genome consists of 3.12 billion paired nucleotides.

The physical maps were first determined using markers (known sequences), and the neighbouring genes in the proximity of markers were studied. Chromosome-specific nucleotide sequences were determined by DNA sequencing methods, such as Sanger's sequencing, and the huge data was computationally sorted to arrive at the final order of sequences. Celera Genomics, a private organization, proceeded with a different approach of directly sequencing without the physical map. Both the methods gave good results and discrepancies were insignificant. The first draft of the human genome was completed early in 2003, when the complete nucleotide sequence of all 23 pairs of chromosomes was determined. The success of this project was that all the words in the human gene 'dictionary' were identified. Sequencing of 99% of euchromatic DNA was finished to 99.99% accuracy.

This project identified that there are approximately 22,000 genes in the human genome. The formal HGP Draft completion announcement was made public on 26 June, 2000, jointly by the two organizations. While benefits in terms of diagnosing diseases and genetic counselling were envisaged, there were also concerns of its misuse for political power and profit.

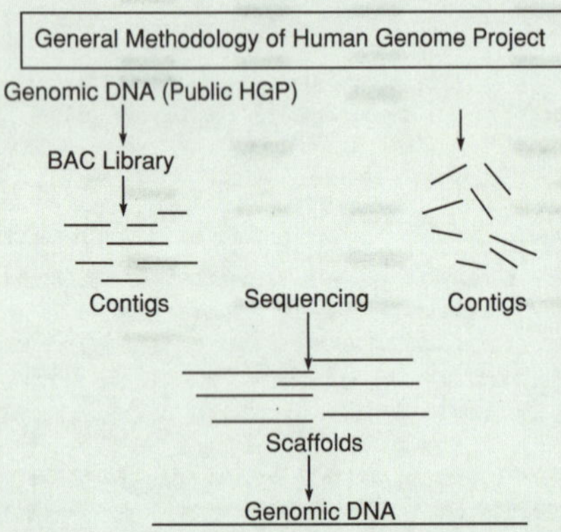

Human Genome Project

- US Department of Energy and NIH
- Celera Genomics corporation
- Wellcome Sanger Institute, Broad Institute/Whitehead Institute/MIT Center for Genome Research, and Washington University School of Medicine Genome Sequencing Center

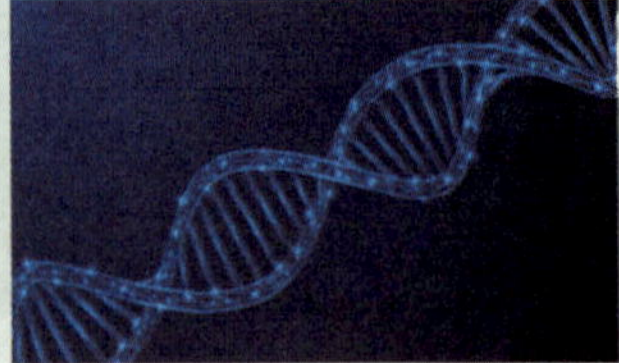

Medical Applications of Genetic Engineering

1. It is used in introduction and expression of genes with high clinical importance in bacteria for bulk production to meet the demand. This approach has been tried for production of insulin, interferon, growth hormone, erythropoietin, atrial peptides, and tissue plasminogen activator. Instead of getting it from other sources, such as closely related animals, these proteins produced by use of bacteria are beneficial because bacteria can be grown in bulk, are inexpensive, and recombinant genes can be used to express large number of proteins. Purification of protein from other bacterial proteins and ensuring proper post-transcriptional modifications are important because prokaryotic cells may lack some mechanisms. Hence, usually cDNA is used where introns are removed to avoid this challenge.
2. Gene therapy can be used to treat single-gene disorders by providing the working copy of the defective gene. This approach has been tried for cystic fibrosis, muscular dystrophy, etc. The transfer of human genes into individuals or affected cells is a positive application of genetic engineering.
3. It is used for creating sub-unit vaccines with the genes for viral coat proteins of herpes simplex virus or hepatitis virus by piggybacking on the harmless vaccinia virus vectors. Such recombinant viruses trigger immune response and help develop immunity.
4. It is used for sequencing of various organism genomes.
5. It is used to knock out animal models for studying gene functions.
6. It is used for gene editing technologies, including the recent CRISPR tools.

Agricultural and Environmental Applications of Genetic Engineering

1. It is used for generating herbicide-resistant plants by overexpressing the target protein or enzyme to detoxify the herbicide: Glyphosate and Round up.
2. It is used by virus-resistant plants by overexpressing viral coat proteins that offer resistance by interfering with the infection process. For example, tomato, pepper, etc., are made virus-resistant this way.
3. Genes for *Bacillus thuringensis* insecticidal toxins are incorporated into plants under appropriate promoters to cause the plants to become insecticidal themselves. Bt cotton is an example of insect-resistant plant.
4. It is used to control ripening of tomatoes by inactivating genes that cause inability to produce ethylene.
5. It is used to enhance nutritional content of important crops. For example, golden rice.
6. Pharming: It is expression of hormones, vaccines, and other pharmaceutical products in plants.
7. It is used in enhancing nitrogen fixation by transferring the *nif* gene from bacteria to plants.
8. Bioremediation: Genetically engineered micro-organisms (GEMs) can break down a wide range of chemical contaminants in soil, groundwater, and activated sludge.
9. It is used in drought-resistant and salinity-resistant plants to adapt to climate change and help in sustainable agriculture.
10. Biofuel production: Genetically modified plants can be used to produce biofuels from engineered algae and yeasts.
11. It is used in genetically modified organisms to tackle oil spill in oceans.

We have studied gene cloning. Can organisms be cloned too? Yes! Somatic cell nuclear transfer is the development in biomedicine that led to development of the first sheep clone called 'Dolly', as depicted in the section 'Interesting Fact 18.2.

Interesting Fact 18.2

Organism cloning—somatic cell nuclear transfer

1. Whole organisms can be cloned too, thus producing genetically identical copies.
2. First successful demonstration of this was DOLLY—the sheep in the year 1996 by a Scotland team.
3. Process used was Somatic Cell Nuclear Transfer. Somatic cell nuclear transfer removes a nucleus from a cell of the organism that will be cloned. After chemical treament, that nucleus is placed into an egg cell from where original nucleus was removed. The egg cell will use genetic information from the new nucleus
4. An electrical shock is used to stimulate the egg to begin to divide as if it were a normal embryo. Then the fertilized egg is implanted into a uterus, and the embryo grows normally. The resulting organism is genetically identical to the organism that donated the nucleus.

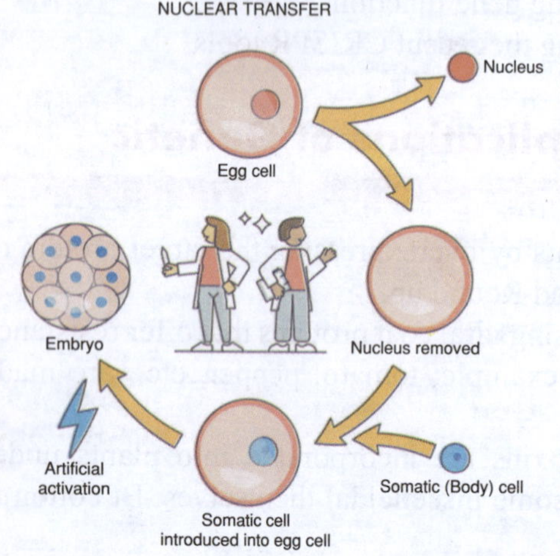

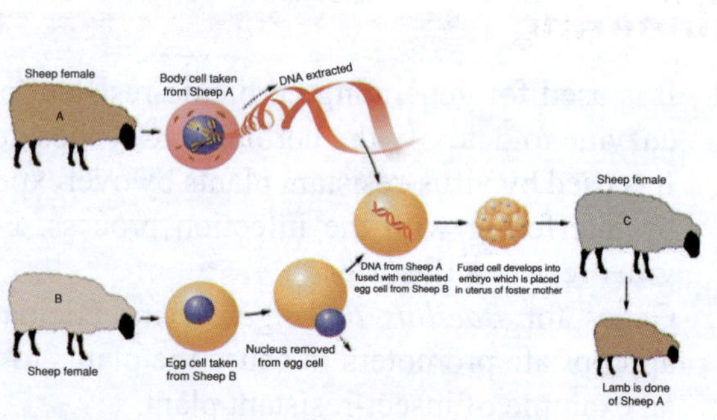

Stem Cells, Characteristics, and Applications

After animal cloning, stem cell research became one of the hot topics in biomedical research. Stem cells are cells that have the ability to divide and regenerate themselves and differentiate into other cell types. Cells with these abilities can be found during early development, where the cells of blastula are totipotent and can develop into complete individual organisms. These are called embryonic stem cells and can differentiate to all cell types of the body. Stem cells in amniotic fluid as well as umbilical cord blood can also form specialized cells and are called prenatal stem cells.

Stem cells are special human cells that can develop into many different types of cells—from muscle cells to brain cells. Stem cells also have the ability to repair damaged cells and have strong healing power. These can evolve into any type of cell. Stem cells are of the following different types:

(i) Embryonic stem cells
(ii) Adult stem cells
(iii) Induced pluripotent stem cells
(iv) Mesenchymal stem cells

Stem cells are relatively 'unspecialized' cells that have the unique potential to develop into 'specialized' cell types in the body (for example, blood cells, muscle cells, or nerve cells). They occur at all stages of human development, from embryo to adult, and in many tissues of the body.

Embryonic stem cells The fertilized egg begins to divide immediately. All the cells in the young embryo are of totipotent cells. These cells form a hollow structure within a few days. Cells in one region group together to form the inner cell mass. This contains pluripotent cells that make up the developing foetus. The embryonic stem cells can be further classified as

(a) *Totipotent stem cells* These can differentiate into all possible types of stem cells.
(b) *Pluripotent stem cells* These are the cells from an early embryo and can differentiate into any cell type.
(c) *Multipotent stem cells* These differentiate into a closely related cell type, e.g., the hematopoietic stem cells, which differentiate into red blood cells and white blood cells.
(d) *Oligopotent stem cells* Adult lymphoid or myeloid cells are oligopotent. These can differentiate into a few different types of cells.
(e) *Unipotent stem cells* These can produce cells only of their own type. Since they have the ability to renew themselves; these are known as unipotent stem cells, e.g., muscle stem cells.

Adult stem cells These stem cells are obtained from developed organs and tissues. They can repair and replace damaged tissues in the region where they are located. For example, hematopoietic stem cells are found in the bone marrow. These stem cells are used in bone marrow transplants to treat specific types of cancers. The differences between embryonic and adult stem cells are given in Table 18.1.

Table 18.1 Difference Between Embryonic and Adult Stem Cells

Embryonic Stem Cells	Adult Stem Cells
Flexible	Less Flexible
Immortal	Mortal
Availability	Limited Quantity
Immunogenic	Non-immunogenic
Tumorigenic	Non-tumorigenic
Isolation Leads to Destruction of Embryo	Relative Ease of Procurement

Induced pluripotent stem cells These cells have been tested and arranged by converting tissue-specific cells into embryonic cells in the lab. These cells are accepted as an important tool to learn about the normal development, onset and progression of the disease, and are also helpful in testing various drugs. These stem cells share the same characteristics as embryonic cells do. These also have the potential to give rise to all the different types of cells in the human body.

Mesenchymal stem cells These cells are mainly formed from the connective tissues surrounding other tissues and organs known as the stroma. These mesenchymal stem cells are accurately called stromal cells. The first mesenchymal stem cells were found in the bone marrow that is capable of developing bones, fat cells, and cartilage.

There are different mesenchymal stem cells that are used to treat various diseases as they have been developed from different tissues of the human body. The characteristics of mesenchymal stem cells depend on the organ from where they originate. Research on stem cells is going on, and it is believed that stem cell therapies can cure ailments like paralysis and Alzheimer's as well. Let us have a detailed look at stem cells, their types, and their functions.

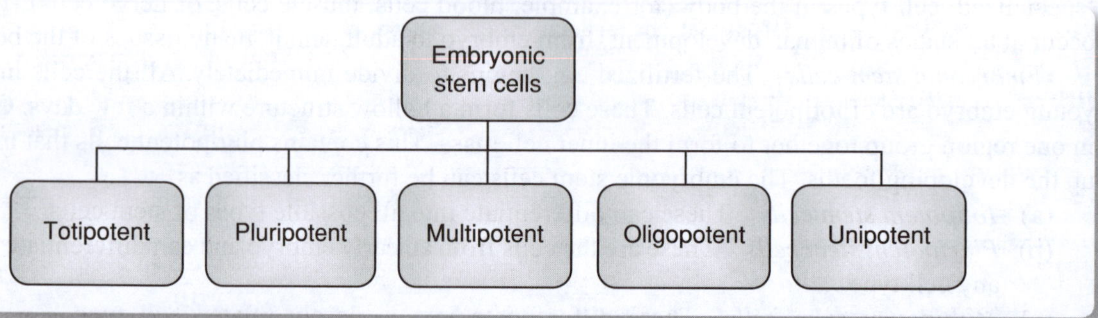

There are also undifferentiated adult stem cells called somatic stem cells in different organs of body, bone marrow, skin, brain, etc., which also have a limited ability to form other cell types and tissues. They are categorized as epidermal stem cells, hematopoietic stem cells, and nerve stem cells. These stem cells have the potential to become tissue factories, can replace lost cells, and help in tissue damage. Some types of differentiated somatic cells can be genetically reprogrammed to form induced pluripotent stem cells (Figure 18.8).

Stem cell applications: The following are the important applications of stem cells:

1. *Tissue regeneration* This constitutes regenerative medicine through stem cell therapy. It generates healthy cells to replace lost cells or cells affected by disease. For example, Parkinson's disease, type I diabetes, aplastic anaemia, immunodeficiencies, leukaemia, etc. This is the most important application of stem cells. The stem cells can be used to grow a specific type of tissue or organ. This can be helpful in kidney and liver transplants. The doctors have already used the stem cells from beneath the epidermis to develop skin tissue that can repair severe burns or other injuries by tissue grafting.
2. *Treatment of cardiovascular disease* A team of researchers has developed blood vessels in mice using human stem cells. Within two weeks of implantation, the blood vessels formed their network and were as efficient as the natural vessels.
3. *Treatment of brain diseases* Stem cells can also treat diseases such as Parkinson's disease and Alzheimer's. These can help to replenish the damaged brain cells. Researchers have tried to differentiate embryonic stem cells into these types of cells and make it possible to treat diseases.
4. *Blood disease treatment* The adult hematopoietic stem cells are used to treat cancers, sickle cell anaemia, and other immunodeficiency diseases. To treat these diseases, the technique used is known as gene therapy. In this technique, insertion of genes into an individual's cells and tissue is made to treat a disease. Defective gene is replaced with a

functional or correct gene. These stem cells can be used to produce red blood cells and white blood cells in the body.

5. Stem cells can be used to test drugs for their safety and effectiveness and assess toxicity to specific cell type and organs, e.g., toxicity to heart cells. Other applications that are possible but not yet successful for humans include therapeutic cloning by somatic cell nuclear transfer for organs.

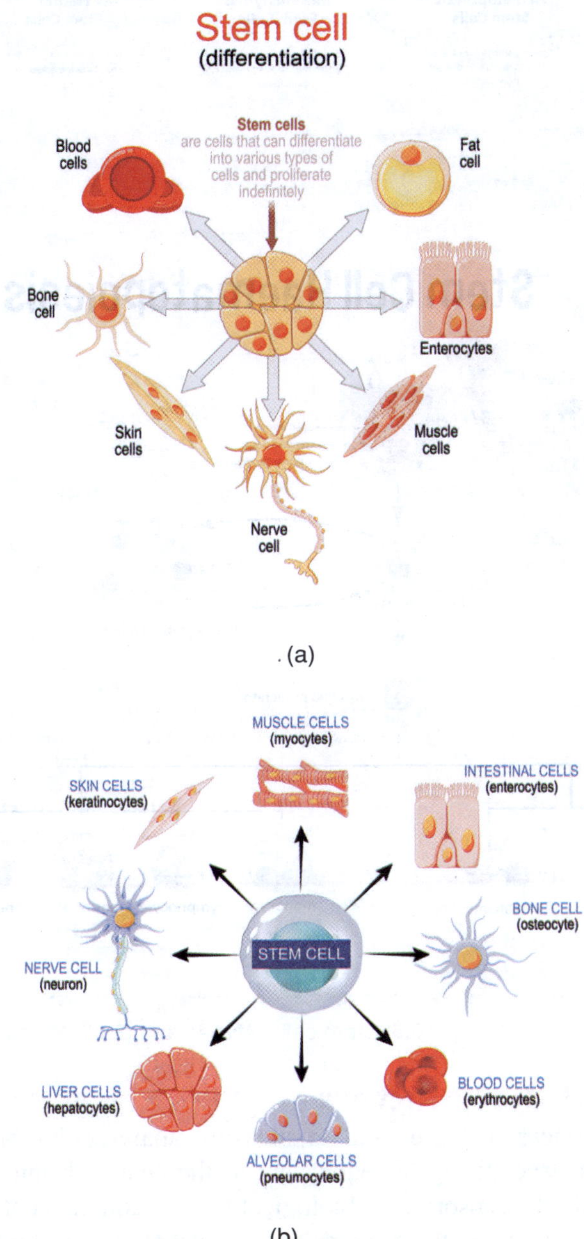

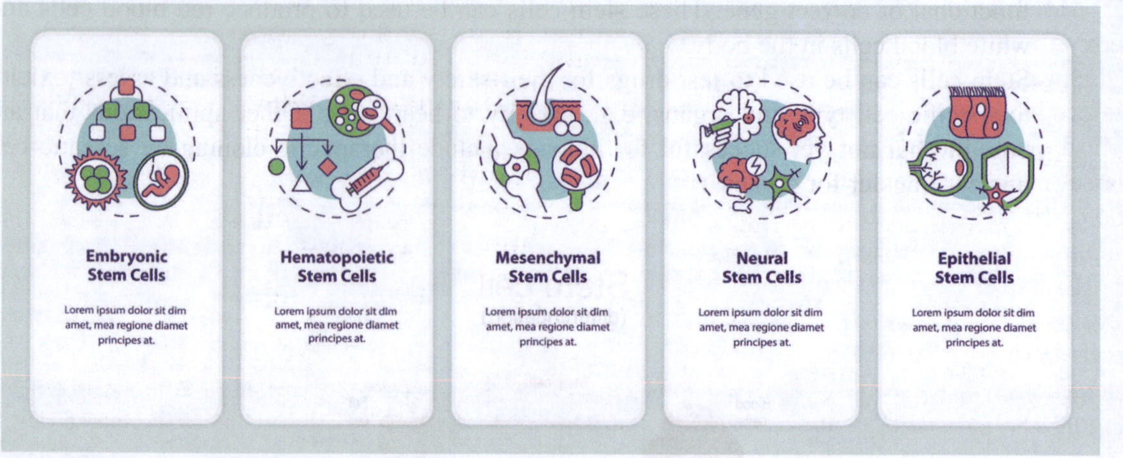

(c)

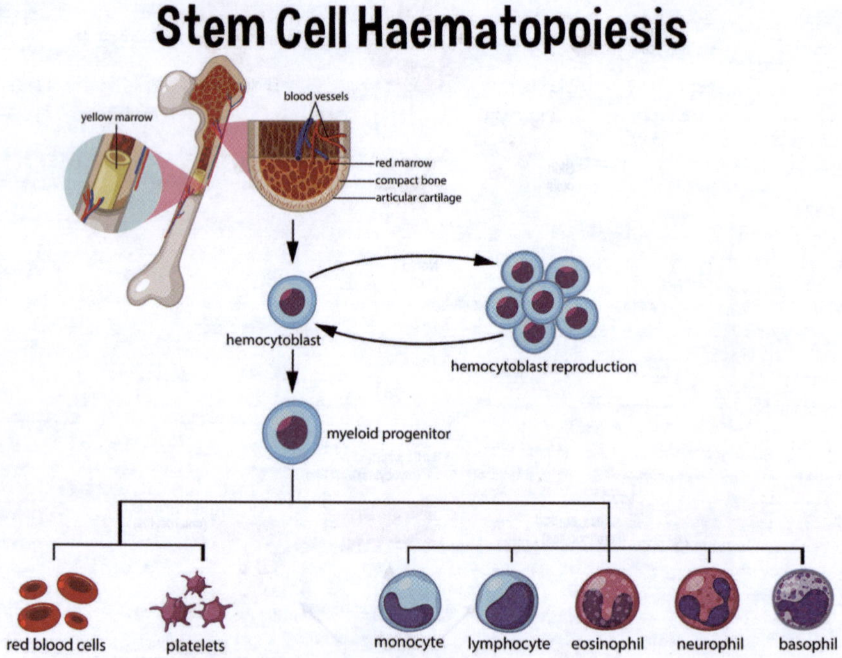

(d)

Fig. 18.8 Stem cells—types and applications

Biosensors and Biodetectors

Biosensors are devices that are used to measure analytes, biological products, or micro-organisms. Biosensors generate signals equivalent to the analyte being measured. There are three essential components of a biosensor: (a) a biological recognition element that specifically identifies the target molecule, (b) transducer that converts the biorecognition event into a signal, and (c) detector or reader signal processing system that converts the signal to readable form.

Molecules that recognize the analyte are called bioreceptors that could be enzymes, cells, aptamers, deoxyribonucleic acid (DNA), and antibodies. The signal generated could be either in the form of light, heat, pH, charge, or mass change, etc. When the analyte is recognized by a bioreceptor causing biorecognition, this binding is converted into an appropriate signal by transducers, which produce optical or electrical signals equivalent to the biorecognition quantum. The five principal transducer classes are electrochemical, optical, thermometric, piezoelectric, and magnetic. Further amplification and conversion of analogue to digital output is achieved by other electronic devices, which is then displayed by the display read out. The output signal on display can be numeric, graphic, or image.

Examples of biosensors: Clark electrode developed by Leland Clark in 1956 was the first biosensor developed to measure oxygen. Glucometer for sensing glucose levels, pulse oximeter for oxygen levels and pulse rates, and smart watches that measure daily physical activity are all examples of our daily interactions with biosensors. Biosensors amalgamate biological knowledge with technological advances (Figure 18.9). These are rapid, easy to use, and have applications across multiple disciplines.

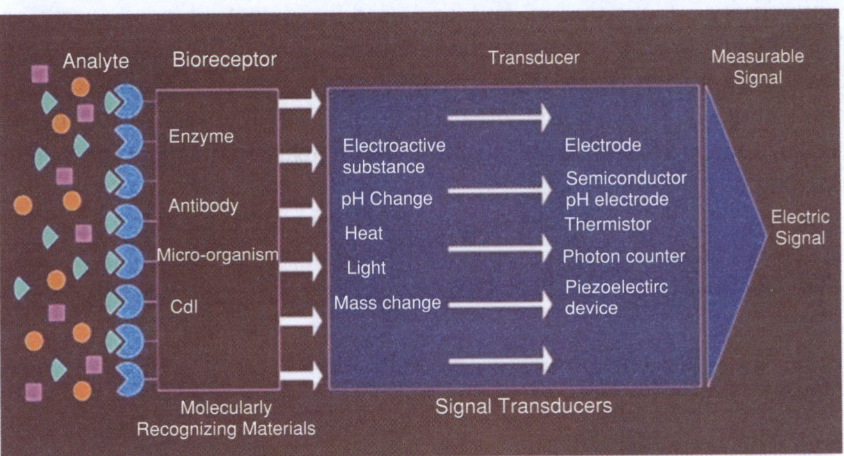

Fig. 18.9 Components and types of biosensors

Source: Adapted from https://www.scienceabc.com/innovation/what-are-biosensors.html

Depending on the nature of transducer used in a biosensor, we can classify them into different types and study their principle and functioning.

(a) **Electrochemical biosensors** Outcomes of biochemical events (like enzyme substrate reactions) are converted to electrochemical signals, and the content or concentration of the analyte is detected and analysed.

The electrochemical biosensor will be called *Amperometric biosensor*, if it measures the current (Ampere) produced as an output. Most glucose biosensors are electrochemical type due to their better sensitivity, reproducibility, low cost, and easy maintenance. For example, in glucose biosensors, immobilized enzyme glucose oxidase (GO*x*) catalyses the oxidation of β-D glucose by oxygen to form gluconic acid (to gluconolactone) and hydrogen peroxide. FAD acts as electron acceptor and gets reduced to $FADH_2$.

The general equations can be depicted as follows:

$$\text{Glucose} + \text{GO}x + \text{FAD} \longrightarrow \text{Glucolactone} + \text{GO}x - \text{FADH}_2$$

$$\text{GO}x - \text{FADH}_2 + \text{Oxygen} \longrightarrow \text{GO}x - \text{FAD} + \text{H}_2\text{O}_2$$

Redox reaction of hydrogen peroxide at the platinum anode helps this working electrode to measure the electron transfer reactions, and this electron flow of GOx, in turn, will be proportional to the concentration of glucose present in blood. The electron flow is measured and depicted as current in Amperes and relates to the concentration of glucose.

An electrochemical sensor is of the *potentiometric type* when the biological sensing element is connected to physico-chemical transducer and helps in electrochemical detection of the analyte. The signal generated is an electrical potential. It measures the analyte activity in a sample by measuring the potential difference between a reference electrode and an indicator electrode. This type of electrochemical biosensors shows advantages of small size, rapid response, ease of use, low cost, and resistance to colour and turbid interferences. These types of biosensors are used in detection of metal ions, small molecules, DNA, proteins, enzymes, toxins, etc. For example, pH meters (Figure 18.10).

Electrochemical biosensors are of the *conductometric type* if they measure changes in conductivity of a solution and are used to detect glucose, urea, and pathogens. Conductance is directly related to the number of ions in a medium, and they are suitable for conductometric biosensors since many enzyme-linked reactions result in a change in ion concentration. For example, urea sensor: It uses enzyme urease to catalyse hydrolysis of urea to produce ionic species like ammonium bicarbonate and hydroxyl ions that change ionic strength and conductivity.

$$\text{Urea} + 3H_2O \longrightarrow 2NH_4^+ + HCO_3^- + OH^-$$

These sensors are inexpensive, reproducible, and have possibility of portable devices.

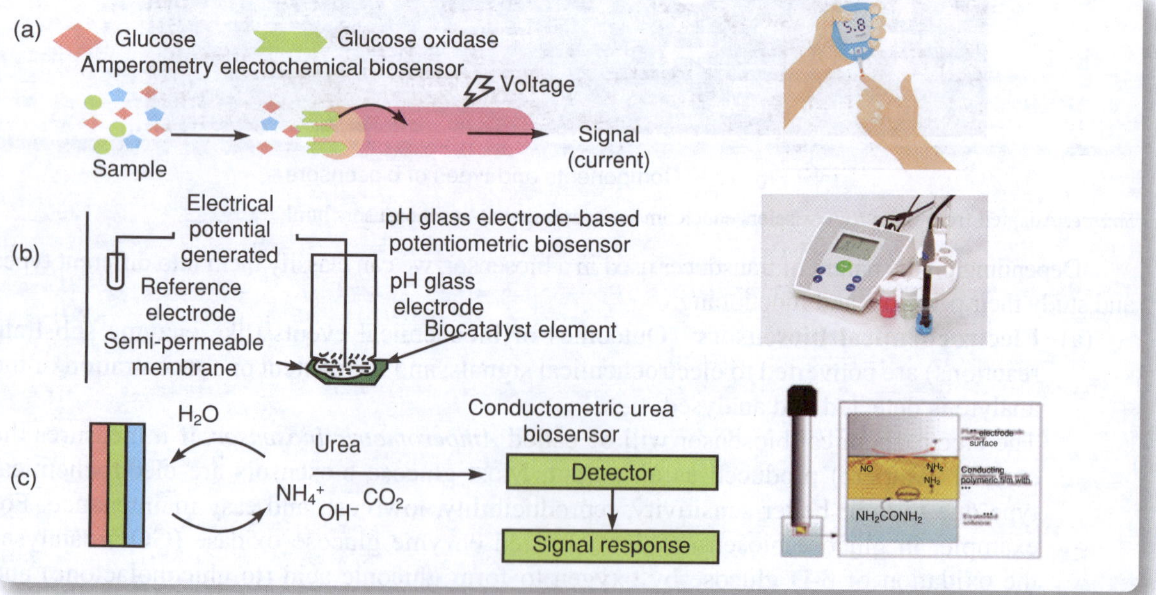

Fig. 18.10 Types of electrochemical biosensors

Source: Adapted from https://www.researchgate.net/figure/Scheme-of-an-amperometric-glucose-biosensor_fig2_343715224
https://www.researchgate.net/figure/pH-glass-electrode-based-potentiometric-biosensor_fig2_364947418
https://www.researchgate.net/figure/Schematic-of-a-typical-urea-biosensor_fig2_222392344;
some Istock images.

Apart from these, others include impedimetric and voltametric biosensors as categories of electrochemical biosensors (Figure 18.11).

(b) **Optical biosensors** They offer great advantages like direct, real-time detection of chemical and biological substances with high specificity, sensitivity, and cost-effectiveness. In these biosensors, the biorecognition element is integrated with an optical transducer to emit an optical signal proportional to the concentration of the analyte. The optical transducers include fibreoptic and optoelectronic transducers such as surface plasmon resonance (SPR), metallic nanostructures base-localized SPRs, refractors, interferometers, resonators, gratings, etc.

(c) **Thermal biosensors** Biological reactions generate/release/absorb heat termed as thermal energy, and these are sensed by thermal biosensors. These biosensors based on thermistors are semiconductor resistors that change resistance in response to temperatures. For example, thermometric biosensor is used to measure or estimate serum cholesterol. As cholesterol obtained is oxidized through the enzyme cholesterol oxidase, heat is produced, which can be calculated. Enzyme thermistors have a role in bioprocess monitoring and are used to monitor biocatalysis, fermentation systems, enzyme catalysed synthesis, food monitoring, etc. For example, glucose, sucrose, lactate, and antibiotics like Penicillin G, etc.

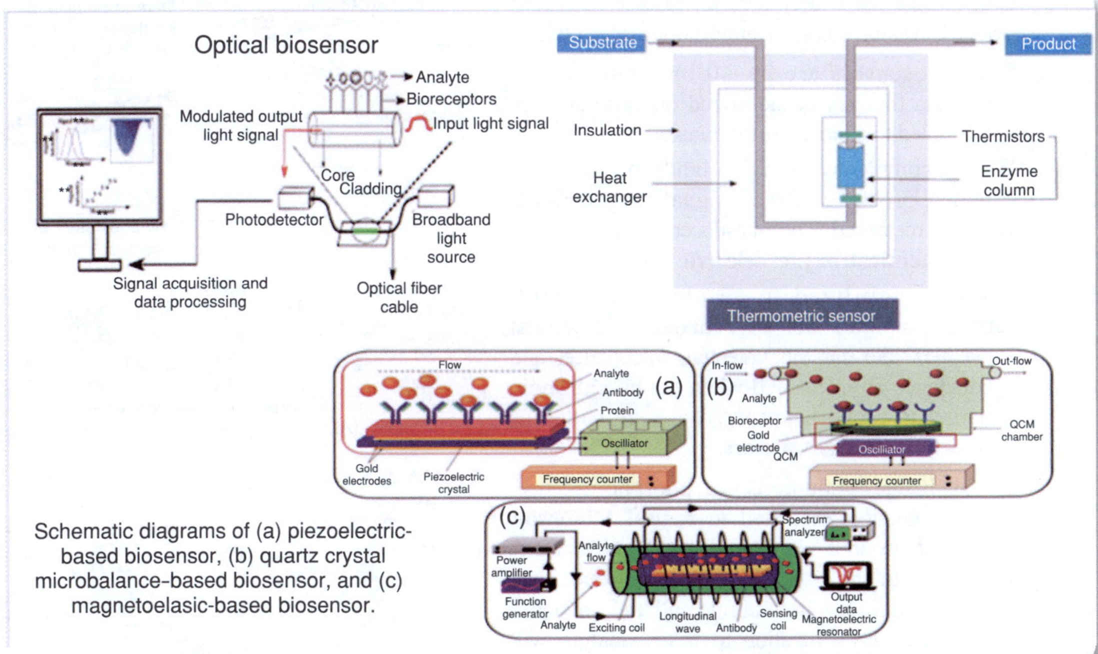

Fig. 18.11 Optical, thermometric, and piezoelectric-based biosensors

Source: Adapted from https://www.researchgate.net/figure/Optical-biosensor-schematic-representation_fig7_380040393
https://www.sciencedirect.com/topics/engineering/thermometric-biosensors
https://www.researchgate.net/figure/Schematic-diagrams-of-a-piezoelectric-based-biosensor-b-quartz-crystal_fig4_349070662

(d) **Piezoelectric biosensor** It uses a piezoelectric material that undergoes alternation when mechanical stress (*piezo refers to press or squeeze*) is applied. Quartz, tourmaline, and

topaz are natural piezoelectric materials, whereas barium titanate, lead zirconate titanate, ammonium dihydrogen phosphate, and lithium sulfate are man-made. Under the influence of mechanical stress, the deformation in the material can decrease the separation between cations and anions, which produces an internal potential difference called piezoelectricity. Peizoelectrics are used as reliable transducers. Examples of these types of biosensors include: (i) the electric cigarette lighter where spring-loaded hammer strikes a piezoelectric molecule causing a spark; (ii) voltage applied to loudspeakers vibrates a piezoelectric polymer creating sound; and (iii) in microbalances, peizoelectrics measure minor changes in mass.

Interesting Fact 18.3

Nucleic Acid biosensors—DNA biosensors

1. Known sequences of nucleic acids like DNA, RNA have a link with biosensors. They can be sensing elements or target analytes, i.e., they can be used to sense or be sensed and are called nucleic acid biosensors.
2. **Nucleic acid biosensors** work on the principle of complementary base pairing between the probe and target. These are widely used for disease diagnosis, gene expression studies using microarrays, etc.,
3. DNA biosensors are 20–40 bp single-strand fragments that are immobilized on solid surface and coupled to a biochemical transducer.
4. When a complementary DNA binds to this single-stranded probe, electrical signal is generated, which is measured. The transducer can be **optical, electrochemical, or piezoelectric** in nature.
5. DNA immobilized to an electrode (gold, carbon, graphite) and any change in electrical parameters like current, potential, impedance, and conductance are measured when the hybridizatoin of target to probe happens. These are called as **electrochemical DNA sensors**.
6. Signal changes in fluorescence, phosphorescence, and chemiluminescence lead to **optical biosensor**. Molecular beacons are examples of optical DNA biosensors. In this, stem loop oligonucleotides are labeled with fluorophore at one end and quencher at the other. Hybridization to target analyte will change fluorescent signal, and it can indicate the number of nucleotides hybridized.

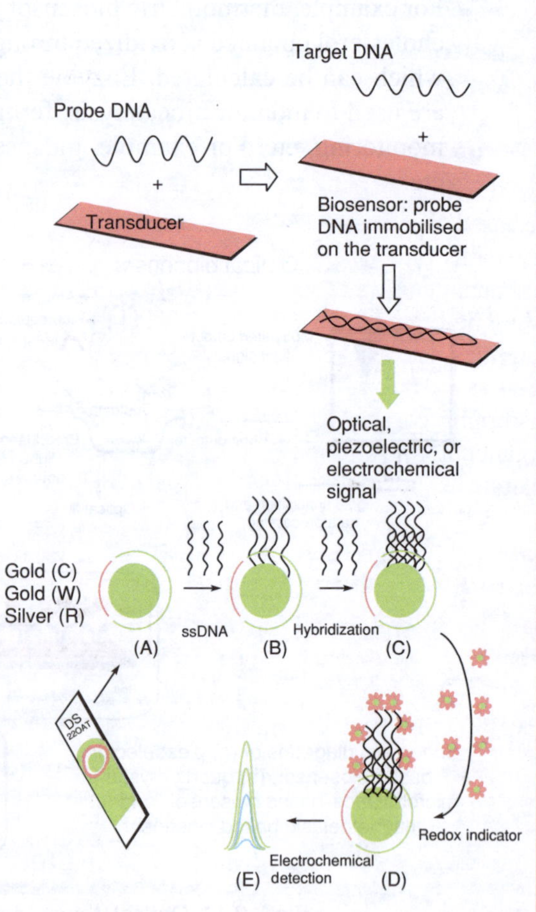

7. A degenerate probe from a conserved genomic region with two specific probes have been used for genotyping of human papilloma virus using piezoelectric measurements—a good example of a **plezoelectric sensor**.

Source: Adapted from https://www.researchgate.net/figure/Schematic-representation-for-nucleic-acid-RNA-DNA-based-Biosensordevelopment-using-Au_fig5_343762195
https://ars.els-cdn.com/content/image/1-s2.0-S2666523921000106-gr22_lrg.jpg

Biosensing applications include drug discovery, disease detection, environmental monitoring, water, soil and food quality monitoring, sensing toxins, etc. Pesticides that are a major environmental concern due to their health impact have been detected even up to trace levels by using electrochemical biosensors. Electrochemical biosensors reported for pesticide-sensing applications include those based on (1) enzymes, (2) whole cells, and (3) antibody–antigen interaction (immunosensors). Heavy metal detection from environment is another significant beneficiary of the biosensor research. Microbial biosensors, protein-based biosensors, and monoclonal antibody-based biosensors are all bioceptor options employed for sensing heavy metals. On the similar lines, enzyme-based biosensors are used for detecting antibiotic pollution from hospital wastes. For example, Laccase enzyme-based biosensor for detecting the anti-hypertensive drug methyldopa.

Biodetectors are also synonymous term used for biosensors and are devices that specifically detect bacteria, viruses, and toxins that could cause potential harm if undetected. This whole phenomenon of detecting and sensing critical substance is a crucial need for patient diagnosis and food safety. They also find application in wastewater treatment to detect pollutants and propose treatment measures. These devices come in multiple types like cell-based, optical detectors, electrochemical biochip-based, laser-induced spectrophotometric, portable, etc.

Pollutant Detection

Environmental pollutants can be monitored using specific biosensors. The detection principle must be based on a suitable physical/chemical transducer integrated with a compatible biological element that reversibly binds the analyte. The detector identifies and converts the resulting reactions into qualitative and quantitative sensing signals for the targeted pollutants from the sample.

The organic and inorganic pollutants are released from industrial, agricultural, and other human activities. Biosensors are used to monitor conditions of soil, water, and air samples to detect pollutants such as pesticides, hazardous chemicals, heavy metals, toxic substances, pathogens, and endocrine-disrupting compounds. The toxicants impacting the environment can be categorized into four types: organochlorine pesticides (aldrin, dieldrin, chlordane, heptachlor), fungicides (hexachlorobenzene), chemicals (polychlorinated biphenyls and their by-products), and heavy metals (Figure 18.12).

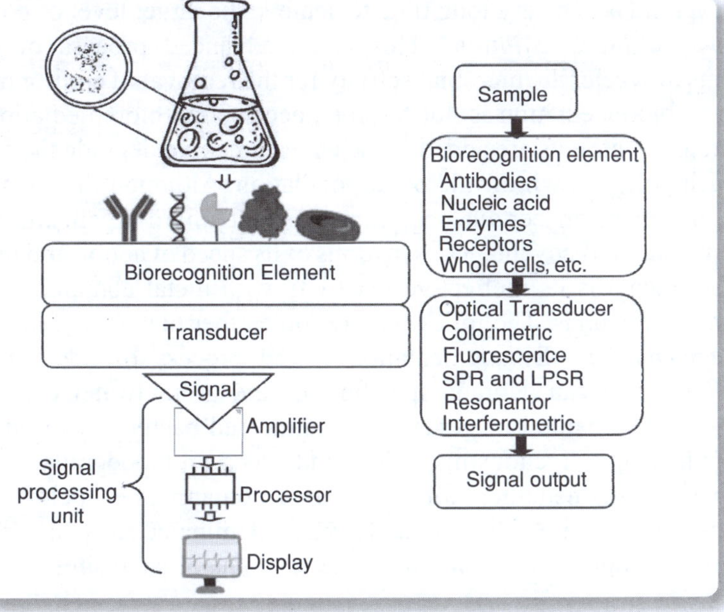

Fig. 18.12 Use of Biosensor in Pollution Detection

Biosensors help in the control of pollution through estimation and monitoring of pollutants. Even though biosensors have proved their abilities to measure air pollutants in various sample types, their efficiency is often poor.

Bioremediation

Detecting and sensing pollutants and harmful agents is just one part of the story. After sensing, What? We need various tools to remediate them. For example, environmental pollutants.

The threat of environmental contamination by pollutants is addressed by various tools, and **bioremediation** is one such promising approach. Bioremediation, by definition, means to involve living organisms, mainly micro-organisms, for degrading the environmental contaminants into non-toxic forms. Bioremediation is very popular in removing contaminants during water treatment and to handle pollutants of groundwater and surface water. Sequestration, adsorption, extraction, and degradation are all ways by which bioremediation works. Various biological agents like bacteria, fungi, and plants are commonly used for bioremediation of various organic and inorganic pollutants in water.

When microbes are taken to contaminated site for detoxification, it is called in-situ bioremediation. It is a direct cleanup mechanism between microbes and the contaminants. In-situ bioremediation is further of two types.

- *Intrinsic bioremediation* This is a passive process of treatment of polluted sites without any human intervention through naturally occurring microbial populations. It utilizes the inherent nature of micro-organisms with no intervention of technology or engineering to augment the process. The major application of intrinsic bioremediation is to control the contaminants at the waste sites. This type of remediation requires monitoring of the role of native micro-organisms in eliminating contaminants via tests performed at field sites or on site-derived samples of soil, sediment, or water. A range of environmental factors such as pH, concentration, temperature, and nutrient availability favour intrinsic bioremediation. One of the advantages of this kind of bioremediation is that it is a cost-effective method. However, it takes a very long time to achieve the target level of pollutant concentration.
- *Engineered bioremediation* This is an advanced process of employing engineered systems to accelerate microbial activity for the removal of environmental pollutants. When intrinsic bioremediation is not feasible, accelerated bioremediation is utilized where the substrate or nutrients are added to the environment to degrade the toxicity of contaminants increasing the growth of microbial population. Although the native micro-organisms are used, however, on certain occasions efficient micro-organisms are added. Engineered bioremediation is advantageous in terms of its speed of action and reliability. The limitation of this process is its ineffective reactivity with metal contaminants that are mixed with organic compounds. The following are some other methods of in-situ bioremediation:
- *Bioaugmentation* Bioaugmentation is the process of addition of culture microbial population that can degrade specific soil and groundwater contaminants. This process uses strains of natural or genetically engineered bacteria with unique metabolic profiles to supplement indigenous microflora and speed up biodegradation. Common application involves bioaugmentation for chlorinated contaminants, petroleum hydrocarbons, etc. Micro-organisms are either isolated from contaminated sites, historical sites, or genetically modified to support remediation process of contaminated sites.

- *Biostimulation* Biostimulation is the process of environmental modification via addition of limiting nutrients and elements like phosphorus, nitrogen, oxygen, or carbon to enhance the growth of existing microbial population. It is the most common remediation approach to remove petroleum pollutants from soil.
- *Bioslurping* Bioslurping is also known as multi-phase extraction and is the process of in-situ aerobic bioremediation of contaminated soils using bioventing and vacuum recovery that extracts light from the capillary fringe. The technique uses dewatering technologies to remediate hydrocarbon-contaminated sites. The volatile organic compounds (VOCs) and semi-volatile organic compounds (SVOCs) are removed by this method.
- *Biosparging* Biosparging is the injection of a gas and nutrients pressure into the saturated zone to promote aerobic biodegradation. It is the most recommended approach for aerobic degradation of sites having petroleum contaminants such as oils, diesel, gasoline, jet-fuels, etc.
- *Bioventing* It involves supplying air and nutrients through wells to contaminated soil for stimulating the growth of natural micro-organisms. This method is applied for remediation of petroleum hydrocarbon contaminants in soil to an unsaturated soil zone through air supply using pumps and blowers for continuous injection of air.
- *Phytoremediation* Phytoremediation is the use of plants and their products for the decontamination or stabilization of contaminants and metals from soil. There are some plant species that can remove heavy metals from the soil through their roots and concentrate them in their stems, shoots, and leaves. These plants possess genes that regulate the quantity of metals taken up from the soil by roots and deposited at other locations within the plant. Phytoremediation is of many types such as:
    - *Phytoextraction/phytoaccumulation* It is the process by which plants accumulate pollutants in their roots, shoots, or leaves above ground.
    - *Phytotransformation* Also known as phytodegradation, it is the transformation of organic pollutants from soil, sediments, or water into a more stable, less hazardous, and less mobile form.
    - *Phytostabilization* It is a process by which plants limit contaminated soil movement and migration.
- *Rhizodegradation* Also known as phytostimulation, it is the breakdown of pollutants through rhizosphere activity.
- *Rhizofiltration* It is a method of removing harmful chemicals and surplus nutrients from water by filtering it through a mass of roots.
- *Mycoremediation* It is a type of bioremediation in which fungi are utilized to clean up a site.
- *Mycofiltration* It is a similar method that filters harmful waste and germs from water in soil using fungus mycelium.

On the other hand, when the contaminated site is taken away for processing and worked upon, it is called ex-situ bioremediation. Also known as off-site bioremediation, it is the removal/excavation of contaminants and pollutants by subsequent transportation of contaminants from one site to another. Like in-situ techniques, this remediation also occurs with the help of micro-organisms. These techniques are based on the nature of contaminants, type of site, degree of pollution, and cost of treatment.

Techniques of ex-situ bioremediation are as follows:

- *Biofiltration* It is the biological treatment process of biodegradable waste, which relies on degrading the microbial populations. It is the process of purification of the air

contaminated with volatile organic compounds. The treatment of contaminants is done using various materials like biofilters, biotrickling filters, bioscrubbers, conventional biofilters, etc. Biofilter is a treatment bed that comprises of compost, soil, or peat in which the micro-organisms act and degrade the pollutants.

- *Biopile* It is a type of remediation process that involves enhancement techniques, viz., above-ground piling of excavated polluted soil, nutrient amendment, and aeration to increase the microbial population and enhance their activity. This method reduces volatilization of low-weight pollutants and effective remediation of extremely polluted environments. It is a cost-effective approach that ensures effective biodegradation.
- *Bioreactor* It is an engineered system involving a series of biological reactions in which pollutants are fed into the bioreactor vessel for their degradation to facilitate the growth of biological mass. Bioreactors maintain a suitable controlled environment and optimum growth conditions for the growth of microbial population.
- *Composting* It is a process of degradation of waste under favourable, controlled conditions with the help of micro-organisms. The microbes act upon the organic matter and reduce the volume, forming compost, which is rich in nutrients.
- *Land Farming* This refers to the excavation of polluted soil above the ground surface leading to aerobic biodegradation of pollutants with the help of micro-organisms.

Bioaugmentation and bioventing techniques can also be used to enhance the speed of degradation of pollutants in environment. Bioaugmentation involves the addition of genetically modified or other relevant microbes to native microbes, thus facilitating pollutant breakdown and removal. Pesticide removal, oil spills, etc., are handled by bioaugmentation.

Bioventing involves the addition to air, which increases pollutant breakdown. Air can be either be actively pumped through a blower, or passive gas exchange due to atmospheric pressure can be introduced. Hydrocarbon-contaminated sites are treated this way to reduce the release of pollutants into the atmosphere.

Biosparging is another variation where the air is injected through a pipe below the water table. This enhances the growth of indigenous microbes in that region due to high oxygen saturation, and slowly the volatile organic compounds move upwards towards the unsaturated area. The treatment is moved in a localized way; it has low construction cost and offers flexibility in design.

Some prominent success stories of bioremediation include remediation of heavy metals in environment and water, remediating oceans after oil spills, landfill clean up and agricultural pesticide clean-up, composting processes, wastewater treatment of urban runoff, etc.

Advantages of Bioremediation The advantages of bioremediation are as follows:

- Removal of harmful contaminants present in the environment
- Cost-effective method with minimal requirements of complicated equipment and expensive infrastructure
- No use of harmful chemicals
- Mostly on-site operations reducing the transportation cost
- Minimum site destruction and disruption
- Less energy requirement
- Reliable and easy approach
- A natural process

Disadvantages of Bioremediation The disadvantages of bioremediation are as follows:

- The use is restricted to biodegradable waste and contaminants.
- Requires extensive monitoring

- Dependent on biological factors such as environmental conditions, nature of micro-organisms, nutrient requirements, and nature of contaminants
- Could be even more toxic than the original contaminant
- Highly specific
- Time-consuming process

A Case Study: Bioremediation of Heavy Metals from Environment

Erin Brocovich (movie released in 2000) depicts the fight against a prominent company due to the ground water contamination by hexavalent chromium leading to health issues. Heavy metal contamination has serious health consequences; hence, remediation is an absolute need. Bioremediation of heavy metals using microbes offers one solution and follows many mechanisms and possibilities for successful remediation (Figure 18.13).

(a) *Biosorption* It is the passive adsorption of metals and other pollutants on the surface of microbe due to physicochemical interactions between the metal and the functional groups on the surface of microbes. This adsorption can be achieved using both live microbes and dead biomass.

(b) *Bioaccumulation* Intracellular sequestering (Bioaccumulation) is a process by which the complex formation of cell–metal occurred inside the cytoplasm. This is metabolism linked and hence only seen in living microbes. In this, accumulation of metals happened by attaching with the cell surface followed by slow penetration to periplasm of the cell cytoplasm by a process that looks like nutrient uptake.

(c) *Biotransformation* relies on the cellular metabolic activity of the microorganism through the redox mechanism; reduction of metals by changing the oxidation number of the metal is common in nature, such as the reduction of chromium, selenium, uranium, and mercury.

(d) *Bioleaching* is the secretion of low-molecular-weight compounds that aid the transformation of a toxic form of metals to a non-toxic form by dissolution or precipitation mechanisms. For example, leaching of Cd is promoted by the secretion of organic acids by some micro-organisms; also, the production of inorganic phosphate by Citrobacter bacilli leads to precipitation of metal phosphate coat.

(e) *Siderophore-mediated biosorption,* also called a chelating agent, in aerobic soils: Some micro-organisms produce siderophores that mediate the ability of the micro-organisms to utilize low water-soluble metals using an energy-dependent process. For example, *Microbacterium flavescens* was found to use siderophore to uptake their nutritional requirements of iron; also the organism uses the siderophore Desferrioxamine (DF) to bind with uranium, plutonium, and iron.

(f) *Methylation of metals (volatilization) by microbial methylation* plays a significant role in metal remediation. For example, *Pseudomonas* spp., *Escherichia* spp., *Clostridium* spp., and *Bacillus* spp. can convert Hg (II), Se, As, and Pb to a gaseous methylated form.

A lot of factors could exhibit significant effects on the bioremediation process; for instance, metal ion concentration, valance state and chemical forms of the metal, the bioavailability of the metal, redox potential, availability of low-molecular-weight organic acids, and environmental factors such as temperature and pH.

Thus, for successful bioremediation of pollutants, both microbial factors, such as identifying microbes capable of remediation, and the environmental factors, such as temperature, pH, moisture, oxygen, nutrients, and toxicity of contaminants, have to be optimized. This ensures that adequate microbial growth is permitted and interaction between the microbe and pollutant is well-facilitated for it to detoxify the same.

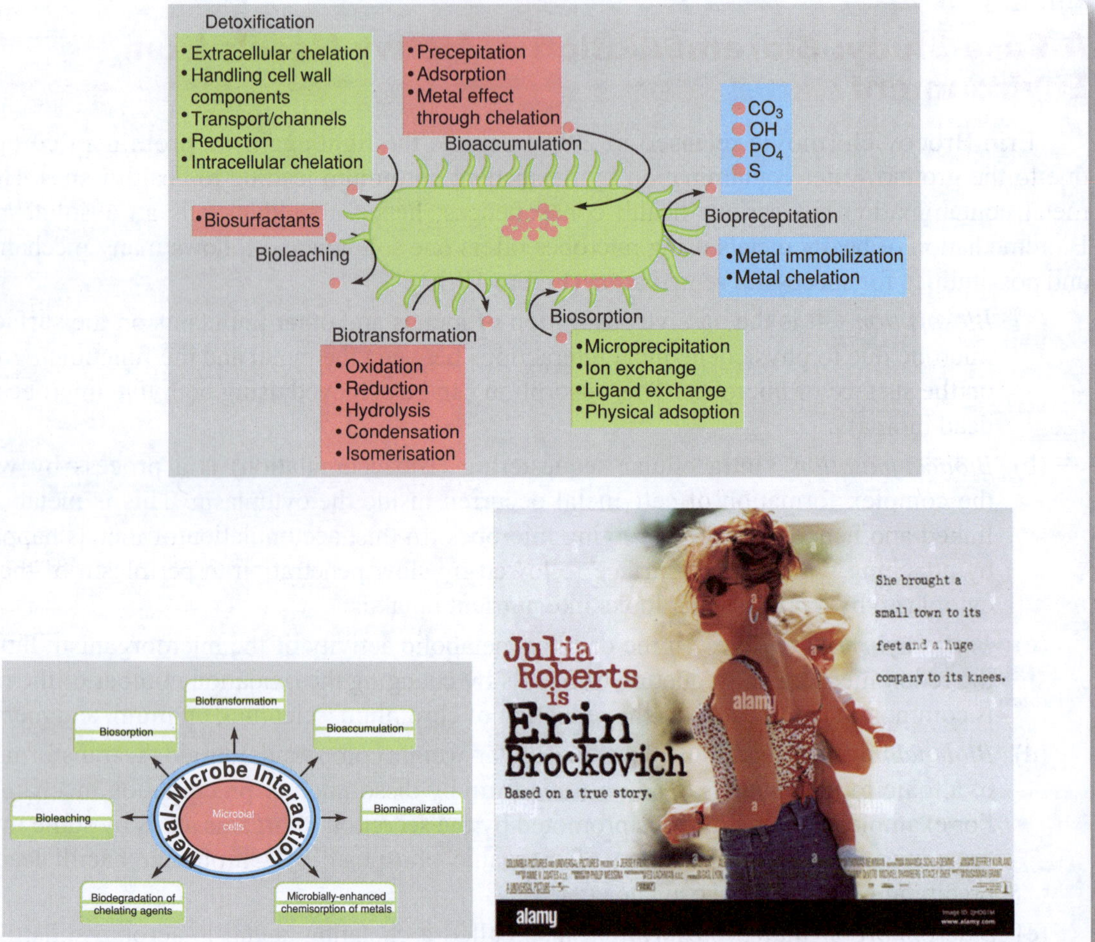

Fig. 18.13 Role of microbes in bioremediation of heavy metals

Source: Adapted from semanticscholar.org/paper/IMPLICATIONS-OF-BACTERIAL-RESISTANCE-AGAINST-HEAVY-Ahemad/ 456b52a6cfe4ac02fe83aabbe1665c2093ac9a00
https://www.frontiersin.org/journals/microbiology/articles/10.3389/fmicb.2022.824084/full

Interesting Fact 18.4

BIOFUELS

1. In the recent past, biofuels have been explored as sustainable alternative to fossil fuels. The idea is to convert the microbial biomass directly into liquid fuel for use in transportation. First-generation biofuels in use today are ethanol and biodiesel.
2. Biofuel formation happens over a short time, unlike the huge time taken for fossil fuels.
3. Biofuels are derived from plant, algae, or animal waste. Fermentation of starch and sugar leads to ethanol production, which is mixed with gasoline to form gasohol. Brazil and US are major producers in this category.
4. Biodiesel is made from oily plants like soyabean and palm oil and has found acceptance in Europe.
5. Third-generation biofuel uses algae and cyanobacteria for biodiesel production.
6. Methane and biogas as biofuels are derived from decomposition of biomass in absence of oxygen, and in this category, methanol, butanol, and diethyl ether are in development.
7. Biofuels make little contribution to global warming, are carbon neutral, hence are sustainable alternatives.
8. New technology called carbon capture and storage, biofuel crops would remove carbon dioxide from air; and hence they are sequestered beneath oceans. Food versus fuel over use of land resources to produce biofuel would also be somewhat acceptable in that case.

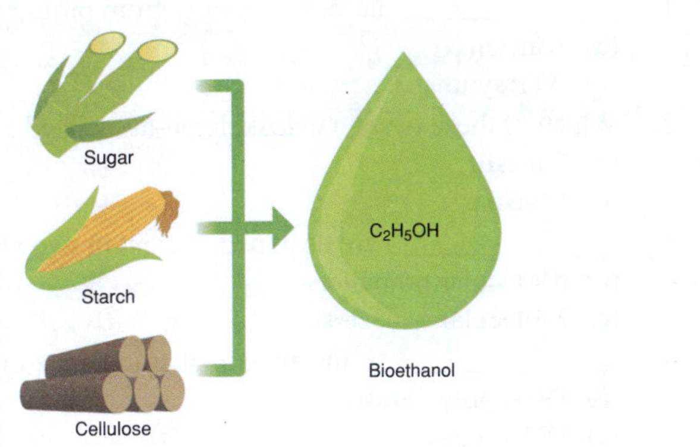

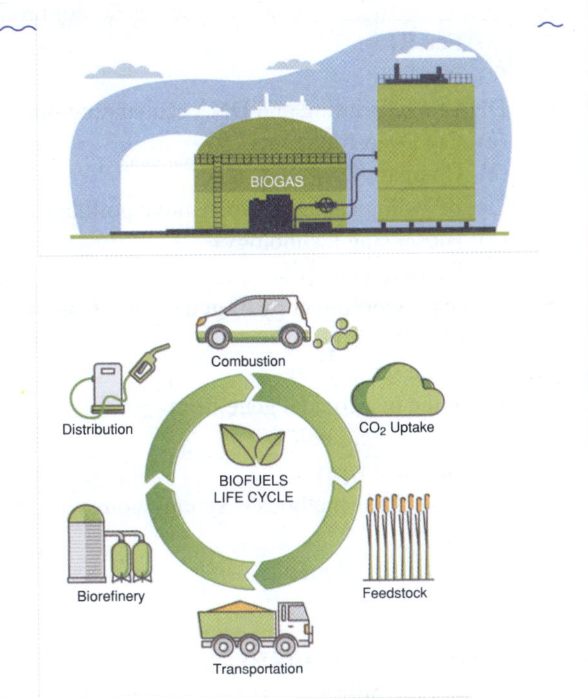

SUMMARY

- Concepts of molecular machines and molecular motors were understood
- It has special focus on ATP synthase and cytoskeletal motor proteins to details.
- The method of genetic engineering to manipulate DNA for beneficial purposes was discussed.
- DNA fingerprinting, DNA sequencing, DNA cloning, and other associated methodologies were detailed.
- Medical, agricultural, and environmental applications of genetic engineering were elaborated on.
- Basic understanding of biosensors, bioremediation, and biodetectors was obtained by lucid and clear explanations.

EXERCISES

A. Objective-Type Questions

1. _______________ harnesses energy from proton gradient to form the energy currency of the cell.
   (a) Kinesin (b) Myosin
   (c) ATP synthase (d) None of the above
2. Which of these is not a cytoskeleton-associated molecular motor?
   (a) Kinesin (b) Dynein
   (c) Myosin (d) Gibberellin
3. _______________ are designed to exist in two alternative stable forms.
   (a) Molecular propellers (b) Molecular switches
   (c) Molecular tweezers (d) Molecular motors
4. _______________ is the enzyme that joins the plasmid with the gene of interest.
   (a) DNA polymerase (b) Taq Polymerase
   (c) DNA ligase (d) Restriction endonuclease
5. _______________ kind of enzymes do not produce sticky ends.
   (a) Eco RI (b) Bam H1
   (c) Alu I (d) Hind III
6. DNA fingerprinting and DNA sequencing both help to:
   (a) Cut DNA fragments (b) Clone DNA
   (c) Compare DNA sequences (d) Process cloning of Dolly
7. Use of biological agents to remove pollutants in environment is ....
   (a) Biosensing technology (b) Biodetection technology
   (c) Bioremediation (d) Biodegradation
8. Adding a working copy of gene to treat single-gene genetic disorders is ....
   (a) Gene therapy (b) Gene editing
   (c) Gene sequencing (d) Gene testing
9. When the insecticidal gene from _______________ is added to plants, they become insect resistant.
   (a) *Escherichia coli* (b) *Bacillus thuringiensis*
   (c) *Vaccinia virus* (d) *Thermus aquaticus*
10. Which of these methods is synonymous with in-vitro replication and helps to increase concentration of DNA?
    (a) Gel electrophoresis (b) DNA sequencing
    (c) Transformation (d) Polymerase Chain Reaction

B. Short-Answer Questions

1. Explain the different steps of gene cloning experiment.
2. What were the perceived benefits and concerns of the Human Genome project? After two decades, how many have been proven true or otherwise?
3. What were the problems associated with Dolly—the sheep cloned by somatic cell nuclear transfer?
4. Read about the safety and ethical concerns of genetic engineering and write four points on each?
5. Is human cloning possible? Justify your answer.

FURTHER READINGS

1. Rotary ATP'ases dynamic molecular machines
   https://www.sciencedirect.com/science/article/pii/S0959440X13002091

2. https://www.opastpublishers.com/open-access-articles/concepts-of-bioremediation-bio-augmentation-and-biosensors.pdf
3. Human Genome Project and After https://www.iamg.in/genetic_clinics/full_text.php?id=194;
4. https://www.fda.gov/food/agricultural-biotechnology/science-and-history-gmos-and-other-food-modification-processes

USEFUL WEBLINKS

1. The Human Genome project https://www.youtube.com/watch?v=-hryHoTIHak
2. Genetic engineering methods https://www.youtube.com/watch?v=CDw4WPng2iE
3. Your body's molecular machines https://www.youtube.com/watch?v=X_tYrnv_o6A;
4. Molecular machine: ATP synthase https://www.youtube.com/watch?v=XI8m6o0gXDY

CHAPTER 20

Technology and Human Impact

LEARNING OBJECTIVES

After reading this chapter, you will be able to:

- Understand the historical perspective of technology and humans.
- Understand the positive aspects of technology in human life.
- Understand the challenges in human life due to technology.
- Get to know about futuristic technology and possible evolution.
- Understand the adaptations to minimize challenges and make the best use of technology.

CHAPTER TAKEAWAY **Impact of technology on human life**

This chapter deals with the various facets of technology influence in our lives and the health aspects of humans that need adjustments and adaptations.

PRELUDE

In today's world, the role of technology in our lives is here to stay. Technology is continuously evolving, morphing into different forms and versions, and is expanding at a pace that is difficult to ignore. Such an omnipresent factor that touches our lives intimately and ubiquitously is bound to impact us in more ways than one. Technology does bring a lot of positive benefits to mankind; it has made communication better with increased productivity and made positive contributions to healthcare and education. So, the promise of technology is evident. However, technology was an external tool or servant for making our lives easier till a few years back; but today, it is scary to note that it has intertwined and our dependence on technology is inescapable. 'Imaginary identities', 'virtual worlds', parallel lives', and 'augmented reality' are common vocabulary of our younger generation. The integration of technology in our lives is a reality that the human race needs to adapt, considering that the 'Beta' generation—kids born after 2025—will not know a world without 'AI'. Human–Computer interaction is an emerging field focusing on how humans interact with computers and design new ways to enrich it. This book cannot be complete without touching the multiple facets of technology on humans, which is the focus of this chapter.

Historical Perspective of Technology and Humans

Societies have a reciprocal relationship with technology. Society produces new technologies and initiates change, but they begin to change societal connotations once technologies have set in.

Technological innovations have been happening from time immemorial. The earliest stone tools and wheel are the beginnings of technological options for humans. From there, it has reached the modern-day advances such as information technology and genetic engineering. New knowledge leading to new tools spurs technological innovations. The framework of technology rests in applied science principles and has always had implications on economic development too. Early anthropologists believed that technological advancement was the primary factor that drove human progress and development.

In the mid-20th century, the definition of technology as 'the means or activity by which man seeks to change or manipulate his environment' emerged. Historical progression of technology can be classified into:

(a) The beginnings: Basic tools of the stone age, assemblages of the head and haft, pottery making, use of metals, textile handling to form woven cloth, fire as power source, and wind for small boats. Spears, bow and arrows, and boomerangs were common objects during this period. Slowly fish traps, animal nets, and bird snares were developed. The latter stages in the new stone age saw developments of structures, tombs, religious edifices, etc. Some primitive beginnings in manufacturing, communication, and transport were also seen in this age. All these progressive changes set the stage for human civilization and growth.

The community development and group colonization marked the beginning of agriculture, metallurgy, and invention of a city. Metals and their utilization for commerce marked the beginning of the metal age. Astronomy, alloying, and irrigation were new developments that followed. Marine ships and military developments, philosophy, and medicines were conceptualized.

(b) The middle: Civilizations grew in strength, and colonization brought the spirit of experimentation in all aspects of life and human behaviour. Inventions begin to appear like windmills and gunpowder out of spontaneous necessity. This was the period of revolution for power sources, and wind power and waterpower was harnessed. Mills, grinding, oils, spinning wheels, soap, architecture, railways, glass—all these technologies emerged during this period. Mechanical clock, ink making, and printing technologies followed. This opened up the domain for possibilities of mechanical devices. All these created an aspiration for a new and dynamic civilization. The outcome was the development of advanced machines such as steam engines, telescopes, blast furnaces, and commodities out of sugar refining and processing of tobacco, with emergence of cotton and new textile industry. Slowly, high standard of living and agricultural revolution ensured an essential prelude to the industrial revolution.

(c) Till date: The 18th and 19th century saw a plethora of inventions and moved it from a pre-industrial to a post-industrial society. Refinements in power technology started vast improvements in water, wind, and steam utilization, and high-powered steam and steam turbines remain viable solutions even today. Electricity brought with it multiple developments. Independent industries started developing, such as metallurgy, mechanical engineering, textiles, chemicals, pharmaceuticals, immunology, civil engineering, and transport and communications. Photography, telephones, and telegraphs emerged. These brought out innumerable possibilities in almost all areas of human existence due to better communication.

The 20th and the 21st centuries cemented the contemporary developments. Detonation of the atomic bomb was a significant event. World wars spiralled many innovations such as the evolution of modern, highly efficient airplanes, alloys, innovations in fuels, jet engines, combat aircrafts, etc. The gas turbine became a subject of experiments in road, rail, and marine transport. Finally, plastics emerged during this period. Cost-effective large-scale electricity production became a reality changing lives. Emergency of atomic power and the detonation of the atomic bomb was a significant event. Synthetic rubber, synthetic fibres, dyes, anaesthetics, surgeries, reinforced concrete, elevators, and innovations in automobiles are all credited to this era.

Boom of electronics and digital technologies: Twentieth century saw the booming of the electronics industry, live voice broadcasting, radio, microscopes, cinema, cathode ray tube, etc. Television became a universal medium of mass communication. Emergence of space age technology and nuclear energy gained momentum. Digital computers were one of the biggest 20th century inventions that grew from an accessory to become one of the essential elements in any enterprise. Computerization and automation became buzz words. Computers impacted all areas of industries, including health and research. Space exploration and all areas of engineering flourished. Mission to moon, rockets, and satellites became modern day success stories as were test tube babies and genetic engineering.

Computers with fibre optics networks and optical amplifiers ushered in the information age. The internet began in 1996, which was a major breakthrough. Wireless technology, Human Genome Project, laser, and fibre optics are all new age developments opening multiple possibilities for the future. Quantum computers, 3D printing, nanotechnology, bioengineering, advanced materials, robotics, alternative fuels, wearable devices, and augmented reality are all set to bring in huge transformations in the way we live and think (Figure 20.1).

The above account is not exhaustive; however, it clearly depicts that technology development is integral to civilizations and is here to stay. It has the potential to change societies both positively and negatively, and it is important that humans meaningfully adapt to these changes and emerge with optimal solutions.

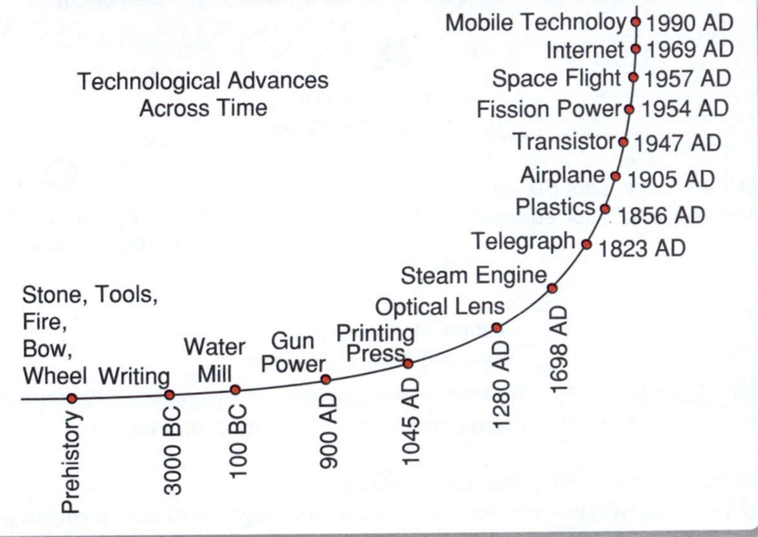

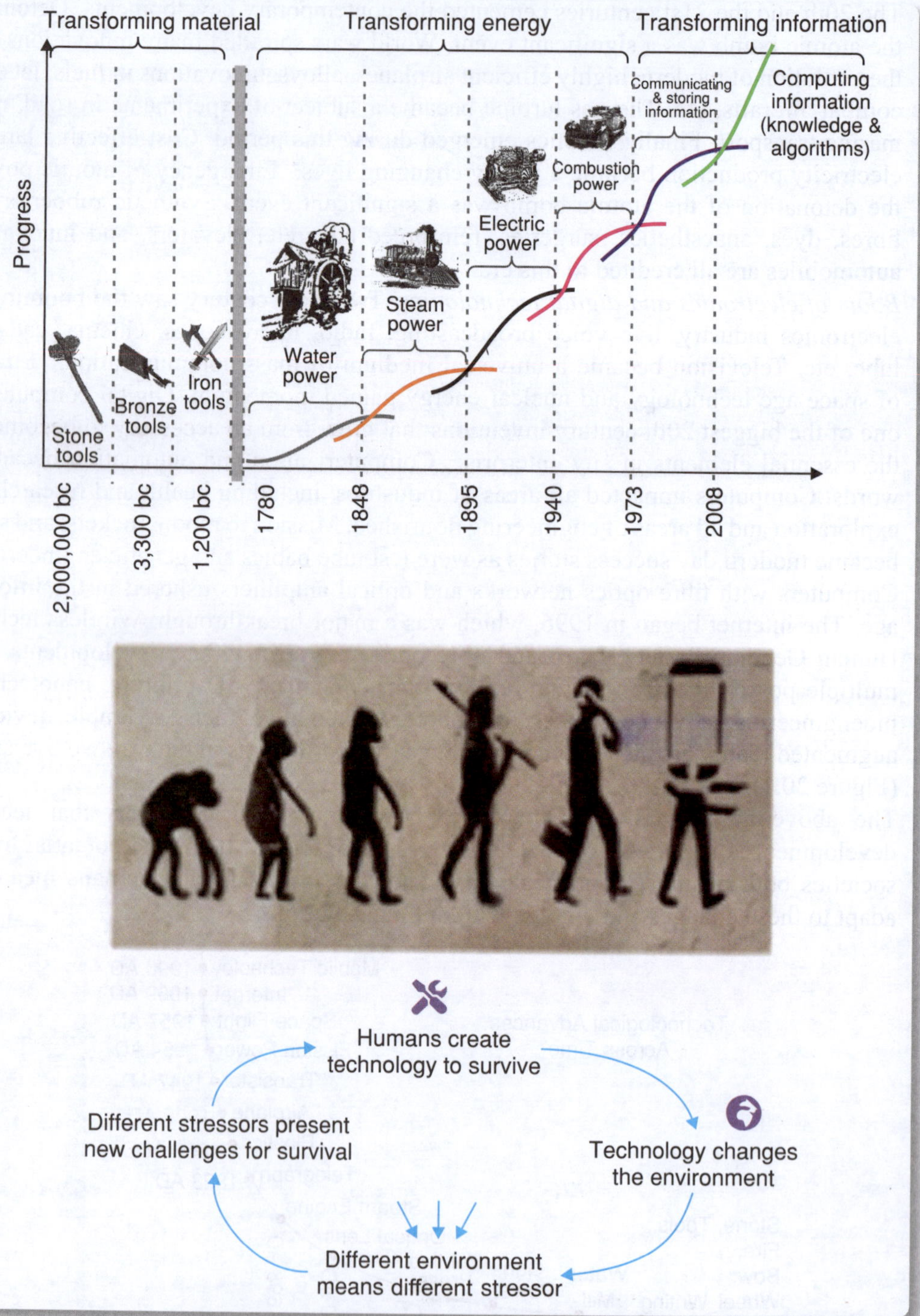

Fig. 20.1 Development of technology over time

Sources: https://seanmcclure.substack.com/p/technology-is-humanity
https://en.wikipedia.org/wiki/Information_Age#/media/File:LongWavesThreeParadigms.jpg
https://energyscholar-20361.medium.com/general-purpose-technology-is-a-thing-5712383d777e

Positive Aspects of Technology in Human Life

Technology is everywhere and a critical component of the world today. Tech tools have supported our lives in many ways: From communication, to manufacturing, to healthcare.

Few very critical positives that have emerged out of technology include:

(a) Technology has eased many manual rigorous tasks, and machines have made life easy and comfortable for humans, e.g., dishwasher, washing machine, etc.

(b) Humans are more connected than ever before. Anyone across the planet is just a call, text, email, or video call away.

(c) Humans can be more productive with time, thanks to technology that helps you to remember, record and clock your routines, write, present, buy, fix appointments, attend classes, shop, play games, stay aware of current events, watch programmes, and attend virtual meetings, all using small gadgets like smart phones, smart watches, etc.

(d) Easy access to information through the internet, search engines, generative AI is another boon.

(e) Technology provides opportunities for social and economic development by providing learning opportunities, connecting employers, online work, etc.

(f) 3D images of organs help surgeons with surgeries. Technology has helped in better diagnosis and improved research outcomes; it has also opened possibility of telemedicine.

(g) Resilient cloud, internet security, and AI technologies help in securing data storage, analysis, and customer satisfaction.

(h) The Internet is leading to greater awareness and shapes the vision and consciousness of society.

Overall, technology has helped in enhancing efficiency, faster decision-making, cost and time saving, building competitiveness, and increasing innovation and quality of life.

Challenges Due to Technology in Human Life

As we have seen in the previous section, modern technology has many benefits, but they all come with a price. There is too much of a good thing, and it appears as if technology is taking over our lives (Figure 20.2).

Some challenges that humans face due to misuse or overuse of technology are highlighted below:

(a) Long attention to screens and devices like tablets, smartphones, and computers leads to eyestrain, blurred vision, and dry eyes.

(b) Long sitting hours with poor posture also causes pain in the head, neck, and shoulders. Text neck and Blackberry Thumb are becoming more common.

(c) Blue light from the screens stimulates the brain and affects sleep if gadgets are used close to bedtime.

(d) These technologies generally promote a sedentary lifestyle and can promote obesity, type II diabetes, and cardiovascular diseases.

(e) Hyperconnected environment makes it difficult to disconnect and restricts possibility and space for unwinding. You could be working, playing, and be on the gadget 24 × 7 due to its accessibility; then, when and where to rewind?

(f) Virtual spaces provide a sense of anonymity and lead to expressing and saying things that we may not otherwise do face to face. Instant sharing of views or information can then lead to uncomfortable or dangerous situations. So, mindful space away from the virtual world is a need that should be included.

(g) Bullying and cyberattacks are common side effects of cyberspace.

(h) Social media can have a dramatic impact on self-worth, success, inclusion, and, in some cases, if they spill over to the real world, can lead to negative consequences like anxiety, depression, and other mental health disorders.

(i) Ability to remember and memorize is on the decline as there is too easy access to information. Students are especially vulnerable and are tempted to use these. It creates learning barriers sometimes, as they find it easier to text a friend, search the Internet, look up answers, or send pictures instead of conceptualizing and visualizing the content and having a complete understanding.

(j) In general terms, there has been a loss of privacy for individuals due to overload of information that is available one click away, so it can be misused by hackers to access personal information.

Overall, it can be said that the experiment of the information age and technological advancement can have implications on people's physical and cognitive capabilities. Declining capacity for analytical thinking, memory, focus, creativity, reflection, and mental resilience are outcomes of digital dangers. Other aspects that cannot be ignored include digital distrust/divisiveness and digital addition and duress. So, a digital detox regimen for maintaining optimal physical and mental faculties is the need of the hour.

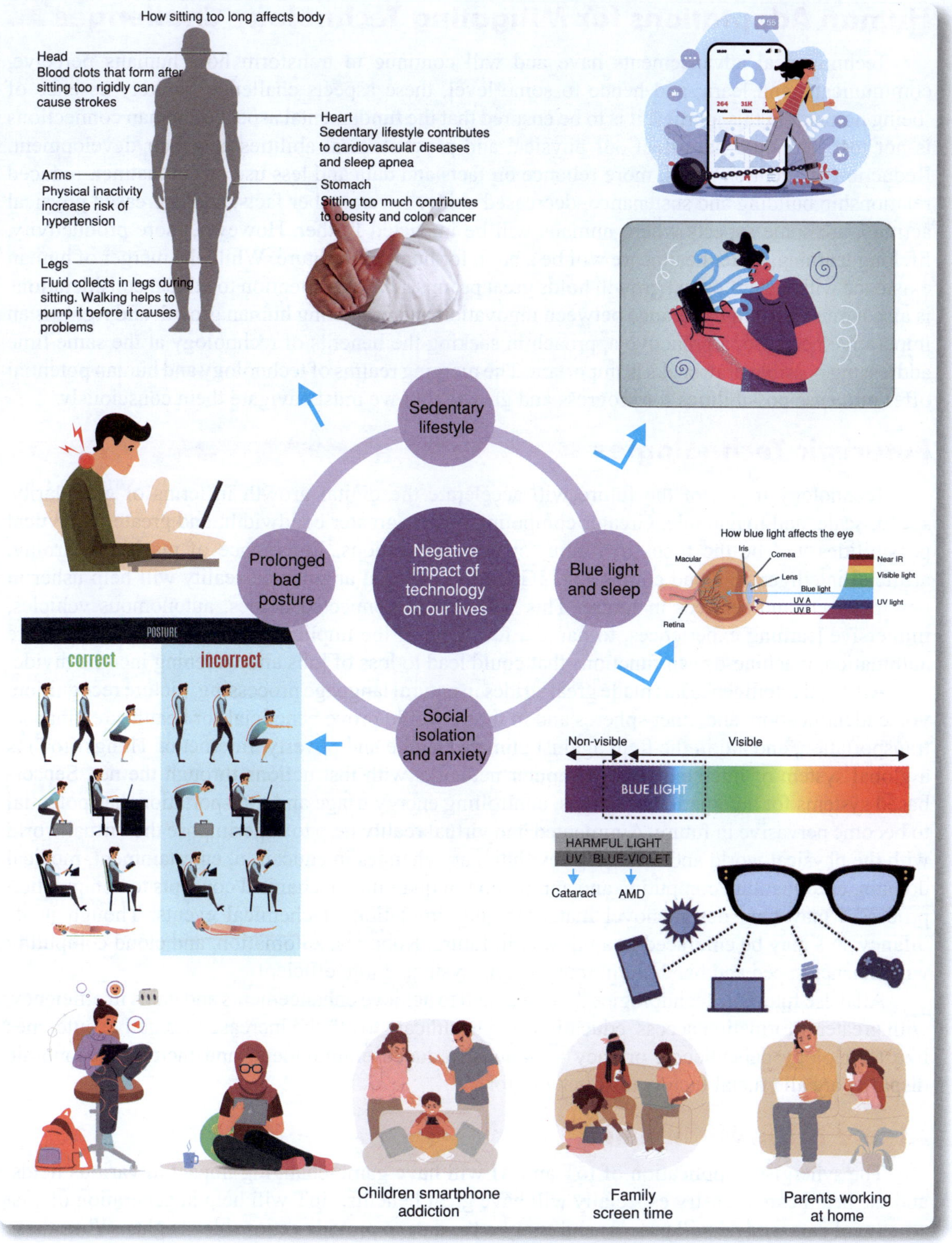

How sitting too long affects body

Head
Blood clots that form after sitting too rigidly can cause strokes

Heart
Sedentary lifestyle contributes to cardiovascular diseases and sleep apnea

Arms
Physical inactivity increase risk of hypertension

Stomach
Sitting too much contributes to obesity and colon cancer

Legs
Fluid collects in legs during sitting. Walking helps to pump it before it causes problems

Negative impact of technology on our lives

Sedentary lifestyle

Prolonged bad posture

Blue light and sleep

Social isolation and anxiety

POSTURE

correct

incorrect

How blue light affects the eye

Macular

Iris

Cornea

Lens

Retina

Blue light

UV A

UV B

Near IR

Visible light

UV light

Non-visible

Visible

BLUE LIGHT

HARMFUL LIGHT

UV BLUE-VIOLET

Cataract

AMD

Children smartphone addiction

Family screen time

Parents working at home

Fig. 20.2 Challenges due to technology

Human Adaptations for Mitigating Technology Challenges

Technological advancements have and will continue to transform how humans perceive, communicate, and learn, and hence to some level, these aspects challenge the very essence of 'being human'. In this journey, it is to be ensured that the fundamental aspect of human connections is not lost. It should not affect our physical and psychological abilities and their development. Reduction in attention spans, more reliance on facts and data and less use of gut instinct, reduced relationship building and sustenance, decreased ability to remember facts, and decreased physical activity, are some aspects where humans will be impacted further. However, more productivity, lifelong learning, better healthcare will be a boon for humans in future. While the merger of human existence with technological growth holds great promise, careful attention to ethical considerations is also a must. Striking a balance between innovation and protecting human growth and rights is an imperative need. So, a proactive approach in seeking the benefits of technology at the same time addressing possible drawbacks is important. The merging realms of technology and human potential offer immense possibilities for progress and growth, but we must navigate them consciously.

Futuristic Technologies

Technology trends of the future will accelerate the digital growth in terms of granularity, speed, scale, and magnitude. Greater computing power, greater bandwidth, and greater analytical possibilities will be the focus in future for new innovations. Emergence of machine learning, artificial intelligence, cloud computing, 5G, and virtual and augmented reality will help usher in new products and services in future. This will lead to connected homes, autonomous vehicles, immersive learning experiences, to name a few. One of the implications of these could be more automation, machine-driven functions that could lead to loss of jobs and widening income divide.

Artificial intelligence has made great strides in natural language processing, picture recognition, voice identification, and other spheres and in future would prove beneficial for healthcare, finance, transportation, and aid in the fight against climate change and poverty. Internet of Things (IoT) is a global system of interconnected computer networks with instructions through the net. Sensor-based systems for automation of homes, controlling energy usage and transportation, has potential to become pervasive in future. Augmented and virtual reality help to superimpose the virtual world with the physical world and can bring revolutionary changes in education, entertainment, medical domain, etc. Quantum computing and application of quantum mechanical concepts to computation processes may help design novel materials and simulation of chemical events. Though in its infancy, this may be embraced in a big way in future. Robotics, automation, and cloud computing will enhance speed and bandwidth, thus making systems more efficient.

All these futuristic technologies though aimed to achieve enhancements and gains in efficiency, with greater information access, education, and healthcare to all, the increase in negative outcomes like technology dependency, privacy violations, reduced employment, and increased economic disparity are all crucial factors to watch out for.

AI and IoT in Healthcare

The synergistic application of IoT and AI will have game-changing impact in various fields, and the healthcare industry especially will have great benefits. IoT will help in generation of vast amount of data, and AI will process and analyse the data to provide actionable insights. When these

two potent technologies come together, it will open up new avenues that will facilitate better health management. Few possibilities that can be envisaged due to this IoT–AI marriage are given below:

(a) Development of IoT wearables can monitor patient health parameters like heart rate, glucose levels, alert healthcare professionals who can analyse the data real time, provide timely interventions, and overall ensure better patient monitoring both within and outside the hospital settings.

(b) The data generated by IoT devices will be huge and overwhelming. Application of AI can lead to thorough analysis of this data and contribute to predictive analytics that can predict future events based on health vitals, and data-driven details can keep the professionals well-informed to take appropriate action.

(c) The edge computing approach of connected devices and sensors can provide real-time support and also improve decision-making in hospital operations. Maintenance of medical equipments, handling patient flow, decentralizing internal data to protect from cyberattacks, manage compliance and privacy requirements, monitor the movements of convalescing patients, etc., will be positive outcomes. Human pose detection AI model detects patient movement in large hospital settings.

(d) Telemedicine and virtual patient–doctor interactions for remote consultations are made more effective due to integration of AI and IoT with real-time data collection and analysis.

Slowly but surely, we are witnessing the evolution of IoT and AI from conceptual frameworks to essential tools in healthcare for a resilient healthcare system of the future. Increase volume of patient management through these technologies also brings in associated economic benefits. However, all these promising developments also open up ethical complexities of AI and IOT in healthcare where exercising caution and care is paramount. Data privacy, digital divide, transparency, and algorithm-induced bias are challenges that advocate the need to emphasize critically on data protection laws, equitable technology adoption, and ethical auditing to circumvent the pitfalls.

SUMMARY

- The historical view of technology growth was studied.
- Positive benefits accrued by humans through technology was enlisted
- The inevitable challenges associated with the technology boom were highlighted.
- Futuristic growth of technology and their impact was discussed.
- Aspects that need attention in the form of human adaptation to manage and sustain technological advancements were touched upon.
- Specific discussion on the futuristic role of IoT and AI in healthcare with their challenges was introduced.

EXERCISES

A. Objective-Type Questions

1. Which of the following technologies is not associated with enhancing access to information?

   (a) Search engines (b) Generative AI

   (c) Internet (d) Blockchain technology

2. Which of the following are on a decline due to technology?
   (i) Attention span (ii) Productivity
   (iii) Ability to remember (iv) Sleep and focus
   (a) All of the above (b) Only (i) and (ii)
   (c) Only (i), (iii), (iv) (d) Only (ii), (iii), and (iv)
3. Which of the following are futuristic technologies of 21st century?
   (i) Refrigeration (ii) Internet of Things (IoT)
   (iii) Artificial intelligence (iv) Quantum computing
   (a) All of the above (b) Only (i) and (ii)
   (c) Only (i), (iii), (iv) (d) Only (ii), (iii), and (iv)
4. Excessive social media consumption could be responsible for
   (a) Social isolation and anxiety (b) Reduced self-esteem
   (c) Sleep deprivation (d) All of the above
5. Which of these is true about augmented reality?
   (a) Involves design, creation, and use of robots for real-world applications
   (b) Composite view of computer-generated image and real world
   (c) A means of crypto currency
   (d) A substitute for cloud computing

B. Short-Answer Questions

1. What could be ethical challenges due to the development of artificial intelligence?
2. How can technology foster economic disparity and divide in the future?
3. What kind of vocations or skills should humans develop for the future?
4. What are the psychological, physical, and cognitive impacts of technology on humans?

FURTHER READINGS

1. History of technology
   https://www.britannica.com/technology/history-of-technology/The-technological-dilemma
2. History of technology https://en.wikipedia.org/wiki/History_of_technology
3. History of technology, (2012) Volume 31; Ian Inkster (anthology editor), Institute of Histological Research, University of London. https://www.bloomsbury.com/us/history-of-technology-volume-31-9781441152794/
4. Technology in 2030: Top 10 big tech predictions
   https://www.pluralsight.com/resources/blog/tech-operations/tech-in-2030

USEFUL WEBLINKS

1. https://www.visualcapitalist.com/history-of-technology-earliest-tools-modern-age/
2. What is technology? https://www.youtube.com/watch?v=INKYtIMBDPQ
3. AI, Machine Learning, Deep Learning and Generative AI Explained
   https://www.youtube.com/watch?v=qYNweeDHiyU
4. Technology is Redefining What it Means to be Human
   https://www.youtube.com/watch?v=bXSd6G7r4Ek

Index